NEW TOPICS IN SUPERCONDUCTIVITY RESEARCH

NEW TOPICS IN SUPERCONDUCTIVITY RESEARCH

BARRY P. MARTINS

EDITOR

Nova Science Publishers, Inc.

New York

LIBRARY OF CONGRESS CATALOGING-IN-PUBLICATION DATA
Available upon request

ISBN 1-59454-985-0

Published by Nova Science Publishers, Inc. ✛ New York

CONTENTS

PREFACE

Superconductivity is the ability of certain materials to conduct electrical current with no resistance and extremely low losses. High temperature superconductors, such as La2-xSrxCuOx (Tc=40K) and YBa2Cu3O7-x (Tc=90K), were discovered in 1987 and have been actively studied since. In spite of an intense, world-wide, research effort during this time, a complete understanding of the copper oxide (cuprate) materials is still lacking. Many fundamental questions are unanswered, particularly the mechanism by which high-Tc superconductivity occurs. More broadly, the cuprates are in a class of solids with strong electron-electron interactions. An understanding of such "strongly correlated" solids is perhaps the major unsolved problem of condensed matter physics with over ten thousand researchers working on this topic. High-Tc superconductors also have significant potential for applications in technologies ranging from electric power generation and transmission to digital electronics. This ability to carry large amounts of current can be applied to electric power devices such as motors and generators, and to electricity transmission in power lines. For example, superconductors can carry as much as 100 times the amount of electricity of ordinary copper or aluminum wires of the same size. Many universities, research institutes and companies are working to develop high-Tc superconductivity applications and considerable progress has been made. This new volume brings together new leading-edge research in the field.

The recently discovered high temperature superconductors (HTSCs) have been found, to be a very suitable class of materials, for the development of THz radiation sensors. They are easy to manufacture and operate, over the efficiently achievable,thermoelectric temperature control interval of, 80-140K. The temperature control over this range, can easily be manipulated by liquid nitrogen as a refrigerant. But the recently developed, low temperature thermoelectric refrigeration materials (e.g. Zr, Hf penatellurides), provide an efficient and a very convenient alternative. The THz, hot electron thermal sensor (HETS) technique, has now reached an advanced stage of development. Here the electrons are heated by a local femtosecond pulsed laser pump, above the lattice temperature. This copious supply of electrons, is then modulated by the chopped, remote THz radiation signal, which is to be detected. The chopped radiation, modulates the non equilibrium, electron-phonon thermal dynamics.

The associated response of the dynamics, of the electrical resistance of the superconductor, on a femtosecond time scale, is the key,to the THz radiation detection. This is facilitated by adding to the sensor an external heterodyne electronic system. Pulsed boost power of the laser pump, provides for the electrons, the AC (alternating current) bias. The DC (direct current) bias is carried out by, a fixed, I (current) through or V (Voltage) across the device. The sensor is operated, in the hot electron mode i.e. near the critical temperature Tc of the superconductor.The THz thermal sensors based on the ' hot electron phenomena',like any other thermal sensor, require cooling of the electrons.This may be an external, and or an internal (self) cooling. A convenient, thermoelectric method of cooling, for the HTSC sensors, is the aim in the development of the sensor.The heated electrons, due to Ac and DC bias, are above the lattice temperature. Hot electrons, are easy to manipulate, at the THz frequencies. The quasiparticle excitations, resulting from the electron-phonon thermaldynamics, provide the desired tool of sensing. These excitations have relaxation time of 10^{-12} - 10^{-15}s, in the HTSCs. The superconducting coherence length in HTSCs, is much shorter in comparison to that found in low temperature superconductors (LTSCs). This puts a much lower limit, on the workable dimensions in space,in the superconducting film, for use as an HTSC sensor.HTSC thermal sensing provides a tool, between temperatures 50-140K,for making a convenient class of THz band of radiation sensors. This window of the electromagnetic radiation spectrum (10^{-5} m to 10^{-3} m wavelength), provides a faster and yet invisible light,for applications, in communication, astronomy and medicine. In order for the electrons, to be an efficient means of sensing the THz radiation, they must loose heat, at a very fast rate.This should happen as close as possible to the point of incidence, of the incident radiation. This means electrons need to travel as short a distance, as possible, in the sensor, before being collected. The devices developed so far, have electron heat dissipation, mainly via the sensor film-substrate, system. In this system, the electron-phonon interactions, with the relaxation time 10^{-12}-10^{-15}s, lasting on a much wider space,control the ultimate speed of the sensor. The phonons are an obstruction to the fast movement of the electrons, away from the sensitive areas of the sensor. They originate, in the sensor film material, and also as a result of the back flow from the substrate. This is after the sensitive area has been heated by the laser beam. The DC boas also adds its share of heating to the sensor,raising the average temperature of the sensor, above that of the substrate.

It is an easy way out, but costly and cumbersome, to use mK range of temperatures, to reduce electron-phonon interactions. The much reduced phonon specific heat, at mK temperatures, provides the much desired environment, for the easy transport of the electrons, but with the added complexity of a cryogenics infrastructure. One can produce scatter free travel, of the electrons in the film, in various alternate ways. One possible way would be, to design, a regiment of superconducting dots, on a low dimensional structure,thermoelectric film. The interface of the, HTSC-thermoelectric junctions, so produced, can be used, as an efficient source of highly mobile electrons.The output can then be collected, as an integrated signal from the dots. May be one can sandwich, an insulating layer, in between, the superconducting and thermoelectric layers, to achieve tunneling of the electrons, with minimum back flow of the heat. The materials considered highly efficient, in thermal sensing, at present, arc the low temperature superconductors (LTSCs).Typical examples are Al, Pb, Nbn, etc. and are operated at mK temperatures.The specific heat of the phonons at mK temperatures, is very low ~ 10^{2}J/K/m^{3}. The specific heat of the electrons may be, slightly higher or lower than that of the phonons. At these low temperatures, electrons travel almost free of interactions with the phonons. In the case of HTSCs, e.g. YBaCuO, BiSrCaCuO, etc., the scenario is quite different. Here, the specific heat of the electrons is high ~ 10^{4}J/K/m^{3}. The specific heat of the phonons is much higher ~10^{6}J/K/m^{3}. The phonons, play a very deleterious role, in the performance of an HTSC thermal sensor. A new ray of hope, in the direction of efficient electron cooling, with minimum involvement of phonons, has now emerged. It is by virtue of the nano structure, thermoelectric cooling, materials and devices, operating between 50 and 200K. Band gap engineering using nano structures, can channel much faster electrontransport in low dimensions. Integrated cooling, can be provided by the already well known thermoelectric cooling materials e.g. Bi-Sb-Te compounds, or the recently developed Zr,Ti penta-tellurides. The low dimensional materials and structures, under investigation world wide, now form a suitable class of materials, for the required thermoelectric cooling. They have the capability, to bridge the cooling gap, 50 to 200K. These new materials, have a strong potential, for developing self (Peltier) cooling devices, for an efficient operation of the HTSC thermal sensors. It is the purpose of Chapter 1 to bring home, the importance of the HTSCs, as a suitable class of thermal sensing materials, with integrated electronic cooling, as compared to the LTSCs. In the case of the HTSCs, it is convenient to use, liquid nitrogen as a refrigerant. A thermoelectric or a combination with liquid nitrogen, if required, as the cooling technique, would also be an achievement.Thermoelectric cooling, is very economical in space, and does not involve, moving parts, thus very quiet in operation. A simple current manipulation of the devices, provides the desired temperature control.

As discussed in Chapter 2, acoustic emission is widely known as a nondestructive method for investigating the dislocation movement and accumulation accompanying plastic deformation as well as the generation and propagation of cracks in solid state materials subjected to mechanical stress. Other extensively studied sources of acoustic emission include martensitic phase transitions in metals and alloys under thermal ramping and martensitic-like structural phase transitions in ferroelectric and ferroelastic materials under both thermally and electric field induced stresses. During the last decade, the acoustic emission method has been successfully applied to studying the physical properties of high-T_c superconductors under variable temperature, electric current and external magnetic field conditions. The most important issues emphasized in the present review are: (i) superconducting and structural phase transitions in a wide temperature range, (ii) kinetics of superconducting ceramics sintering and oxygenation, (iii) dislocation mechanisms of mechanical work hardening during long term thermal cycling and (iv) magnetic flux penetration into the superconductor and flux lines pinning and interaction. Most of the results have been obtained with YBCO ($YBa_2Cu_3O_x$) ceramics, yet some properties of BISCCO ($Bi_2Sr_2CaCu_2O_x$) high-T_c superconducting composite tapes have been also addressed. The authors show that by monitoring the acoustic emission bursts it is possible to measure the temperature hysteresis of phase transitions and to reveal their order, to determine the temperature of maximal oxygen absorption (and calculate the absorption kinetic coefficient) as well as to measure the lower critical magnetic field H_{c1} and the full penetration field H^* under electrical current transport. The cumulative results demonstrate that acoustic emission method is an indispensable tool for studying the high-T_c superconducting phenomena.

In Chapter 3, the authors give a general description of their approach which explains many physical properties in the superconducting and normal states of almost all 2D high T_c superconductors (HTSC). This 2D character leads to the existence of Van Hove singularities (VHs) or saddle points in the band structure of these compounds. The presence of VHs near the Fermi level in HTSC is now well established. They review some physical properties of these materials which can be explained by this scenario, in particular: the critical temperature

T_c, the anomalous isotope effect, the superconducting gap and its anisotropy, and thermodynamic and transport properties (eg: Hall effect). The effects of doping and temperature are also studied, and they are directly dependent of the position of the Fermi level relative to the VHs position. They show that these compounds present a topological transition for a critical hole doping p ≈ 0.21 hole per CuO_2 plane. Most of these compounds are disordered metals in the normal state, they think that the Coulomb repulsion is responsible for the loss of electronic states at the Fermi level, leading to a dip, or the so-called "pseudo-gap".

Melt textured $YBa_2Cu_3O_{7-\delta}$ superconductor has been widely used in the field of high temperature superconductor (HTS) Maglev technology, such as the flywheel energy storage system and the transportation system. The induced (shielded) current may flow at large density without loss, circulating in large single-grained superconductors. So they can be used as permanent magnet, but with much higher magnetic fields. However, before good engineering designs for these applications can be derived, a deeper understanding of the magnetic behavior of YBCO superconductor must be obtained. Therefore, the studies on the electromagnetic properties of HTS YBCO bulks are reported for Maglev technology in Chapter 4. Both experimental and computational results have been discussed in terms of Electromagnetic Properties of Bulk High Temperature Superconductor for HTS Maglev Technology. It was found that not only growth sector boundaries (GSB) between the five growth sectors (GS) but also superconduction property variations in these growth sectors contribute to inhomogeneities of bulk YBCO. Experiments were designed to investigate the macroscopic anisotropy of magnetization critical current density of bulk YBCO. While the field is kept constant at 1.0 T, the ratio increases as the temperature decreasing from 85 K to 20 K. Although levitation force has linear relationship with the applied field in the case of symmetrical, such a linear relationship disappears once the applied field becomes unsymmetrical. However, levitation stiffness has linear relationship with the associated levitation force, whether the applied field is symmetrical or unsymmetrical. The multiple seeded melt growth (MSMG) bulk has grain boundary (GB), but it still can be regarded as single larger grain bulk in the perpendicular mode due to the inter-grain critical currents flowing across GBs, and it has much larger levitation force than the stacked bulk array. During the lateral movement, the decay of levitation force is dependent on both the maximum lateral displacement and the movement cycle times, while the guidance force hysteresis curve does not change after the first cycle. Moreover, A variational approach was presented for the studies on the field dependence of the critical current density in YBCO Superconductor. When the anisotropy ratio into account in the HTS computation modelling, the calculated levitation forces between superconductor and magnet agree with the experimental ones. This work may be helpful to the system optimization and may provide scientific analysis for the HTS Maglev system design.

Chapter 5 addresses five important issues:

1. **Anomalous transport characteristics of high temperature superconductors and Josephson currents**

   The electric currents of superconductor and electrical field are relation of direct proportion; the currents and magnetic field are relation of inverse ratio. In a special condition, the Josephon currents has anomalous characteristic.

2. **Thermodynamic properties of high temperature superconductor**

   A new systematic calculation of the specific heat contributions of vortex liquids and solids is presented. Three derivatives of the free energy with respect to the temperature of superconductor, the entropy, the specific heat, the temperature of superconductor derivative of the specific heat are continuous across the phase transition.

3. **The study of characteristics of superconductive rings**

   The current of superconductive rings is change with jump in theory. The magnetic field of superconductive rings is quantization. If increasing magnetic field, the order parameter is gradually decreasing, leads to a decrease of the size of the jump of the flux in the vorticity. In a special condition, if the outer magnetic field is gathering, the sign of supercurrent can reversal.

4. **Study of thermodynamic properties of type I superconductive films**

   The specific heat of the type I superconductive films

$$C_V = 2a_0^3 \left\{ 5\left(ba_0 - \frac{2k_B\rho_0^{3/2}}{3L_0} \right)T^3 + 6\left(1 - 2ba_0T_{c0} + \frac{k_BT_{C0}\rho_0^{3/2}}{L_0} \right)T^2 \right.$$

$$\left. -9\left(T_{C0} - ba_0T_{C0}^2 + \frac{k_BT_{C0}^2\rho_0^{3/2}}{3L_0} \right)T + 3T_{C0}^2 - 2ba_0T_{C0}^3 + \frac{k_BT_{C0}^3\rho_0^{3/2}}{3L_0} \right\}.$$

5. **Study of high temperature superconductor under pressure**

   When outer pressure is a constant on superconductor, the pressure intensity with the temperature is the relation of quadratic curve. The temperature is increasing with the pressure intensity. When outer pressure on superconductor is not a constant, the external pressure intensity has a relation of partial differential equation with the temperature of superconductivity. As increasing the external pressure intensity, the temperature is rising. The critical temperature is decreasing quasi-linearly with applied hydrostatic pressure for superconductor, and observed negative pressure coefficient of the critical temperature of superconductor. In another special case, the authors obtain the critical temperature increases quasi-linearly with applied pressure on superconductor.

In chapter 6 an attempt has been taken to study the variation of positron annihilation parameters, specially those which are probing the electron momentum distributions, due to superconducting transition in three different high T_c superconducting oxides (single crystalline $Bi_2Sr_2CaCu_2O_{8+\delta}$, single crystalline $SmBa_2Cu_3O_{7+x}$ and polycrystalline $La_{0.7}Y_{0.3}Ca_{0.5}Ba_{1.5}Cu_3O_z$) and also to identify the core electrons with which positrons are annihilating in these cuprate HTSC systems. This will help to understand the reasons of the variation of positron annihilation parameters due to superconducting transition in these HTSC systems in a better way. The anisotropy of the EMD in different crystallographic orientations in the layered structured HTSC system has also been studied by using the positron annihilation technique. The two detector coincidence Doppler broadening of the electron positron annihilation radiation (CDBEPAR) measurement, having peak to background ratio better than 14000 : 1, have been used to study the temperature dependent (300 K to 30 K) electron momentum distributions in these high T_c superconducting oxides. The CDBEPAR data are analysed both by conventional lineshape analysis and the ratio curve analysis.

It is well known that the system of Cooper's pair is described by boson symmetric wave functions, but Cooper's pair operators are bosons only when the moments **k** are different and they are fermions for equal **k**. The analysis of trilinear commutation relations for the Cooper pair (pairon) operators reveals that they correspond to the modified parafermi statistics of rank p = 1. Two different expressions for the Cooper pair number operator are presented in Chapter 7. The authors demonstrate that the calculations with a Hamiltonian expressed via pairon operators is more convenient using the commutation properties of these operators without presenting them as a product of fermion operators. This allows to study problems in which the interactions between Cooper's pairs are also included. The problem with two interacting Cooper's pairs is resolved and its generalization in the case of large systems is discussed. It is shown that in site representation, the hole-pair operators obey the same commutation relations (paulion) as the Cooper pair operators in impulse representation, although the latter describe delocalized quasiparticles. In quasi-impulse representation, the hole-pair operators are also delocalized and their exact commutation relations correspond to a modified parafermi statistics of rankM (M is the number of sites in a "superlattice" formed by the centers of mass of each hole pair). From this follows that one state can be occupied by up to M pairs. Even in the absence of dynamic interaction, the system of hole pairs is characterized by some immanent interaction, named after Dyson as kinematic interaction. This interaction appears because of the deviation of the quasiparticle statistics from the Bose (Fermi) statistics and its magnitude depends on the concentration of hole pairs. In spite of the non-bosonic behavior, there is no statistical prohibition on the Bose-Einstein condensation of coupled hole pairs.

Recent scanning tunneling microscope (STM) experiments on Bi2212 have shed new light on the nature of superconducting state in high-Tc cuprates and have emphasized the important role played by inhomogeneities of superconductivity and energy gap in the CuO_2 plane of the high-Tc cuprates. Summarizing all related observations, they find that there are nine features altogether for the inhomogeneities. Chapter 8 demonstrates that the thermal perturbation leads to the fluctuation of antiferromagnetic short-range coherence length (AFSRCL) in the CuO_2 plane, and further leads to the fluctuation of pairing potential. The latter can cause the inhomogeneities of the gap and the superconductivity. This chapter gives a unified explanation for the nine features of the inhomogeneities. The physical picture of the inhomogeneities of superconductivity and gap in the CuO2 plane is as follows. The values of the gap and the critical temperature Tc in bulk measurements are determined by the most probable value of AFSRCL. At $T = T_c$, a superconducting percolation channel is established by the locations with the most probable AFSRCL and the locations with AFSRCL larger than the most probable one. The proximity effect and pair tunneling effect exist in the locations with lower values of Tc. However, both effects are not important for the inhomogeneities. The authors think that the mobile $O_{p\sigma}$ holes in the CuO_2 plane are of homogeneous distribution. The gap and the superconductivity themselves are stable, and the stability does not need the help of nodal Cooper pair. This chapter also reconciles Lang *et al.*'s experimental observations with the basic concept of superconductivity.

A General Theory of Superconductivity with points of view differing from those of the BCS Theory is presented. In Chapter 9 The formation of electron pairs in a conductor material is investigated upon arriving to the critical temperature where the conductor-superconductor transition occurs. A general equation for the superconductivity is obtained based on the stable pairing of two electrons bound by a phonon for any type of superconductor material. This equation comes from a self-consistent field calculation with a screening, which is temperature dependent, showing that the total energy of the electron pairs is constant and the local energy of the paired electrons is equal to that of the phonon in the range 0 to T_C. A specific condition for the existence of the superconducting state is established, allowing the prediction of the critical temperature. The dispersion law of the elementary excitements produced by the superconductivity is obtained and correctly interpreted. The method is based on represent to the operators of Bose that characterize to phonons and to the electron-phonon interaction as a combination of products of Fermi operators corresponding to the electrons that form the pairs. The expression obtained for the critical temperature is compatible with those obtained by G.M. Eliashberg and W.L. McMillan. An expression for the bond energy of the pairs, or better known as superconductor gap, is also obtained as a function of the temperature and the critical temperature, resulting very similar to that formulated by Buckingham. This theory is reached in the frame of self-consistent field equations for any natural or artificial solid where free electrons exist. The necessity of the electrons must be coupled by phonons for the existence of the superconducting state is also justified, arriving to a general conclusion: the superconductivity theory is based only on the theory used to carry out the electron-phonon interaction and more concretely of the phonons (harmonic or anharmonic theory, low, intermediate and high temperature). The theory is applied to the particular case of low temperature superconductors, obtaining an excellent agreement with the results of other theories (phenomenological and microscopic) as well as with experimental data. An application of the general equation obtained for low critical temperature superconductors utilizing a phononic theory is developed. Then, the authors arrive to a specific expression for the bounding energy as a function of temperature. The density of states of the electron pairs is calculated and used to obtain an equation for the critical magnetic field. This result is needed to determine the electrodynamical properties. Finally, they obtain the specific heat as a function of temperature, they compare it to experimental data for Sn, and they calculate its jump at T_C for eight superconductors. The authors have also determined the variation of the energy gap or bond energy with the temperature of the MgB_2 superconductor and they have compared the results with another theoretical and experimental results reported in the literature, obtaining an excellent agreement with the experimental results.

In: New Topics in Superconductivity Research
Editor: Barry P. Martins, pp. 1-44

ISBN: 1-59454-985-0
© 2006 Nova Science Publishers, Inc.

Chapter 1

HOT ELECTRON NON EQUILIBRIUM HIGH TEMPERATURE SUPERCONDUCTOR THZ RADIATION SENSING AND THE INTEGRATED ELECTRON COOLING

***M.M. Kaila*[*]**

School of Physics, University of New South Wales, Sydney, NSW 2052, Australia

..curiosity is the mother of all inventions .
....necessity is the mother of all battles...

It is the curiosity, that creates the urge in humans,
to do the impossible, e.g. to go to the Mars.
This my be a small step, for a few individuals,
who are privileged to be involved to do the task.
But a giant step, by the mankind, at large.
They are prepared to pay the price, nevertheless.

Dr. Madan M. Kaila

Abstract

The recently discovered high temperature superconductors (HTSCs) have been found, to be a very suitable class of materials, for the development of THz radiation sensors. They are easy to manufacture and operate, over the efficiently achievable,thermoelectric temperature control interval of, 80-140K. The temperature control over this range, can easily be manipulated by liquid nitrogen as a refrigerant. But the recently developed, low temperature thermoelectric refrigeration materials (e.g. Zr, Hf penatellurides), provide an efficient and a very convenient alternative. The THz, hot electron thermal sensor (HETS) technique, has now reached an advanced stage of development. Here the electrons are heated by a local femtosecond pulsed laser pump, above the lattice temperature. This copious supply of electrons, is then modulated by the chopped, remote THz radiation signal, which is to be

[*] E-mail address : m.kaila@unsw.edu.au, Ph: 612-93854561, Fx .: 612-93856060

detected. The chopped radiation, modulates the non equilibrium, electron-phonon thermal dynamics.

The associated response of the dynamics, of the electrical resistance of the superconductor, on a femtosecond time scale, is the key,to the THz radiation detection. This is facilitated by adding to the sensor an external heterodyne electronic system. Pulsed boost power of the laser pump, provides for the electrons, the AC (alternating current) bias. The DC (direct current) bias is carried out by, a fixed, I (current) through or V (Voltage) across the device. The sensor is operated, in the hot electron mode i.e. near the critical temperature Tc of the superconductor.The THz thermal sensors based on the ' hot electron phenomena',like any other thermal sensor, require cooling of the electrons.This may be an external, and or an internal (self) cooling. A convenient, thermoelectric method of cooling, for the HTSC sensors, is the aim in the development of the sensor.The heated electrons, due to Ac and DC bias, are above the lattice temperature. Hot electrons, are easy to manipulate, at the THz frequencies. The quasiparticle excitations, resulting from the electron-phonon thermaldynamics, provide the desired tool of sensing. These excitations have relaxation time of 10^{-12} - 10^{-15}s, in the HTSCs. The superconducting coherence length in HTSCs, is much shorter in comparison to that found in low temperature superconductors (LTSCs). This puts a much lower limit, on the workable dimensions in space,in the superconducting film, for use as an HTSC sensor.HTSC thermal sensing provides a tool, between temperatures 50-140K,for making a convenient class of THz band of radiation sensors. This window of the electromagnetic radiation spectrum (10^{-5} m to 10^{-3} m wavelength), provides a faster and yet invisible light,for applications, in communication, astronomy and medicine. In order for the electrons, to be an efficient means of sensing the THz radiation, they must loose heat, at a very fast rate.This should happen as close as possible to the point of incidence, of the incident radiation. This means electrons need to travel as short a distance, as possible, in the sensor, before being collected. The devices developed so far, have electron heat dissipation, mainly via the sensor film-substrate, system. In this system, the electron-phonon interactions, with the relaxation time 10^{-12}-10^{-15}s, lasting on a much wider space,control the ultimate speed of the sensor. The phonons are an obstruction to the fast movement of the electrons, away from the sensitive areas of the sensor. They originate, in the sensor film material, and also as a result of the back flow from the substrate. This is after the sensitive area has been heated by the laser beam. The DC boas also adds its share of heating to the sensor,raising the average temperature of the sensor, above that of the substrate.

It is an easy way out, but costly and cumbersome, to use mK range of temperatures, to reduce electron-phonon interactions. The much reduced phonon specific heat, at mK temperatures, provides the much desired environment, for the easy transport of the electrons, but with the added complexity of a cryogenics infrastructure. One can produce scatter free travel, of the electrons in the film, in various alternate ways. One possible way would be, to design, a regiment of superconducting dots, on a low dimensional structure,thermoelectric film. The interface of the, HTSC-thermoelectric junctions, so produced, can be used, as an efficient source of highly mobile electrons.The output can then be collected, as an integrated signal from the dots. May be one can sandwich, an insulating layer, in between, the superconducting and thermoelectric layers, to achieve tunneling of the electrons, with minimum back flow of the heat. The materials considered highly efficient, in thermal sensing, at present, are the low temperature superconductors (LTSCs).Typical examples are Al, Pb, Nbn, etc. and are operated at mK temperatures.The specific heat of the phonons at mK temperatures, is very low ~ 10^{2}J/K/m^{3}. The specific heat of the electrons may be, slightly higher or lower than that of the phonons. At these low temperatures, electrons travel almost free of interactions with the phonons.

In the case of HTSCs, e.g. YBaCuO, BiSrCaCuO, etc., the scenario is quite different. Here, the specific heat of the electrons is high ~ 10^{4}J/K/m^{3}. The specific heat of the phonons is much higher ~10^{6}J/K/m^{3}. The phonons, play a very deleterious role, in the performance of an HTSC thermal sensor. A new ray of hope, in the direction of efficient electron cooling, with minimum involvement of phonons, has now emerged. It is by virtue of the nano structure, thermoelectric cooling, materials and devices, operating between 50 and 200K. Band gap engineering using nano structures, can channel much faster electrontransport in low dimensions. Integrated cooling, can be provided by the already well known thermoelectric cooling materials e.g. Bi-Sb-Te compounds, or the recently developed Zr,Ti penta-tellurides. The low dimensional materials and structures, under investigation world wide, now form a suitable class of materials, for the required thermoelectric cooling. They have the capability, to bridge the cooling gap, 50 to 200K. These new materials, have a strong potential, for developing self (Peltier) cooling devices, for an efficient operation of the HTSC thermal sensors.

It is the purpose of this work to bring home, the importance of the HTSCs, as a suitable class of thermal sensing materials, with integrated electronic cooling, as compared to the LTSCs. In the case of the HTSCs, it is convenient to use, liquid nitrogen as a refrigerant. A thermoelectric or a combination with liquid nitrogen, if required, as the cooling technique, would also be an achievement.Thermoelectric cooling, is very economical in space, and does not involve, moving parts, thus very quiet in operation. A simple current manipulation of the devices, provides the desired temperature control.

Foreword

It is my intention, in this work, to provide the reader, with a broad base of physics, to appraise oneself, with the present status, in the area of non equilibrium hot electron thermal sensing and electronic cooling.The non equilibrium electron-phonon thermal dynamics, in the sensor film, is the result of the power boost by a femtosecond laser pump. In a hot electron operation, an HTSC is operated close to its critical temperature Tc (transition from superconductor to normal state), for developing thermal sensors. I have included, in this work, a comparison of the HTSC and the LTSC materials. But the work, is particularly aimed at the HTSC sensors. It is the efficient cooling of the sensor, that is a necessary step,in the overall performance of the sensor. One can today, develop, a low dimensional materials based, THz (10^{12}, cycles/second) radiation thermal sensor, with an integrated electronic cooling. This work, creates the direction, in the knowledge required,so as to enable the application, of the developed low temperature bulk thermoelectric cooling materials, to the HTSCs thin film thermal sensors.

Room temperature bulk thermoelectric cooling materials, were developed during the sixties and seventies. There has been little development since then. One can see today, a rapid surge in research and development efforts, towards the low dimensional (nano structure) thermoelectrics. There is now an urgent demand, to create,

a conveniently operated, low temperature electronic refrigeration system, for the hot electron thermal sensor (HETS).All thermal sensors basically, have to be cooled. This is to keep their noise equivalent power, NEP (the minimum power of a signal that can be detected, below which it is dominated by the noise), very low, and the responsively (voltage or current output / power of the input signal) very high. A sensor based on an LTSC (e.g. Al, NbN, etc), enjoys a very good performance. But the price paid, in terms of cryogenics, is very high. They need to be operated at mK temperatures. Semiconductor Ge, has equivalent performance, but requires close to 1K, as the operating temperature.

The III-V strain tuned, super lattice semiconductors, HgCdTe, InGaAs, have limited performance capabilities. This is due to their narrow band gap, resulting in easy generation of thermal noise.But the conveniently operated, thermoelectric temperature control, over the range, 300K and 200K, makes them commercially, very popular at present.Semiconductors, in general, due to their band gap restriction, are not suitable for the THz band of radiation. HTSC thermal sensors operate, over the achievable electronic cooling range of 50-140K. The new materials (Zr,Hf pentatellurides) are a suitable candidate, for developing, HTSC-integrated or otherwise, electronic cooling devices, over this range. The liquid helium, mechanical cryo-coolers, etc. are an option to use milli Kelvin temperatures. But they require a large infrastructure and involve noisy equipment. Thermoelectric cooling, is a solid-state phenomena. It offers the technology, with potential of developing, miniaturized cooling devices. Micro, Peltier coolers, can be in integrated with the sensors. These coolers can be designed, using quantum wells, wires and dots. LTSC materials, have a much sharper resistive transition, as compared to the HTSC materials, and thus are much faster in operation. But the added complication of the cryogenics involved, makes them less attractive.

The possible extension of thermnoelectric cooling down to 50K, using the old (Bi-Sb-Te compounds) and or the new (Zr, Hf pentatellurides) materials, using low dimensional designs, is challenging, as much as tempting. Electrons used, in the hot electron thermal sensing technique,in HTSC materials, are derived, from the top end, of the superconductor- normal state transition, close to the critical temperature Tc.They require, a very small excitation (nW-μW) power and are capable of performing same function as a normal electron does, in a semiconductor. Their ability to revert back to the cold state (cooper pairs), makes them more efficient and less susceptible, to thermal noise. Thus they are more suitable as thermal sensors.Some information, in preparing this work, has been used from other sources. The author gratefully acknowledges those sources.Information, in some cases, is reproduced, in the appendices, at the end, as an illustration. I express my highest gratitude, to those origins of information. This work, I am sure would be very valuable to young scientists, who wish to pursue research and development career, in he area of thermal sensors. It will be equally valuable, I am sure, to the scientific community at large.

I   Introduction

The semiconductors so far, e.g. GaAS[1], have provided, as a suitable class of materials, for the development of the THz radiation sources and sensors. But over the last decade, the attention has shifted to the HTSCs, as a better alternative[2-5]. In order to be able to make the THz radiation sources or sensors, using HTSCs, it is essential to understand first, the basics of the physical processes involved, i.e., the electron and lattice (phonon) interactions, band gap, quasi particle excitations, etc., etc[Appendix A1]. There results, a rapid increase in the electrical resistance, of a superconducting thin film, when it is irradiated by a femtosecond,optical or thermal laser pulses. A suitable choice of current or voltage bias, can be used to operate the superconductor, near the upper end of the resistive transition (superconductor to normal) width, in a hot electron mixerarrangement[Appendix A2. A3]. When the superconductor is biased, to operate, at a temperature, much below the critical temperature, the sensing is in a different mode of operation. In that case, the excitation of the supercurrents (circulating electrical currents over small areas) are used for sensing.

These currents can be externally initiated in the superconductor, by a momentary application, of a small magnetic field, and then removed. What results is a mixed (superconducting electrons and normal electrons) state. This is the familiar Meissner effect state. A cooper pair, is a bound state of two electrons, with zero spin and momentum. Large number of these pairs, constitute, the supercurrents. Some paired electrons, can be made to oscillate, between free and bound state, by cyclic heating, of the pairs. The result is a in an oscillatory electric filed, being generated in the film. The superconductor in this way, can be so devised, as to have, a beam of THz electromagnetic waves emanating, from the back of, a suitable thin film superconductor/substrate system[2]. In this case, the electron-phonon thermal dynamic processes happening, in an HTSC, can be read, by means, other than the usual, I-V heterodyne electronic read out system.This is the time Domain Terahertz Transmission Spectroscopy (TDTTS) technique [3]. Here the transmitted beam through the superconductor, provides a convenient tool to study the electron-phonon thermal dynamics.This transmission, if modulated by the signatures of an unknown THz signal, can be unscrambled, by a video technique[Appendix A4, A5].

The TDRTS (time domain reflection THz spectroscopy), uses reflection from the superconductor surface instead, and has been found equally valuable[4].Thus one can use, the electron-phonon dynamics, via the reflected beam, for thermal imaging. The readout may be through a low temperature grown Gallium Arsenide sensor or an another HTSC sensor. Other useful superconducting properties of the HTSC thin films, e.g., the penetration of the electromagnetic waves through the surface, which leads to a change its surface resistance, the kinetic inductance (resulting from the oscillations of the super current carriers), etc., etc., can be used to design a THz radiation sensor. The hot electron phenomena, has already been extensively and successfully used, in the design of THz thermal sensors. In the hot electron technique, the heated electrons, being more mobile, can be manipulated easily, for a desired operation.High performance THz sensors, have been produced, among the HTSC materials[5]. At present the LTSCs e.g. Nbn (response time ~ ps, responsivity~ 10^4 V/W) are the materials, considered up to the mark, for designing fast and sensitive, hot electron thermal sensors. An efficient LTSC nano structure thermal sensor, with integrated

Peltier cooling has already been developed. This is the **S(Superconductor) I(Insulator) N(Normal Metal) I S (SINIS)** pair junction, operated at 300mK

One can effectively realize, the self thermoelectric (Peltier) cooling, of the metal electrode, which is, the sensing element. This is achieved, provided one uses, a bias voltage across the sensor, close to the band gap energy, of the superconductor.A temperature drop from 300mK to 100 mK, was easy to produce[6].Peltier cooling is only current controlled. This simple factor makes it an ideal technique, of temperature control, particularly, when a small wattage of heat, and a small area is involved. The nano-engineering design, of the thermoelectric cooling materials, is now gaining a fast momentum world wide.

It is anticipated, this will lead to the development of fast and efficient, much needed cooling devices. The low dimensional space, could be, the quantum wells, where the space is a two dimensional plane in which electrons and phonons move. The other structures are, quantum wires, the one dimensional structure and the quantum dots, providing the zero dimensional space. Electronic cooling, using semiconductors, so far has provided a very beneficial technique for the range of temperatures, 300K-200K. The materials developed for this purpose were, Bi_2Te_3 - Sb_2Te_3 alloys. Commercial cooling devices based on these materials have found applications in medicine, infrared sensors, etc.

The efficiency of a thermoelectric material, cooling or generation, is measured, through the coefficient of performance, called as the thermoelectric figure of merit (**TEFM**).In its simplicity, one can writ, for a single material leg, the **TEFM,** as **ZT = (α^2 σ / K)**. Here **Z** is the dimensional figure of merit, **ZT** the dimensional less figure of merit, **T** the temperature in degrees Kelvin, α the thermoelectric power (Seebeck coefficient) and σ and $\mathbf{K}$ are the electrical and thermal conductivity respectively.of the leg. The thermal conductivity, $\mathbf{K = K_e}$ $\mathbf{+ K_{ph}}$, where $\mathbf{K_e}$ and $\mathbf{K_{ph}}$, are respectively the electronic and thermal (phonon) parts of the thermal conductivity. In any material suitable for applications, one need to optimize, α, σ and $\mathbf{K}$. in such a way that $\mathbf{ZT >>1}$.It is now possible, to engineer, the phonon-electron propagation through a material, so as to achieve the highest **TEFM**. Recently, there have been extensive studies for the design of thermoelectric nano structures. These studies have been performed, towards the materials previously well known,in the bulk form[Appendices B1-B4]. It is due to the easy access, to the experimental data available on those materials. These studies can equally well be applied to the Zr, Hf pentatelurides[Appendix B5].

The typical low dimensional structures studied are, quantum wells, PbTe/PbEuTe[7, Appendix B6, B7], quantum wires, PbTe, GaAs[8], Bi_2Te_3[9, Appendix B8], Bi[10, Appendix B9], quantum dots, Ge on Si[11, Appendix B10]. Detailed experimental studies on these new devices, have yet to be performed. The new class of thermoelectric materials, e.g. the quasi-crystals $A_lP_dM_n$[12], the Skutterudites C_oAs_3, etc[13], etc., may develop into efficient thermal sensors, when designed as low dimensional structures. The penta-tellurides $H_f,Z_r(Ti)T_e(Se)$, on the other hand, need special mention, due to their efficient performanceas cooling materials, over the temperature interval 200-50K[14]. One can engineer quantum wells, wires, etc. of the Zr/Hf pentatellurides, and develop thermoelectric cooling devices, with an efficient operation. An HTSC/ Zr/HfTe$_5$/HTSC, heterostructure, could be developed as a self cooling thermal sensor. The non equilibrium dynamic superconductivity in HTSCs, has electron-phonon relaxation times of 10^{-12}-10^{-15}s. The HTSCs and the new thermoelectric materials, have a strong potential to develop THz radiation sensors with integrated cooling[15].

It has been recently demonstrated, that one can use femtosecond laser pulses to excite thermally, the bound antiferromagnetic (anti parallel) spins in a rare earth, e.g. theorthoferrite TmFeO$_3$. The iron moments order antiferromagnetically, but with a small canting of the spins on different sublattices. This small anisotropy is very sensitive to temperature. It can oscillate, through a maximum and minimum, over the temperature interval, 80-90K, in a picosecond[16].Spin oscillation phenomena, in conjunction with an HTSC,can be an incentive for developing, a THz radiation sensor. The study of dynamic superconducting-antiferromagnetic interface, on the time scale[17], would provide a good base to understand, spin-lattice relaxations, in HTSCs.In the area of ultra high density magnetic recording media, the nano structures of the ferromagnetic materials[18, 19], present good candidates, for the study of optical and thermal phonon-spin relaxations. The study of antiferromagnetic-ferromagnetic phase in conjunction with HTSC materials, may open, another interesting area of exploration, in the field of THz radiation[20].

II Theory and Experiment

II A Hot Electron Thermal Sensor (HETS)

II A.1 Non Equilibrium Dynamic High Temperature Superconductivity : Theoretical Modeling

One should note that there is a critical difference between a superconducting Hot Electron Thermal Sensor and a conventional superconduicting bolometer.They are both operated in the transition-edge (superconductor to normal)regime. In the bolometer, thermal equilibrium between electrons and phonon is established instantly. In the **HETS**, these two systems are not in equilibrium. A two temperature, electron (t_e) and phonon (t_p)dynamics, is in operation. Electron-phonon interactions, in the case of [HTSCs[5], play a crucial role in thermal sensing.. This is in contrast to the LTSCs, where the electrons, have little involvement with the phonons. The sensors therefore perform much better. The author has recently carried out detailed modeling studies, among HTSCs, applied to the THz radiation sensing. I have particularly endeavored, to obtain responsivity and conversion gain, of the device, particularly, in the form of analytic, mathematical equations.

The present work, should provide the reader, with the platitude, of the problems involved, and the possible solutions, in the design of a THz thermal integrated sensing-cooling device. Modeling, in the performance studies of the THz sources and sensors in HTSCs, is a difficult task. This is basically, due to the lack of understanding and not enough experimental data available, on the properties of the materials involved. It is the right time now, to make nano HTSC- THz devices,with integrated electronic cooling, and test their performance. In developing the theoretical modeling, of an HTSC thermal sensor, I have used lower case letter, symbols, in the mathematical equations.This may be considered as contradictory, to the normal practice of the use of the capital letters. The author apologizes for that. But there is advantage in doing so when, one is using ' Mathmatica Computer Pacakage ' for the computations. In the ' Mathematica' program, capital letters are reserved for standard mathematical functions. Thus the use of lower case letters allows, easy back and forth movement, between the development of mathematics and carrying out the computations. One can write the heat dissipation in a thin film device, under the two temperature (electron and phonon), model in the form of two coupled linear differential equations, as follows[5, 21].

$$c_{el}\, v_s\, d\, t_e\, /\, dt = -\, l_{ep}\, v_s\, (t_e^{\,3} - t_p^{\,3)} + I\, v_b + a_{bs}\, p_{rad} \tag{1}$$

$$c_{ph}\, v_s\, dt_p\, /\, dt = l_{ep}\, v_s(t_e^{\,3} - t_p^{\,3)} - (s_{flm}\, /\, r_b)\, (t_p\text{-}t_s) - 8\, K_{ph}\, v_s\, (t_p\text{-}t_s)\, /\, l_b^{\,2} \tag{2}$$

The diffusion term, i.e. the third on the right hand side of the equation (2),is taken as negligible in this study. This is what happens, if the sensorlength l_b, is greater than, the diffusion length; this is the distance traveled by an electron before scattered by a phonon. A normal meander (length $>>$ l_b)sensor, would automatically satisfy this condition. In order to keep the length of the analytic equations, within limit, one should reduce the number of independent variables, in the equations. In the heat transfer by the phonons in the sensor, to the substrate, in the equation (2), for the second term on the RHS, I have replaced,t_s (substrate temperature) by the electron temperature t_e. This means, electrons and phonons maintain, a non equilibriumheat exchange, within the electron-phonon thermal dynamics.This condition can be considered equivalent to the case, where the sensor remains thermally isolated from the substrate. A thermally insulating layer e.g. Silicon Nitride (Si_3N_4)[22], Yttria stabilized Zirconia[23],etc., in between the sensor and the substrate, would enable to achieve the desired environment, for the electron transport. The meaning of various terms and symbols is as follows.

l_{ep} = $\gamma / (3\, t_{mes}\, t_s)$ = electron-phonon interaction (cooling) parameter (units $W / m^3\, K^3$)

γ =   Somerfield constant (units $W / m^3\, K^2$)

t_s =   the temperature of the substrate

k_{ph} =   electron phonon diffusion thermal conductivity

l_b =   length from the sensitive (middle of the bow tie antenna) area to the metal contacts at the end

c_{el} =   electron specific heat per unit volume

c_{ph} =   phonon specific heat per unit volume

t_{mes} =   phonon escape time to the substrate

a_{bs} =   the coupling factor between incident radiation and the detector

c_{el} =   electron specific heat per unit volume

c_{ph} =   phonon specific heat per unit volume

v_s =   volume of the sensor = $t_{flm}\, s_{flm}$

t_{flm} =   thickness of the sensor film

s_{flm} =   area of the sensor film

t_{mes} =   phonon escape time in to the substrate

r_b =   thermal boundary resistance = $t_{mes} / c_{ph} / t_{flm}$

l_{ep} =   electron-phonon coupling constant

t_e =   electron temperature = $t_{e0} + t_{e1}\, \exp\,(I\,\omega\,t_{minst})$

t_p =   phonon temperature = $t_{p0} + t_{p1}\, \exp\,(I\,\omega\,t_{minst})$

p_{rad} =   total (signal + local pump) = Dc power + AC power = $p_{rad0} + p_{rad1}\, \exp(I\,\omega\,t_{minst})$

v_b =   source voltage; = $v_{b0} + v_{b1}\, \exp\,(I\,\omega\,t_{minst})$

i =   total current bias through the bolometer = $i_0 + i_1\, \exp\,(I\,\omega\,t_{minst})$

ω_1 =   frequency of the fast elctro-optic switch (chopper) through which the remote THz signal passes

ω_2 =   frequency of the local laser boost pump

ω = $\omega_1 - \omega_2$ = the intermediate frequency (**IF**)

r_1 =   AC resistance of the load at IF = v_{b1} / i_1

r_0 =   dc resistance of the device (bolometer + load) = v_{b0} / i_0

Carrying out the solution of the equations (1) and (2) simultaneously,and retaining terms to the order of ω^4 only, one leads to the following analytical result, for the responsivity v_{rspf}[15].

$$v_{rspf} = \{\ \sqrt{[\{(c_{el}{}^2\, s_{flm}{}^2\, t_{e1}{}^2\, t_{flm}{}^{2)} + (c_{el}{}^2 s_{flm}{}^2 t_{e1}{}^2 t_{flm}{}^{2)}\ t_{mes}{}^2 \omega^2\}\omega^2\ ]}/(i_0\, p_{rad1})$$
$$/\sqrt{[(6(l_{ep}/c_{ph})\, t_{mes}\, t_{p0}{}^2 + 9\,(l_{ep}/c_{ph}){}^2\, t_{mes}{}^2\, t_{p0}{}^4 + (1 + t_{mes}{}^2\,\omega^2)]} \tag{3}$$

The conversion gain g_{cnvrsn}, in the THz dynamic case, is evaluated from the equation (3) by using, the following expression[15].

$$g_{cnvrsn} = (1/10)Log_{10}[\ 2\ (v_{rspf}^{\ 2}\ p_{lol}\ i_{1)}\ /\ v_{b1}\ ] \tag{4}$$

The resulting expression is long, and it would be out of space to reproduce here. I feel, the simplifications below, under the extreme conditions, $\omega t_{mes} \gg 1$ and $\omega t_{mes} \ll 1$, for the responsivity, will provide the reader, with an interesting overview, of the importance of the various processes involved.

In the approximation, $t_{mes}\ \omega \ll 1$, equation (3) would reduce to

$$v_{rspf} = (c_{el}\ s_{flm}\ t_{flm}\ \omega)\ t_{el}\ /\ (i_0\ p_{rad1}) \tag{5}$$

In this simplification process, the first and second term in the denominator within the square brackets, have been taken as $\ll 1$, and thus neglected. Various outputs can be achieved by suitably choosing the material properties, c_{ph}, t_{mes}, etc., and the operating parameters, to a desired effect. The above simplified voltage responsivity, can be expressed as follows.

$$v_{rspf} = \{\ (USP)\ /\ (p_{rad1})\ \}\ [\ t_{el}\ /\ i_0\ ] \tag{6}$$

$$USP = c_{el}\ s_{flm}\ t_{flm}\ \omega \tag{7}$$

The equation (6), exposes, the physics of the THz sensing, in a much clearer manner. **USP** can be regarded as, useful sensing power, per unit increase, of the electron temperature, t_{el}. Change in the electron temperature alone, would then drive the sensing technique. This is exactly, one is trying to do, by taking an easier path, i.e. going to mK temperatures. A tacit assumption has been made, about the thermal resistance, between the electrons and the phonons, in the sensor film, in this model.It is realized, through the second term on the RHS, in the equation (2). Electron-phonon thermal resistance within the sensor film, is taken as the highest limit r_b; this is the thermal resistance (phonon-phonon)between the sensor and the substrate.

A suitable insulating layer over the substrate, makes the substrate relatively unimportant. In actual practice however, there will always be some heat transferred to the substrate.

One can consider, this as a limiting case, where heat transfer within the film, is just little bit better, than the heat transfer to the substrate. The following simplifications are included for a useful insight.

$$v_{rspf} = [\ c_{el}\ s_{flm}\ t_{el}\ t_{flm}\ t_{mes}\ \omega^2\ ]\ /\ (i_0\ p_{rad1})\ /\ [\sqrt(6(l_{ep}\ /\ c_{ph})\ t_{mes}\ t_{p0}^{\ 2} + t_{mes}^{\ 2}\ \omega^2)]$$

$$\omega t_{mes} \gg 1;\ 6(l_{ep}/c_{ph})t_{mes}t_{p0}^{\ 2)} \gg 1 \sim \omega t_{mes} \tag{8}$$

$$[\ c_{el}\ s_{flm}\ t_{el}\ t_{flm}\ \omega\ ]\ /\ (i_0\ p_{rad1)}$$

$$\omega t_{mes} \gg 1;\ 6(l_{ep}/c_{ph})t_{mes}t_{p0}^{\ 2)} \ll \omega t_{mes} \tag{9}$$

$$v_{rspf} = [\ c_{el}\, s_{flm}\, (t_{e1})\, t_{flm}\, \omega^2\, (\sqrt{}\ t_{mes})\ ]\ /$$
$$\sqrt{}\ [\ (6\ (l_{ep}/c_{ph)})\ ]\ /\ t_{p0}\ /\ (i_0\, p_{rad1})$$
$$\omega t_{mes} \gg 1;\ 6(l_{ep}/c_{ph})t_{mes}\, t_{p0}{}^{2)} \gg \omega t_{mes} \qquad (10)$$

$$v_{rspf} = [\ c_{el}\, s_{flm}\, t_{e1}\, t_{flm}\, \omega\ ]\ /\ (\sqrt{}t_{mes}\ /)$$
$$/\ [\sqrt{}(6(l_{ep}/c_{ph)})\ ]\ /\ (t_{p0}{}^{)}]\ /\ (i_0\, p_{rad1})$$
$$\omega t_{mes} \ll 1,\ 1;\ 6(l_{ep}/c_{ph})t_{mes}t_{p0}{}^{2)} \gg \omega t_{mes} \qquad (11)$$

The various parametric manipulations, of the responivity in the above equations, provides the designer, with a wide range of possibilities, those can be achieved.In the equation (11), one sees that, the responsivity is inversely proportional to $(\sqrt{}t_{mes})$. This is a character, particularly exhibited, in this chosen model. An improvement (a decrease)in t_{mes}, also improves the responsivity.This is just opposite to what is observed in a normal static superconductivity sensor i.e. a bolometer. In that situation, t_{mes}, is the response time of the device.It would be interesting to fabricate, a THz sensor, using one lobe of the bow tie antenna (normally used in THz source, sensor technology), as a thermoelectric material. This, will result in, addition of a Peltier heat term in the equation (1). This term will be $-\,I\,(\Pi)$. I is the current bias and Π the Peltier coefficient.

In order to be able to cope with the high speed of the response of the sensor, the thermoelectric materials should provide for the electrons an easiercooling route of transport[7-11]. This may just be possible, by using the nano fabrication technology. In the above analysis, it is found that the term $6(l_{ep}/c_{ph})t_{mes}\, t_{p0}{}^{2})$ plays a very crucial role. Solving this term for t_{mes}, by putting it $= 1$, taking $c_{ph} = 10^6$ J/K/m^3, $t_{p0} = 85$K, and $l_{ep} = 10^{10}$ W/m^{3}K^3(the case of a YBCO sensor), one gets $t_{mes} \sim 10^{-9}$s. The time of escape of the phonons to the substrate ($t_{mes} = r_b\, c_{ph}\, t_{flm}$)should be interpreted as the upper limit on the electron-phonon relaxation time.The lower limit, which will extend the frequency band, towards the higher THz frequencies, will come from, the quickest possible removal, of the electrons for conduction. One should realize that the thermal resistance $r_b \sim 5\times10^{-8}$K m^2 / W, between the YBaCuO sensor and the MgO substrate, should be higher than that, between the electron-phonon cooling process, within the film. If the speed of transmission of the electrons, to the readout system, is fast enough, then this model can be realized in practice. How fast the electrons are transported, is limited, by the low dimensional structure designed, through which the electrons move. In the limit, however, the speed of the sensor, is controlled by the speed of the read out system.

II Figures 1 and 2 (below), included here are from a more detailed study [15]. They are reproduced here for the sake of illustration.Figure 1, depicts, the variation of the conversion gain vs intermediate frequency (**IF**), and the electron specific heat. One can notice, that ωt_{mes} is between 1 and 10. Under this condition, the rate of change of the electron temperature, due to the incident radiation, is much faster, than, the rate of heat transfer, from electrons to the phonons in the film. One should note, the phonon specific heat, is much higher (by 10^2), than the electron specific heat. This results in a slower heat exchange between electrons and phonons,thus, a quicker saturation, in the conversion gain vs ω (Figure 1). Figure 2 is another interesting illustration. Here the conversion gain is plotted vs l_{ep} and c_{ph}. The higher the phonon specific heat, the greater the thermal mass of the phonons, the slower the heat exchange between the electrons and the phonons. One can see that the effect, starts around c_{ph}

~ 10^4 J/K/m^3. From here on, the conversion gain is independent of $\mathbf{l_{ep}}$ (Figure 2). Thus electron-phonon cooling ($\mathbf{l_{ep}}$), in an HTSC sensor, is much less effective, than in an LTSC. Not much change in conversion gain is observed for $\mathbf{c_{ph}} > 10^4$ J/K/m^3.One should compare this $\mathbf{c_{ph}}$, with that for an LTSC i.e., 10^2 J/K/m^3.The theory of the thermal sensing remains practically the same, whether it is the LTSC or the HTSC. It is the magnitude of the specific heats of the electrons and the phonons in these materials, which makes them so different.

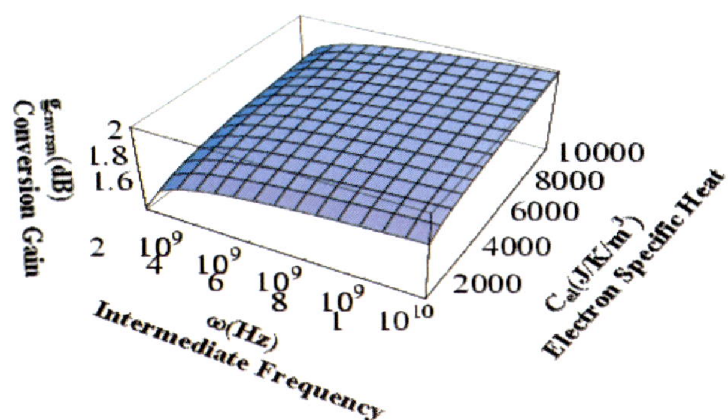

II A Figure 1. Plot of Conversion Gain $\mathbf{g_{cnvrsn}}$ (dB) vs, Intermediate Frequency ω (Hz) and Electron Specific Heat C_{el} (J/K/m^3). The Fixed parameters are $\mathbf{C_{ph}} = 1\text{x}10^6$ J/K/m^3, $\mathbf{l_{ep}} = 1\text{x}10^4$ (W/K^3/m^3), $t_{mes} = 1\text{x}10^{-9}$ s, $\mathbf{t_{flm}} = 1\text{x}10^{-6}$ m, $\mathbf{s_{flm}} = 1\text{x}10^{-6}$ m^2, $\mathbf{t_{el}} = 5\text{K}$, $\mathbf{t_{e0}} = \mathbf{t_{p0}} = 77$ K, $\mathbf{a_{bs}} = 0.5$, $\mathbf{p_{rad1}} = 1\text{x}10^{-6}$ W, $\mathbf{i_0} = 1\text{x}10^{-6}$ A, $\mathbf{i_1} = 1\text{x}10^{-9}$ A, $\mathbf{v_{b0}} = 1\text{x}10^{-5}$ V, $\mathbf{v_{b1}} = 1\text{x}10^{-6}$ V.

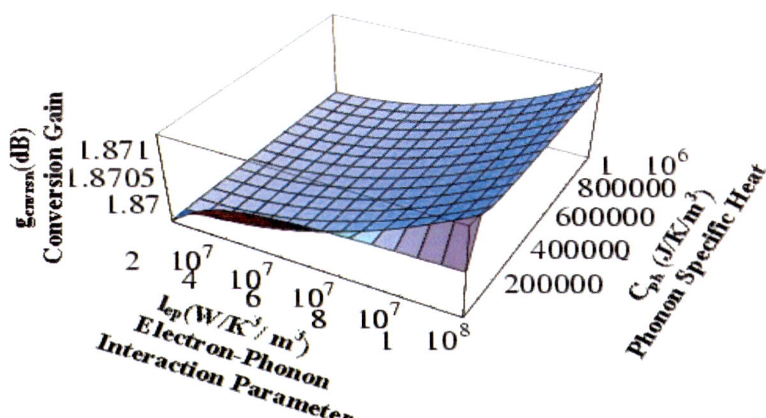

II A Figure 2. Plot of Conversion Gain $\mathbf{g_{cnvrsn}}$ (dB) vs Electron-Phonon Interaction Parameter $\mathbf{l_{ep}}$(W/K^3/ m^3) and Phonon Specific Heat. $\mathbf{C_{ph}}$ (J/K/m^3). The Fixed parameters are $\mathbf{t_{mes}} = 1\text{x}10^{-9}$ s, $\mathbf{t_{flm}} = 1\text{x}10^{-6}$ m, $\mathbf{s_{flm}} = 1\text{x}10^{-6}$ m^2, $\mathbf{i_0} = 1\text{x}10^{-6}$ A, $\mathbf{t_{el}} = 5\text{K}$, $\mathbf{t_{e0}} = \mathbf{t_{p0}} = 77$ K, $\mathbf{a_{bs}} = 0.5$, $\mathbf{p_{rad1}} = 1\text{x}10^{-6}$ W, $\omega = 1\text{x}10^9$ Hz, $\mathbf{v_{b1}} = 1\text{x}10^{-6}$ V, $\mathbf{C_{el}} = 1\text{x}10^4$ (J/K/m^3).

II B   Thermoelectric Cooling for a HETS : Bulk Materials

II B.1   Developments Past and Present

Research and development in the area of thermoelectric refrigeration during sixties and seventies, was very much limited to the bulk materials[Appendices B1-B4]. Cooling devices developed were restricted in use, close to room temperature applications only. The best materials found were Bi_2-Sb_2-Te_3 compounds. The author fabricated and tested a bulk material three stage cascaded cooler and obtained a temperature drop of 100K at 300K.The cooling area was $1cm^2$ across and had a heat load of half a Watt[24]. There has been little progress, in the area of thermoelectric cooling,since then. But the field has recently experienced, a complete turnaround. This is particularly so in the area of low temperature control applications.The thermal sensors of interest today are 10^{-12} cm^2 acrossand dissipate around a micro Watt of power.This film thermoelectric cooling, external or internal, thus makes a sense.Low dimensional thermoelectric materials and structures e.g. superlattice naao two dimensional heterostructures, quantum wires and quantum dots have raised the hope of designing, highly efficient cooling devices for the low temperature control applications.

Quantum well devices of, Si-Ge, Bi_2Te_3, etc., have actually been fabricated and the dimensionless thermoelectric figure of merit, **ZT** $>> 1$, has been demonstrated[25, Appendix B11]. It is anticipated, that the nano fabrication developments in the field of the hot electron thermal sensing, together with convenient thermoelectric control of temperature, would revolutionize, the field of thermal sensing. The new class of bulk thermoelectric materials, Z_rH_f Te_5, have the right material parameters i.e. the thermoelectric power (Seebeck coefficient), lattice thermal conductivity and the electrical conductivity, for the development of 50 - 140K range, efficient refrigeration devices[14, Appendix B5]. Low dimensional thermoelectric materials–structures, raise much hope for developing the desired thin film thermoelectric cooling devices[25]. But a researcher in the field need to have a knowledge thresh hold, in the area ofthe thermoelectric phenomena[26-28]. Unfortunately the development of the text, in this area, has not kept pace with the recent developments. The task for a beginner, in the field thus become all the more difficult. This work, I feel, will make the task easier, foe a new researcher, as well as, for the scientific community, on the whole.

II B.2   Efficiency of a Thermoelectric Refrigerator

Most efficient, thermoelectric materials are semiconductors.At the low dimensional materials level, a high level of theoretical and experimental understanding, of the thermoelectric phenomenawould be necessary. A simplified relationship between theory and experiment, for the bulk materials, developed mainly during seventies, is available e.g. in [28, Appendices B1- B4].A good knowledge of the Band Theory of Solids, is a prerequisite, to make a good progress in the field[29]. It is essential to comprehend the steps involved in the systematic theoretical development,of he mathematical relations. What is reproduced here, in the sections IIB 3.1 and IB 3I.2 below, is only a a summary. Details can be found in a standard text[26-28]. They are included here, to show a direct relevance,to the low dimensional thermoelectric situation. The most basic component of a thermoelectric heat pump, is a thermo-junctionfabricated form one, n-type (negative) and the other, p-type (positive) leg of a semiconductor material. The diagram shown in Appendix B1 is for a thermoelectric generator. In a cooler, one need to replace the load R_L by an external source of DC power.

The thermo junction is reverse biased by an electrical current. The electrons (in n leg) and holes (in p leg), carry heat away from the junction, and thus the cooling produced. The hotter ends of the legs are electrically insulated from, a thermally conducting heat sink. An equilibrium temperature difference $\Delta T = T_2$ (temperature of the hot end) $-T_1$ (temperature of the cold end) will ultimatelybe established, between the cold end and the heat sink.Thermoelectric refrigerators, are designed to operate at the maximum heat pumping. The coefficient of performance of the heat pump (**COP**), can be written as follows.

$$\mathbf{COP} = \text{Cooling Power (Heat Extracted)/Input Power (Electrical Energy Consumed)} \quad (1)$$

The current through the device for maximum **COP** is given as

$$\mathbf{I}_{max\,(COP)} = \lambda\,(\mathbf{T_2} - \mathbf{T_1})\,[\,\sqrt{(1 + \mathbf{Z_cT})} + 1)\,]\,/\,(\alpha_{np}\,\mathbf{T_{av}}) \quad (2)$$

Here $\mathbf{T_{av}} = (\mathbf{T_1} + \mathbf{T_2})\,/\,2$, is the average temperature of the pump,λ = is the average thermal conductance (in parallel) of the two leg across cooler-sink interface. $\alpha_{np} = (\alpha_n - \alpha_p)$, α_n and α_p being the thermoelectric power of the n and p leg respectively. $\mathbf{Z_cT}$ = the dimensionless thermoelectric figure of merit of the couple. The maximum possible value for the **COP** is

$$\mathbf{COP_{max}} = \eta\,\gamma \quad (3)$$

where $\eta = \mathbf{T_1}\,/\,(\mathbf{T_1} - \mathbf{T_2})$ is the normal carnot cycle efficiency and

$$\gamma = [\sqrt{(1 + \mathbf{Z_cT})} - (\mathbf{T_2}/\mathbf{T_1})\,]\,/\,[\,\sqrt{(1 + \mathbf{Z_cT})} + 1)\,] \quad (4)$$

is the thermoelectric efficiency. $\mathbf{Z_cT}$ for a couple, in terms of the material properties of the two legs of the couple can be written as follows

$$\mathbf{Z_cT} = (\alpha^2{}_{np})\,/\,[\,(\lambda_n\,/\,\sigma_n)^{1/2} + (\lambda_p\,/\,\sigma_p)^{1/2}\,] \quad (5)$$

It is convenient to address **ZT** as for a single type of material, as follows.

$$\mathbf{Z} = (\alpha^2\,\sigma)\,/\,\mathbf{K} \quad (6)$$

α = the Seebeck Coefficient, σ = the electrical conductivity, and
$\mathbf{K}$ = the thermal conductivity.

In an integrated HTSC sensor-thermoelectric situation, one would in fact be interested in a single type of material, being involved. Electrons with higher mobility (i.e. an n type material), would be a preferred choice.

II B.3 Optimization of ZT
II B.3.1 Single Band (Conduction) Approach
A more realistic approach to the theory of optimization of **ZT** requires,a two band conduction model, for carrier (electrons and holes) transport. The two bands are conduction(CB), where

electrons freely move and valence bands(VB), where holes freely move. Consideration of the carriers within a single band over simplifies the calculation for the transport parameters.

This may be the ideal model for the design of an efficient electronic cooler. It will provide for the beginner, a smoother transition, to adapt oneself later, to the understanding required for the more complex physics involved, in real models. In a single band approach, the conduction band, considered here, only electrons (no holes) participate in the transport. For a complete mathematical treatment, i.e. taking into account, multi valley (more than one electron energy ellipsoid participating), multiple types of electron scattering, the reader should follow a more rigorous text. The following, is only an outline for, a single valley conduction, acoustic, ionized impurity and optic phonon scattering. The calculation of charge and heat transport, in any device,involves first finding, the density (number per unit volume) of the free carriers (electrons here) in a band. This is done by multiplying the density of states **D(E)** (energy states available per unit energy and per unit volume of the material, in the band), within a small interval of energy, **dE,** the probability, of occupation (**f**) of the energy levels **E,** and integrating over various energies.The statistics controlling the occupation of energy levels, is the familiar Fermi Dirac statistics **f**.

The important mathematical expressions involved in the transport processes are.

$$\mathbf{D(E)} = (4\pi / \mathbf{h}^{3)}(2\ \mathbf{m}^{*}_{e})^{3/2}\ \mathbf{E}^{1/2}\ (\mathbf{E} > 0)\ (\text{conduction band edge}) \qquad (7)$$

$$= (4\pi / \mathbf{h}^{3)}(2\ \mathbf{m}^{*}_{e})^{3/2}\ (\mathbf{E} - \mathbf{E_g})^{1/2}\ (\mathbf{E} < -\ \mathbf{E_g}) \qquad (8)$$

$$\mathbf{f} = 1 / (\mathbf{E} - \mathbf{E_f}/\ \mathbf{kT}) \qquad (9)$$

$\mathbf{E_f}$ = Fremi Level, $\mathbf{k}$ = Boltzmann Constant, $\mathbf{T}$ = temperature in degree Kelvin
$\mathbf{Eg}$ = energy band gap, between valence and conduction band,
$\mathbf{m}^{*}_{e}$ = the effective (expressed in terms of the free electron mass $\mathbf{m_e}$ and
$\mathbf{h}$ = the Planck's constant.

$\mathbf{Eg}$ = the gap of energy, which the electrons have to overcome,before being in the conduction state. Refer to adjacent **IIB.3 Figure 1**, below, for the positions of the various energy levels, in the two band perspective.When the material is pure (intrinsic), the Fermi level (FL) (with respect to which the electron energy is measured) is in the middle of the gap. While when the material is impure (doped), the FL, is closer to the conduction ban edge, for the electron, and to the valence band edge, for the holes, if present[Appendix B14].E is the spread of energy levels, available in a band, which the carriers can occupy. The above **D(E)** is peculiar to a bulk material.

In a low dimensional situation, it would have a different expression and a discrete structure. The reader is referred to a standard text on Solid State Physics e.g. [29] for a good understanding of the Band Theory of Solids and the scattering dynamics of the electrons.The evaluation of the transport parameters, α (Seebeck Coefficient), σ (electrical conductivity) and **K** (thermal conductivity), is first carried out.

Then on substitution in the equation (6) one gets **ZT,** as follows. **II B.3 Figure 1 .** Energy Levels in a Semiconductor Model, showing Valence and Conduction Bands, separated by an Energy Gap

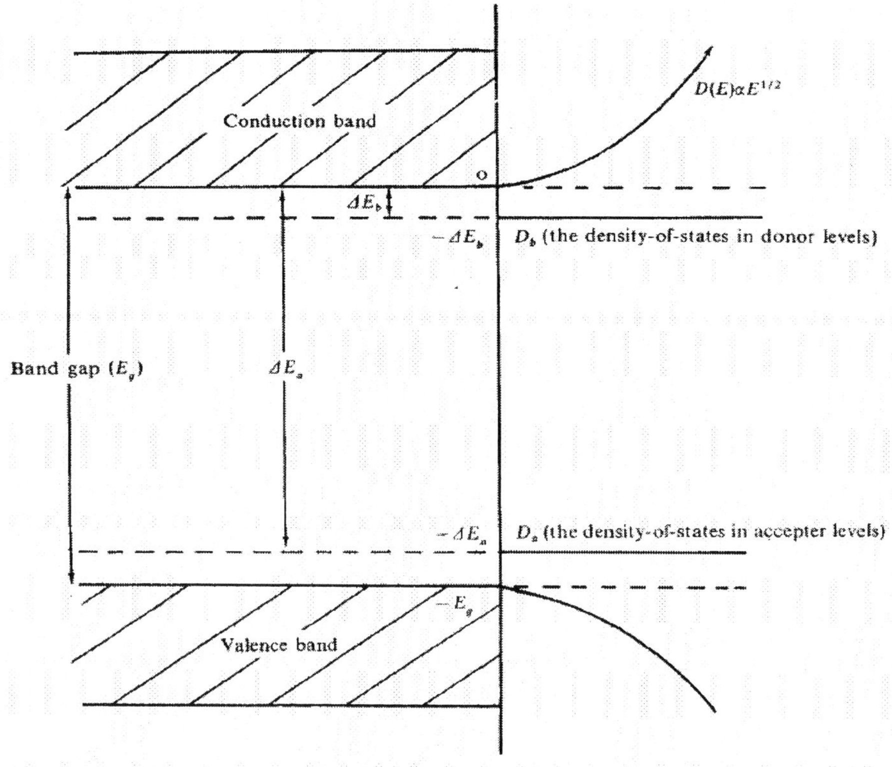

II B.3 Figure 1. Energy Levels in a Semiconductor Model, showing Valence and Conduction Bands, separated by an Energy Gap

$$\mathbf{ZT} = (\delta - \xi)^2 / (l + 1 / \beta_e) \tag{10}$$

$$\delta = [\ (S + 5/2)\ \mathbf{F}_{S+3/2}(\xi)\ ] / [\ (S + 3/2)\ \mathbf{F}_{S+1/2}(\xi)\ ] \tag{11}$$

The meaning of various terms and symbols is as follows.

$$\mathbf{F_l} = {}_0\int^{o} (\mathbf{x^l\ dx} / [\ (1 + \exp(\mathbf{x} - \xi\ /\ kT)\ ]\ (\text{Fermi Integral of order } \mathbf{l}) \tag{12}$$
$$(x = E\ /\ kT = \text{integral variable}\ ,\ \xi = \mathbf{E_f}\ /\ kT,\ l = S + 1/2,\ S + 3/2.\ \text{etc.}),$$

$$l = \mathbf{K_e}\ /\ (\sigma\ \mathbf{T})\ (= \text{the Lorentz Number},\ \mathbf{K_e} = \text{Electronic Thermal Conductivity})$$
$$= \{\ [\ (S + 7/2)\mathbf{F}_{S+5/2}(\xi)\ ] / [\ (S + 3/2)\mathbf{F}_{S+1/2}(\xi)\ ]$$
$$-\ [\ (S + 5/2)\mathbf{F}_{S+3/2}(\xi)\ ] / [\ (S + 3/2)\mathbf{F}_{S+1/2}(\xi)\ ]^2 \tag{13}$$

$$\alpha(\text{Seebeck Coefficient}) = -/+\ (\mathbf{k}\ /\ \mathbf{e})\ (\delta - \xi),\ -\ \text{for electrons and}\ +\ \text{for holes} \tag{14}$$

The calculations for the transport parameters, can based on the relaxation time approximation approach. What it means is that as the carriers move through the solid, various scattering processes, modify the occupation of the higher energy levels, for the electrons. The electrons move, between the excited and the groundstates of energy, according to a resultant relaxation time. This relaxation time corresponds to the various scattering mechanisms involved.The change in occupation, in time, as a result of scattering, in the statistics **f**, can be written as, δ**f** / δt = [(**f** – **f$_0$**) / τ)], Here **f** is the scattered and **f$_0$** is the unscathed, statistics of the occupation of the electrons. Where, one takes the relaxation time, as energy dependent, according to the relation

$$\tau(\mathbf{E}) = \mathbf{a}\ \mathbf{E^s} \tag{15}$$

a is a parameter independent of energy but may depend upon temperature.
τ = the energy dependent relaxation tome of the scattering of the carriers,
S = -1/2, for acoustic phonon scattering, = +3/2, for ionized impurity scattering, = +1/2, for optical phonon scattering, etc. The meaning of more terms and symbols is as follows.

$$\sigma = \mathbf{n}\,\mathbf{e}\,\mu = \sigma_0\,\varepsilon \tag{16}$$

$$\sigma_0 = 2\,(2\,\pi\,\mathbf{m^*}\,\mathbf{k}\,\mathbf{T}\,/\,\mathbf{h^2})^{3/2}\,\mathbf{e}\,\mu_\mathbf{c} \tag{17}$$

Here $\mu_\mathbf{c}$ = carrier mobility, **n** = the carrier density and **m*** = the density of states effective mass

$$\varepsilon = \mathbf{F_{S+1/2}}(\xi)\,/\,(\mathbf{S}+1/2)! \tag{18}$$

$$\mathbf{n} = (4\,/\,\sqrt{\pi})\,(2\,\pi\,\mathbf{m^*}\,\mathbf{k}\,\mathbf{T}\,/\,\mathbf{h^2})^{3/2}\,\mathbf{F_{1/2}}\,(\xi) \tag{19}$$

$$\mu = (2/3)\,\mathbf{a}\,(\mathbf{kT})^5\,(\mathbf{e}\,/\,\mathbf{m_c})\,[\,(\mathbf{F_{S+1/2}})\,/\,\mathbf{F_{1/2}}\,] \tag{20}$$

Here $\mu_\mathbf{c}$ = the carrier mobility, **m$_\mathbf{c}$** = the carrier mass and

$$\mu_\mathbf{c} = (\mathbf{e}\,/\,\mathbf{m_c})\,<\tau> \tag{21}$$

$$<\tau> = {_0}{^{\circ\equiv}}\,[\,\tau(\mathbf{E})\,\mathbf{E}^{3/2}\,(\,\beta\,\mathbf{f_0}\,/\,\beta\,\mathbf{E})\,\mathbf{dE}\,]\,/$$
$$[\,\mathbf{E}^{3/2}\,(\,\beta\,\mathbf{f_0}\,/\,\beta\,\mathbf{E})\,\mathbf{dE}\,] \tag{22}$$

τ(**E**) would have different energy dependence for different scatterings.Taking various scattering mechanisms, as independent of each other, one can write $(1\,/\,<\tau>)\,> = 1\,/\,<\tau_\mathbf{ac}$ (acoustic phonon) $>\,+\,1\,/\,<\tau_\mathbf{I}$ (ionized impurity) $>\,+$ etc. Detailed mathematical expressions for $\tau_\mathbf{ac}$ (acoustic or lattice scattering), $\tau_\mathbf{I}$ (impurity scattering), etc., can be found in a standard text. In the literature a **special** material parameter, β has been used to gauge, the effects of various material properties, on **ZT** (Appendix B2). The parameter can be written as follows.

$$\beta = [\,(\mathbf{k}\,/\,\mathbf{e})^2\,\mathbf{T}\,\sigma_0\,]\,/\,(\mathbf{K_{ph}}),\ \mathbf{K_{ph}} = \text{lattice thermal conductivity} \tag{23}$$

$$= (\mathbf{k}\,/\,\mathbf{e})^2\,2\,(2\,\pi\,\mathbf{k}\,\mathbf{T}\,/\,\mathbf{h^2})^{3/2}\,\mathbf{e}\,(\mathbf{m^{*3/2}}\,\mu_\mathbf{c})\,/\,\mathbf{K_L} \tag{24}$$

The above mathematical summary is only an oversimplified picture.One need to take the following into account for a complete treatment.Non parabolic, multiple valley carrier conduction pockets. This leads to higher carrier concentration and is in favor of **ZT**.2. Mixed scattering mechanisms rather than one or tow, dominant ones. This will reduce the mobility of the carriers and thus the conductivity. It may however enhance kinetic energy transport in a particular situation. The material parameter **β**, depends upon the basic material properties e.g. the effective mass of the carriers, mobility, etc. 3. The two band approach (including minority carriers, bipolar conduction, etc.)

The higher **ZT** values will be found in materials, with **(s)** deep in the band gap E_g, **((s) < 0)**, smaller carrier effective masses, higher carrier mobility, small K_L, etc. There is no single material which will meet all the optimization requirements. It is by performing experiments on selected materials, comparing with their theoretical modeling, one really can find, what exactly is in favor of **ZT**. This was done for the materials in their bulk form, during sixties and seventies. Extensive data is available for materials like S-Ge alloys, Bi-Sb-Te compounds, etc. A summary of the results can be seen in the literature [Appendices B3-B4]. Mathematical description of the phenomenology, of the two band approach, is included below, for the sake of completion.

B 3.2   Two Band (Conduction and Valence) Approach

Assuming that the thermal conductivity K results exclusively from the freecarriers, one can write

$$K = K_c = (k/e)^2 \, T \, \Sigma^N_{i=1} \big|_I \, \sigma_I \tag{25}$$

The suffix i runs over N extrema (energy ellipsoids) in the conduction (I = 1) and valence (I = 2) band. The total electrical conductivity is

$$\Sigma_i \, \sigma_i = N_e \, \sigma_{1e} + N_h \, \sigma_{1h} \tag{26}$$

where N_e and N_h are the number of equivalent extrema in the conduction and valence bands (non degenerate) and σ_1 is the contribution to electrical conductivity from one extrema. The other relations, as part of the **ZT** formulation are

$$\alpha = (k/e) \, \Sigma^N_{i=1} \, [\, (A_i \sigma_I) \, / \, \sigma_i \,] \tag{27}$$

where $A_i = \delta_i - \xi_i$, and involves the equations (11) – (14).

For simplicity one assumes, that the electrons obey classical statisticsand that equal contribution from scattering parameters,are in operation, for the electrons and holes i.e. r_e = r_h = (s + 1/2). Here **r** comes from the energy dependence of the mean free path (l)approach for the transport of the carriers, i.e. $l = l_0 \, E^r$. The relaxation time, equation (15), and the mean free path approaches, are equivalent.

On substitution of the respective electrical conductivities, one gets

$$\sigma_e = N_e \, \sigma_{1e} \text{ and } \sigma_h = N_h \, \sigma_{1h} \text{ in (26) one gets}$$

$$\alpha = (k/e) \left[A_e / (1 + \sigma_h / \sigma_e) + A_h / (1 + \sigma_e / \sigma_h) \right] \tag{28}$$

$$\sigma_e / \sigma_h = \gamma^2 \left[F_{re} (\xi_e) / re ! \right] / \left[F_{rh} (\xi_h) / rh! \right] \tag{29}$$

$$Fr_i (\xi_i) = {}_o\int^{o\equiv} (x)^{r}i \left[1 + \exp (x - \xi_i) \right]^{-1} dx \tag{30}$$

$$\gamma = \left[(m^*_e / m^*_h)^{3/2} (\mu_{ce} / \mu_{ch}) \right]^{1/2} \tag{31}$$

Here μ_c, represents the mobility in the low carrier (classical) limit. m^*_e, m^*_h are the density of sates effective mass for electrons and holes respectively.

$$m^* = N^{2/3} (m^*_1 m^*_2 m^*_3)^{1/3} \tag{32}$$

m^*_1, m^*_2, m^*_3 are the masses along the principal directions of ellipsoids of energy. Information on the upper bound of **ZT** can be more easily worked out from the model

$$Ai = \left[r + 2 - \xi_i \right] \tag{33}$$

$$|_i = (r + 1/2) \tag{34}$$

$$\sigma_e / \sigma_h = \gamma^2 \exp (\xi_g + 2 \xi) \tag{35}$$

Here $\xi_g = E_g / kT$, is the reduced energy gap. From these equations, it is apparent that in this model, the Lorentz number, $K_e / (\sigma T)$, i.e. $|_i$, is independent of ξ, ξ_g and γ, and therefore the variation in these parameters will affect **ZT,** only through the Seebeck coefficient α (equation (27)). It is worth noting in this model, that the upper bound in **ZT**, crucially depends upon γ. For lattice (acoustic) scattering

$$\gamma_L = \chi_L (m^*_h / m^*_e)^{1/2} \tag{36}$$

$$\chi_L = (N_e / N_h)^{5/6} (\sigma_h / \sigma_e)^{1/2} (\varepsilon_h / \varepsilon_e) \quad (\varepsilon \text{ refers to the deformation potentials}) \tag{37}$$

For ionized impurity scattering

$$\gamma_I = \chi_I (m^*_e / m^*_h)^{1/2} \tag{38}$$

$$\chi_I = (N_e / N_h)^{1/6} (a_h / a_e)^{1/4} \tag{39}$$

The mass anisotropy **a** is given by

$$a = 3 / \left[(m_2 m_3 / m_1^2)^{1/3} + (m_1 m_3 / m_2^2)^{1/3} + (m_1 m_2 / m_3^2)^{1/3} \right] \tag{40}$$

One finds that for higher **ZT,** a high value for the ratio m^*_h / m^*_e is required and a small m^*_e / m^*_0 (m^*_0 = the free electron mass). In addition, acoustic lattice scattering should be the dominant scattering process in the negative branch of the thermocouple and impurity scattering in the positive branch. The conduction band should also have a largenumber of extrema with a small number in the valence band. A large mass anisotropy for the electrons, but not for the holes, is another requirement for a large **ZT**. The energy gap also matters. For a reduced energy gap (E_g / **kT**) of 4, ZT_{max} = 8, while for energy gap of 16, a value = 70 is predicted, assuming a solubility of $10^{27} m^{-3}$.

II C Generalized Thermoelectric Theory Applied to a Junction

II C.I Peltier Cooling, Ref[28, Bulk Device Structure, Appendix B14]

Ideally the Peltier heat transported per second, dQ_P, away from a junction, constructed from two materials a and b, should result exclusively (reversible effect)from the charge transport. The phonons should not interfere with electrons. This heat is proportional, to the magnitude of the external current applied, and the duration dt, over which it is applied. Mathematically it can be expressed as

$$dQ_P \sim I \, dt \tag{1}$$

$$= \Pi_{a,b} \, I_{a,b} \tag{2}$$

$$= \Pi_{a,b} \, (de \, / \, dt) \tag{3}$$

Here $\Pi_{a,b}$ is called as the Peltier coefficient, or the Peltier voltage, and (de/dt)is the rate of charge transported. It is positive, if the current flows from a to b (Appendix B14). $dQ_P > 0$ means, that the heat is absorbed, at the junction.From the reversible thermodynamic consideration, one can also write,the effect, in reverse to (1), i.e. the thermoelectric power, as

$$\alpha_{a,b} = \Pi_{a,b} \, / \, T \tag{4}$$

Here, $\alpha_{a,b}$ is the Thermolectric power (or the Seebeck coefficient), and is the Voltage generated per unit temperature difference, across the junction (material a and b), as a result of the heat absorbed at the hot junction. A third effect, called the Thomson effect, is also part of the heat equilibrium in the circuit. There will also be an evolution or absorption of heat, whenever, an electric current, passes through a single homogenous conductor, along which a temperature gradient is maintained. This, is called as the Thomson heat, is generated or absorbed (throughout the length of a single conductor)and is in proportion to the current **I** passing, for a time **dt** i.e.

$$dQ_T \sim I \, dt \, dT \tag{5}$$

$$= \tau \, I \, dt \, dT \tag{6}$$

$$= \tau \, e \, dT \tag{7}$$

τ is referred to as the Thomson coefficient. It is positive, if the heat is absorbed, when current flows to the hotter region.All the three, thermoelectric effects, are expressed as the reversible effects,In the equations (1) - (7).

The Joule heat, which forms a part of the thermal effects in the circuit,is irreversible and is not included in the equations, just for the sake of simplicity. As a result, of the temperature gradient between the hot and cold the junction, an irreversible loss of heat (conduction, radiation, etc.) also occurs.Disregarding, the irreversible effects, one can write, the net rate of absorption of heat, required to maintain equilibrium, in the ab circuit as follows.

$$Q = [\Pi_{a,b}(T_H) - \Pi_{a,b}(T_c) + \int(\tau_a - \tau_b)\,dT\,] \qquad (8)$$

By combining first and second law of thermodynamics, one can write[28]

$$d(\alpha_{a,b})/dT = (\tau_a - \tau_b)/T \qquad (9)$$

$$\tau_a = d(\alpha_a)/dT, \text{ for a single couple} \qquad (10)$$

This means, a standard reference conductor can be defined as having $\tau_a = 0$, at all temperatures, down to absolute zero ($T = 0$).A superconductor, is taken as a reference material ($\alpha = 0$),in the thermoelectric phenomena.

Electrons as they travel, get scattered by different targets e.g. phonons, impurities, etc. All these scatterings are dependent on the electron energy.The relation between mean free path l and energy E is expressed as follows.

$$l = l_0 E^r \qquad (11)$$

$$\alpha = (k/e)(A + \xi^{*}) \qquad (12)$$

$\xi^{*} = E_f/kT$, is the reduced Fermi Energy.

$$A = [\,(<v \blacksquare l>)\,]/[\,(<v\,l>)<(k\,T)>\,] \qquad (13)$$

$\blacksquare$ is the electric field and v **is** the velocity of the charge carrier$A = r + 2$, is referred to as the average kinetic energy transported by the carriers. One can view, that Seebeck voltage arises, from two sources, thermo EMF, Π, at the junction, and a distributed array of sources (Thomson), along each of the conductors, $= [\,d(\tau)/dx\,]\,dx$. The total Seebeck voltage arises, solely from the change with temperature, of the continuous Fermi level E_f (electrochemical potential, Appendix B14).

$$\alpha = (1/e)\,[d(E_f)/dT)\,\} \qquad (14)$$

It would be a nice illustration, to see the meaning, of the equation (1 4), in terms of the application of the new low temperature materials, ZrHf tellurides[appendix B5], to thermoelectric refrigeration.The two stationary values of α, with respect to temperature,

Happen to be at 80 and 200K respectively. These two $d(\mathbf{E_f})/d\mathbf{T}$ values, produce the highest values, of the power factor.

$$\mathbf{P_f} = \alpha^2\sigma = \mathbf{K(ZT)} \tag{15}$$

The calculations, amount to, $\mathbf{P_f}$ = 0.5 (W/mK) at 80K and 1.5 (W/mK) at 250K. In between these two limits, $\mathbf{P_f}$, is higher than that of Bi_2Te_3, over 150 to 200K, and less than that of Bi_2Te_3, over 80 to 100K. The $\mathbf{ZT}$ magnitude, would be the same as $\mathbf{P_f}$, as the thermal conductivity is around 1W/mK, in these materials. This is the situation, for the case of,a bulk material. In the case of low dimensional materials, the situation is expected to be much more favorable[Appendices B6-B11]. It is important to pint out, the difference between the Fermi energy and the Fermi level. The Fermi energy is measured from the conducting band edge, whereas, the Fermi level, is measured from some arbitrary fixed energy level (Appendix B14). One can express, the Peltier voltage as

$$\Pi_{\mathbf{a,b}}\,(\mathbf{T}) = \mathbf{T}\,\alpha_{\mathbf{a,b}} = [\,(1\,/\,\mathbf{e})\,(\eta_{\mathbf{a}} - \eta_{\mathbf{b}})\,] - [(1\,/\,\mathbf{e})\,\{\,(\mathbf{r}+2)_{\mathbf{a}} - (\mathbf{r}+2)_{\mathbf{b}}\,\}\,] \tag{16}$$

Here the first term, is the change in potential energy, when a charge crossesthe junction, and the second term is the average kinetic energy transported.The Thomson coefficient can be similarly expressed as

$$\tau = \mathbf{T}\,(d\alpha\,/\,d\mathbf{T}) = (1\,/\,\mathbf{e})\,(d\eta/d\mathbf{T} - \eta\,/\,\mathbf{T}) \tag{17}$$

II C.2 Thermoelectric Theory Applied to the Mixed State in an HTSC

In conventional superconductors, dominant contribution to the heat flow is that due to the moving vortices, at least at temperatures sufficiently far below Tc, and magnetic fields H $\ll$ H_{c2} (upper critical magnetic field). On the other hand, in the HTSCs, both quasiparticle excitations and vortices (to a lesser extent) carry heat. Peltier effect has been observed in the mixed stateof the HTSCs. For the electric field flux ($\mathbf{E_x}$, electric field per unit area, A) and the heat current flux ($\mathbf{J_{hx}}$, heat current per unit area A), along x direction, one can write

$$\blacksquare_{\mathbf{x}} = \rho\,\mathbf{J_{ex}} + \alpha\,\nabla\mathbf{T_x} + \text{transverse effects} \tag{18}$$

Here, $\nabla\mathbf{T_x} = \Delta\mathbf{T}/\Delta\mathbf{X}$ is the temperature gradient, internally created

$$\mathbf{J_{h,x}} = \Pi\,\mathbf{J_{ex}} + \mathbf{K}\,\nabla\mathbf{T_x} + \text{transverse effects} \tag{19}$$

where, Π = Peltier Voltage, internally created, i.e. no external heat input.

$$\Pi = \alpha\,\mathbf{T} \tag{20}$$

When no external heat is present ($\mathbf{J_{h,x}} = 0$), from (19), one gets

$$\Pi = (\mathbf{K}\,/\,\mathbf{J_{ex}})\,\nabla\mathbf{T_x}\;(\mathbf{J_{hx}} = 0,\ \text{adiabatic condition, no external heat input}) \tag{21}$$

In order to measure the Peltier coefficient (Π), one has to determine, the temperature gradient ∇T_x (no external heat, $J_{hx} = 0$), induced by the electric current, J_{ex}. This is a difficult task, trying to do directly.An indirect approach is considered more convenient. First the measurement of ΔT; this is done by finding, the thermal voltage generated, across the sample, for each direction (reversal), of the electric current. The two voltage values, will have opposite (+ and -) signs. Subtracting and dividing by two gives the Peltier voltage.

This is then converted into temperature difference using the Seebeck coefficient, of the HTSC, measured in the normal state. Also there is need to,measure the thermal conductivity **K.** This done by an independent, exclusive thermal ($J_{ex} = 0$) experiment. In the thermal conductivity experiment, a differential thermocouple, connected across two points, little away from the ends of the sample (to avoid contact effect), is used to measure the temperature difference and hence the gradient, across the sample, using, the length of the sample. The thermal conductivity is then calculated, from the familiar relation, Q (External Heat Input) = (K A ($\Delta T/\Delta X$)]. Then from the equation (20), the Peltier voltage for an HTSC- metal junction, can be found.

III Optimization of ZT: A.Low Dimensional Materials/Structures

The following are some interesting results, worth noting, found in the recent studies, on the low dimensional materials. The materials used in the modeling studies, were the conventional room temperature bulk materials. This is due to the well documented details of the material properties being available. One finds that there is a substantial improvement **ZT**. It makes, a very strong incentive, to develop nano structure devices, for cooling applications. The improvements found in the low dimensional material studies;the following included as examples, should be interpreted, in view of thetheoretical formulation, outlined in the section II above.

III A.1 Quantum Wells

II A. 1 n- type PbTe/PbEuTe–planar multi layer Structure, Ref[7, ZT results :
Appendix B6 and B7]

Special Features : Narrow gap class of semiconductors, effects of energy band non-parabolicity, increased carrier density of states, decreased in-plane phonon thermal conductivity, anisotropic effective masses, the multi valley characteristics.**PbTe (ell) :** m_{II}^{W} (electron mass parallel to well plane) = 0.35 m_0, m_0 = mass of free electron, m_L^{W} (electron mass perpendicular to well plane)= 0.034 m_0, E_g^{W} (Band Gap energy) = 321 meV, U (z) (well confinement potential) = 173 meV, d = well width**PbEuTe (Barrier) :** m_{II}^{B} = 0.495 m_0, m_L^{B} = 0. 049 m_0, E_g^{B} = 635 meVIn the case of (100) orientation all the four ellipsoids of constant energy of the bulk PbTe are equivalent and the energy subbands in them coincide.In the case of (11 1) oriented QW, the valley degeneracy is partially lifted.

Two sets of sub bands arising from different valleys appear. One set is from the longitudinal ellipsoid perpendicular to the QW and is situated below the second one from the oblique ellipsoids. The values of **ZT** were found greater than in the bulk case and grow with the decrease of well width. uch a behavior is connected with the effect of density of states increase. At large d (> 2 nm), the **ZT** depends on QW orientation and its maximum value o.4 is achieved at carrier concentration $\sim 10^{18} cm^{-3}$, as in bulk. For small **d (~ 2 nm)** the optimal carrier concentration is higher. $\sim 10^{19} cm^{-3}$ and **ZT ~ 0.8** is achieved.

III B Quantum Wires

III B. 1 One Dimensional Wire, Bi₂Te₃ Ref[9, ZT Results : Appendix B8]

Special Features : Anisotropic one band material, constant relaxation time, parabolic bands in the direction of conduction, free electron like motion in the wire direction, bound state (infinite potential barriers) in the perpendicular directions, only lowest band considered. **Bi₂Te₃** has **ZT** = 0.7 at 300K. It has a trigonal structure, lattice parameters (expressed in terms of a hexagonal unit cell) are a_0 = 4.3 A, c_o = 30.5 A, anisotropic effective mass components are m_x = 0.02 m_0, m_y = 0.08 m_0, m_z = 0.32 m_0, K_L = 1.5 $Wm^{-1}K^{-1}$ and mobility along a_0 (x direction) = 1200 $cm^2V^{-1}s^{-1}$. There are has 6 carrier pockets each with a slightly different orientation in the Brillouin Zone (BZ). The increase in **ZT** found, is due to the change in density of states, but an additional factor is the reduced thermal conductivity due to the increased phonon scattering.

III B.2 Cylindrical Bi Nanowires, Ref[10, ZT Results, Appendix B9],

Special Features : Bi is a semimetals with one anisotropic hole pocket at the T point of the Brillouin zone and three highly anisotropic and non parabolic electron ellipsoids at he L points, has small electron effective mass and the highly anisotropic Fermi surface is a special attraction, the model takes into account anisotropic carrier effective mass tensor, i.e., non parabolic features of the L-point , conduction and valence bands, and the multiple carrier pockets, in **ZT**, there is a rapid increase, with decreasing wire diameter (< 10 nm), for semiconducting Bi nanowire with Fermi energies close to the optimal levelthe system can be approximately described by a one band model at low temperatures, in which the thermal energy **kT** is much smaller than the band gap and adjacent subband separations, the Seebeck coefficient in a one band system is fairly independent of the band character and is determined by the position of the Fermi energy only, the dependence of **ZT** on the carrier effective mass is only influenced by the electrical connectivity and the electronic contribution to the thermal conductivity.

III B.3 Quantum Dots, Ge Dots on Si, Ref[11, ZT Results : Appendix B10]

Special Features : A configuration of regimented quantum dots, with strong coupling among the dots, the carrier transport is facilitated by the extended 3D mini band formation, rather than the localized QD states,the analysis is restricted to heavy holes, the single valley effective mass approximation thus becomes valid, since a single energy maximum in the valence band is located in Γ point, the light hole subband in compressed (superlattice) Ge is well separated from the heavy hole subband, and have much smaller effective mass, the curves show **ZT** for p type Ge/Si QDS, normalized to the Si values,the constant relaxation time of 10^{-12} s is used, there is relatively large region of Fermi energies where **ZT**, is one or two orders of magnitude larger than the bulk Si value.experimental value for bulk Si is 0.05 at 300K and of the bulk p type Si $_{0.95}$ Ge$_{0.05}$ ~ 0.06.

IV Conclusion

IV A General Considerations

Thermoelectric cooling, at the low dimensional material level, is the much wanted temperature control, for the nano HTSC, THz devices. The ultimate speed of the of the device is determined by the fastest possible heat dissipation by the electrons, without a back flow.Self cooling of the electrons by the Peltier technique, in the HTSC sensorsis the right approach to follow. Making a junction between Zr,Hf Te$_5$ and YBCO, as a nano structure would,make a good sensor-cooler to investigate. In a hot electron situation, a full description of the electron-phonon thermal dynamics, in an HTSC, requires inclusion of coexisting systems, i.e. cooper pairs, quasiparticles (electrons from broken cooper pairs), phonons in the film, and phonons in the substrate, etc. When there is, thermal equilibrium, all of these can be described by equilibrium functions with same temperature.That is the case, of a bolometer. If a distribution does not satisfy these conditions, the situation is considered as nonequilibrium. That is what an **HETS** is about. A treatment of a nonequilibrium state requires, a solution of the space and time dependent,thermal distribution functions equations..

The assumption of a uniform non equilibrium state, spread over the entire volume of the film, is applicable only, when the sensor is operated close to the critical temperature Tc. Below Tc, the electron specific heat exhibits an exponential temperature dependence. That requires non linear heat transfer equation for even small deviations from the equilibrium. Near Tc, the superconducting energy gap is strongly suppressed. The concentration of cooper pairs is very small, and the unpaired electrons exhibit no significant superconduting peculiarities. They are regarded as normal electrons, and obey the normal Fermi distribution function. One should notice that there are a wide varieties, of similarities and dissimilarities, among the LTSCs (e.g. NbN) and the HTSCs (e.g. YBaCuo), [Appendices A1-A3]. The thermalization dynamics, in the case of an HTSC is an order of magnitude faster. In YBaCuO, C_p/C_e ~40, while in NbN, it is ~10. On the femtosecond time scale, the non-thermal (hot electron), and thermal, bolometric (phonon) processes are practically, de-coupled in an HTSC. Thus the former, totally dominates the early stages of electron relaxation. In a case, when an HTSC is operated much below Tc,the hot electron approximation is not adequate.

Several alternative models, of using the non equilibrium state of electron and phonons, in an HTSC, for thermal sensing, have been suggested in the literature. It is worth mentioning the virtues of, a hot spot **HETS**.In this approach, a particular selected area on the film, under irradiation, evolves as a hot spot in the film. What this hot spot means is, that the material within the hot spot behaves like a normal material,surrounded by superconducting volume all around. The spatial extent of the spot, increases with the energy received in the spot. The spot, which is in the normal state, thus has a variable interface with the surrounding superconducting phase. The ratio of the normal electrons, inside the spot, as compared to the supreconducting cooper pairs, outside, can be adjusted, by using a suitable bias current and the power of the laser pump. A study can be carried out on the response of the sensor, by gradually reducing the photon flux, to almost zero. The response for a single photon, can thus be manipulated. There may be a one spot, or several sots, working in coherence. A practical device has been developed, as a single photon, hot electron thermal sensor[5]. It is possible to design a low dimensional structure e.g. a regiment of quantum dots, of a thermoelectric material, on a superconducting lattice, or vice a versa.The ratio of the thermoelectric (**T**)-Superconducting (**S**) phase, can be varied. It would make a good study, to investigate a current or a voltage biased, **ST** sensing-cooling integrated device.

One can use a current bias, in the supercionduting film, for the destruction, of a certain number of cooper pairs. The remaining pairs, accelerate to carry the same current. Because of nonzero inertia of pairs, acceleration requires an electric field. This intrinsically generates a voltage V $_{kin}$ ~ L$_{kin}$ (kinetic inductance), in he exterior of the film. Thus this can be an alternative, and may be a better way, of THz radiation detection. One should also realize that in an HTSC thin film sensor, there is a strong potential, to develop, a video technique for the THz radiation sensing , i.e. the thermal imaging [Appendix A4, A5].

IV B Concepts from the LTSCs, Those Can Be Applied to the HTSCs

LTSC Electron Tunneling-Cooling in NIS(Normal Metal / Superconductor / Insulator/) Sensors

Peltier studies in LTSCs, at mK temperatures have brought out interesting results, Ref[6, Appendices B12, 13]. The physics of the Peltier heat, in superconductors, in general, becomes much clearer, by examining the much studied LTSC (mK), **NIS** sensors. The tunneling of the electrons from **N** to **S**, warms up, the quasiparticles in the **S** electrode. If it does not backtunnel **S** to **N**, each quasiparticle carriesan energy = Δ - **eV**, (**V** = bias voltage and Δ = superconductor band gap).

The number of cooper pairs breaking per unit time, per unit area, **Ns,**in the superconductor, is determined, by a balancing current. This is the current of the quasipartcles, in the **S** electrode, due to their decay or out diffusion. If (**k T$_n$**), the thermal heat of the electrons, **T$_n$** = normal metal temperature, $<<\Delta$ - **eV**, the relation between, electron number current number flux (**j$_e$**) and the quasiparticle number flux (**j$_q$**), is **j$_q$** = **j$_e$** - (1 / $\tau_{bt)}$ **Ns**. The cooling power **Pc** can be written approximately as **Pc** ~ [Δ (1 − τ_s / τ_{bt}) − **eV**] **j$_e$**, τ_s = the relaxation time of the quasi particle (superconductor) excitations, and τ_{bt} = back scattering relaxation time.The back flow of the quasipartcles has to be reduced to minimum i.e. one should have $\tau_{bt} \gg \tau_s$. The phonons emitted in the process of self recombination, can also reduce cooling power (**Pc**) by being absorbed in the **N** electrode, when thy deposit energy 2Δ. In competition with **Pc**, is the phonon electron heat exchange, in the **N** material. The achieved cooling power in the LTSC case was approximately 2 pW/μm^2.

IV C Achievements HTSCs

Electronic Cooling : Mixed State : Bi $_{1.7}$Pb $_{0.24}$Sr$_2$ ca$_2$Cu$_3$O$_8$ (BSCO,Tc = 120K), (Bulk Material) Ref[30]

Thermoelectric-thermomagnetic phenomena in the mixed state of the HTSC materials has been extensively studied. The BSCO-Cu junction, is good example to consider[30]. In a bulk material superconductor situation, a direct measurement of the Peltier coefficient (due to its small magnitude), is very difficult. There is need to take into account, effects of both the electrical and thermal currents, across the junction.The mixed state of the superconductor, has a fraction of the volume,where electrons are free and behave like normal electrons.When heat is transported, due to an electric current **I** (Peltier heat), through the superconductor, a temperature gradient is established,across the ends of the superconductor.Knowing the thermal conductivity (**K**) of the superconductor, and measuring the temperature gradient across the sample, produced by the Peltier heat (refer section **II C.2**), one can find the Peltier coefficient Π, from the relation

$$\Pi = (K / Je) \, (\Delta T / \Delta x) \tag{1}$$

In an experiment on the $(Bi, Pb)_2Sr_2Ca_2Cu_3O_\delta$, (Tc = 120K)-Cu junction, Π was found to be 0.5 mV at 100K, at the middle of the transition width [30]. This corresponds to a temperature drop ($\Delta T = \Pi / \alpha$) of 10^{-5}K,across the sample. This electronic cooling, is much smaller, than that (0.25K) observed in a thick film situation, in the YBCO-Bi junction[23]. One should compare these results with that of the case of a nano film LTSC sensor i.e. the **SIN** (superconductor / insulator / normal metal) **IS** sensor[6, Appendices B12, B13]. A cooling of 200mK (at 300mK) was produced, in that device. It is seen in the above BSCO, mixed state HTSC experiment, that the Peltier heat current is, due mainly to the excitations, over the nergy gap of the superconductor. These are he so called, quasiparticle (**QP**)excitations. Since the electric field (disregarding the small Hall angle) is mainly parallel to the electric current **J** , the **QPs** move parallel to the electric current and contribute to the longitudinal heat current., i.e. the Peltier heat. The Peltier coefficient, due to total current **J**, including contributions fromthe supercurrent (**J$_s$**) and the **QP** current (**J$_{QP}$**) , can be written as under. It I believed that the QP excitations carry all the heat.

$$\Pi = \Pi_{QP}\,(J_{QP} / J) + \Pi_s\,(J_s / J) \tag{21}$$

$$\sim \Pi_{QP}\,(J_{QP} / J) \tag{22}$$

IV D Detectivity

Noise Equivalent Power / Temperature : Hot Electron Thermal Sensor (HETS) : [Appendices C1-C3]
The detectivity of a sensor, is a parameter, which determines, the ultimate sensitivity of a sensor. It is inversely proportional to the noise equivalent power (**NEP**) of the sensor.The **NEP**, in a sensor, is taken as the square root, of the sum of the squaresof the **NEP**s, from all possible sources, e.g. the thermal noise (**TN**), the resistance or the Johnson Noise (**JN**), etc. Detectvity modeling in the case of a YBaCuO-Bi junction was the first oneto involve thermeoelctric-voltage responsivity, though it was for a bolometric case[32]. The same technique can be applied to model**NEP** in an **HETS** environment[33]. Remote sensing detectivityanalysis performed in the YBaCuO-BiSb bolmeter case, can similarly be easily applied to an **HETS**.

In a practical device, the specific detectivity **D***, rather than detectvity **D**,is considered as an appropriate parameter. The **D*, D** and the **NEP** are related as follows.

$$D = 1 / NEP \tag{1}$$

$$D* = D\ A^{1/2}\ \Delta f^{1/2} \qquad (2)$$

A = the sensitive area, and Δf = the frequency band of interest and

$$NEP = NEV\ /\ Vrsp(\omega) \qquad (3)$$

Here **NEV** is the noise equivalent voltage and **Vrsp(ω)** is the frequency dependent voltage responsivity of the sensor.In a particular, the open circuit (no current bias) case, for the YBCO-BiSb junction, a specific detectivity, **D*,** of 10^{11} Cm Hz$^{1/2}$/W was estimated [34]. A detailed noise performance analysis for the **YBaCuO-HETS** case, has been carried out before[35]. Noise equivalent temperature, **NET,** rather than **NEP**, is considered as more appropriate parameter. The following mathematical expressions, involved in the noise analysis ofan **HETS**, are worth reproducing here[35, 36, Appendices C3, C4].

$$NET = (NEV)^2\ /\ [\ 2\ a\ k_B\ R_L\ g_{cnvrsn}(\omega)\ ] \qquad (4)$$

$$NET = (NEP)^2\ /\ [\ 2\ a\ k_B\ P_{LO}\ ] \qquad (5)$$

$$g_{cnvrsn}(\omega) = [\ 2\ a\ V^2_{rsp}\ (\omega)\ P_{LO}\ ]\ /\ R_L \qquad (6)$$

Here **a** = the coupling facto between the incident radiation and the sensor,**k$_B$** = the Boltzmn constnt, **R$_L$** = the external (to sensor) load resistance and **g$_{cnvrsn}$(ω)** = the frequency dependent conversion gain of the sensor. In a single side band (SSB) situation, for the case of thermal noise **TN**, the noise equivalent temperature, **NET** can be written as

$$NET = 2\ t^2_{el}\ g_e\ /\ (a_{tcr}{}^2\ p_{lo}) \qquad (7)$$

Here t$_{el}$, is the electron temperature the substrate temperature, **a$_{tcr}$,** is the dimensionaless temperature coefficient of resistance of the sensor, **g$_e$** is the thermal conductance between electrons and phonons. In a complete treatment, noises due to, the thermal resistance between sensorand the substrate, electrical resistance or the **JN** noise, etc, should all be taken into account.

In the case of a model where electro-phonon thermal conductance,is the dominant one in the heat transfer process, the **g$_e$,** is expressed as follows

$$g_e = 3\ l_{ep}\ V_s\ t^2_{el} \qquad (8)$$

where **V$_s$** is the volume of the sensor, **t$_{el}$** is the amplitude of the electron temperature pulse, **l$_{ep}$** = γ / (3 **t$_{mep}$ t$_s$**),**t$_s$** is the temperature of the substrate, **t$_{mep}$** is the electron phonon temperature relaxation time and γ is the Somerfield constant. For a 10 nm thin and a 0.1 μ wide, YBaCuO sensor, a **NET** of 3000K is obtained. An optimum bias of **P$_{LO}$** (local laser pump boost power) $\sim$ 10 μW and I = 50 μA, makes **JN** much less than the **TN** noise.Where phonon are in equilibrium with the substrate, so that the electrons,can easily diffuse to the end electrical contacts, is called as the diffusion limit.In this limit, one can write

$$\mathbf{NET\ (TN)} = 3\ \mathbf{l_{ep}}\ \mathbf{t_e}^4\ \mathbf{L_D}^2\ /\ [\ 4\ \mathbf{K_p}\ (Tc - t_e)\ ] \tag{9}$$

Here $\mathbf{K_p}$ = is the phonon thermal conductivity and $\mathbf{L_D}$ = the diffusion length of the electrons. The phonon diffusivity cab be written as

$$\mathbf{D_p} = \mathbf{k_p}\ /\ \mathbf{C_p} \tag{10}$$

$\sim 0.15\ Cm^2/s$, in the YBaCuO case[35, Appendix C3].

In order to make the device operate under this condition, the lengthof the sensor has to be < 70 nm and need to be < 2 nm across.The electron temperature relaxation time, $\mathbf{t_{me}}$, is limited by the electron-phonon relaxation time, $\mathbf{t_{mep}}$ and the phonon escape time $\mathbf{t_{mes}}$, as follows

$$\mathbf{t_{me}} = \mathbf{t_{mep}} + \mathbf{t_{mes}}\ (\mathbf{c_e}\ /\ \mathbf{c_{ph}}) \tag{11}$$

When a detector can not operate under the diffusion limit, but rather involve, the substrate, then $\mathbf{t_{mes}}$has to be small enough, so that, $\mathbf{t_{mep}} > \mathbf{t_{mes}}\ (\mathbf{c_e}\ /\ \mathbf{c_{ph}})$. Then the band width of the sensor will be[36]. In the lowest noise temperature limit, one can write[36, Appendix C4].

$$\mathbf{NET\ (TN)} = [\ 4\ Tc/\ (\mathbf{a_{tcr}})\ ] \tag{12}$$

Under this limit the band width of the sensor becomes

$$\mathbf{\Delta f\ (TN)} = (1\ /\ \mathbf{t_{mep}})\ [\ \sqrt{}\ (1 + (Tc\ /\ \Delta Tc)\ ] \tag{13}$$

This is he case when the substrate is highly thermally conducting, with conductance much better than between electrons and phonons. With Tc = 100K and ΔTc = 10K (**HTSC**), $\Delta f = \sqrt{11}$ THz ($\mathbf{t_{mep}} \sim \mathbf{10^{-12}s}$). In an **LTSC,** on the other hand, **Tc = 10K, ΔTc = 1K,** one gets $\Delta f = \sqrt{11}$ THz.

Thus in the low noise limit, the **HTSC** results in the sane possible band width, as the **LTSC**. One should notice however that the restrictions on the dimensions of an HTSC sensor, are much more critical, to achieve same level of performance. In the HTSCs, the electrons, have a much shorter mean free path, and thus have to be collected over a much shorter space, for conduction. I feel, a, **S** (superconductor)**I** (insulator) **T** (thermoelectric material) – **I S**, i.e. a **SITIS,** pair junction, as a sensor, analogous to the LTSC, **SINIS** sensor, should form a good study, for an experimental and theoretical analysis. This should lead. to the much desired design, of an HTSC junction sensor, with an integrated electronic cooling[Appendix B13]. One can then extend it, to a hot electron HTSC-Peltier cooling,integrated THz thermal sensor. This can be done by using both, old[Appendices, B3, B4] and new [Appendix B5], class of thermoelectric materials.

Acknowledgements

Family

This work is written in respect of the sufferings of my mother Mrs P.W. Kaila, late, father Dr. M. R. Kaila, brother Dr. K. L. Kaila, and many millions, living or dead, who became innocent victims, least to mention, as refugees, at the independence of India, in 1947.This carnage, resulted from the division of India, which accompanied the freedom.Our family, and many other millions, during 1947-1948, were made refugees twice. We first moved from our home at Lahore, and created another one at Jammu (Kashmir). Then later due to the unsettled fate of Kashmir, moved from Jammu to Delhi. I ma grateful to my wife Mrs Veena Kaila, for her interest and encouragement.My gratitude also goes to my daughter, Dr. Rakhi Kaila. She showed lot of interest in my research work, during her stay at,the University of New South Wales (UNSW), while completing her studies, towards her MBBS degree. I am also thankful to my son, Rohit Kaila, for his support, during family movements, interstate within Australia, and also overseas. The movements became an unnecessary evil, to keep myself, in job and my profession.

Scientific Community

I am grateful to the authors of the publications, both virtual and real, from where some of the information has been used, to prepare this work. The information available, at present is from the specialists in the area.This is disseminated, at isolated spots, and intelligible, by non specialists.I am grateful to those who have made available such publications.This has enabled me to prepare this work. The preparation is in a in a manner, so that a beginner, in the field, has a good starting point. On the other hand, scientific community, at large, would have, at their disposal, a directed information, for research,development, education and training. The research community, realizes, that it there is need for a thermal sensor, which will have an integrated cooling, for its best performance. Thermeolectric cooling, by the researchers in field,is believed, to be the right direction to follow. The way the information is as available at present, is beyond the comprehension, of many non specialist members of the community. There has been, little research and development effort, in creating a suitable knowledge bank, over the last three decades, particularlyin the area of thin film themoelectrics.

The inspiration to take up this task has, originated from the research experience, I gathered, in the area of high temperature superconductors, while working,in association with late A/ Professor, G, J. Russell, at the UNSW, during the nineties.Useful research articles, by several workers in the field, at isolated locations, have been very useful to me. They have helped me to prepare this work, in comprehensible format. I hope this work will stimulate participation, in the development, from a much wider section, of the scientific community. The small step I have endeavored here, I feel, will become a giant step, with the enthusiasm, of the community at large.My association with Emeritus Professor H. J. Goldsmid, during seventies,and part of eighties, at the UNSW, provided me with training, in the area of thermoelectric materials and devices, which helped me to make,this work, a modest one. I much appreciate that association.I am also grateful to Professor J. W. V. Storey, for his continuous support andinterest, including, in the field of thermal sensors. The best ' present ' nature has given to humans, is it mind. The newborn, if nurtured in the right direction, can do wonders.It is the right environment, which evolves the right progress. The wrong one takes the society backward, and produces wastage and destruction.

Appendix A 1 Courtesy Ref[5] Supercond Sc. Technol, 2002, 15, R1

INSTITUTE OF PHYSICS PUBLISHING

Supercond. Sci. Technol. 15 (2002) R1-R16

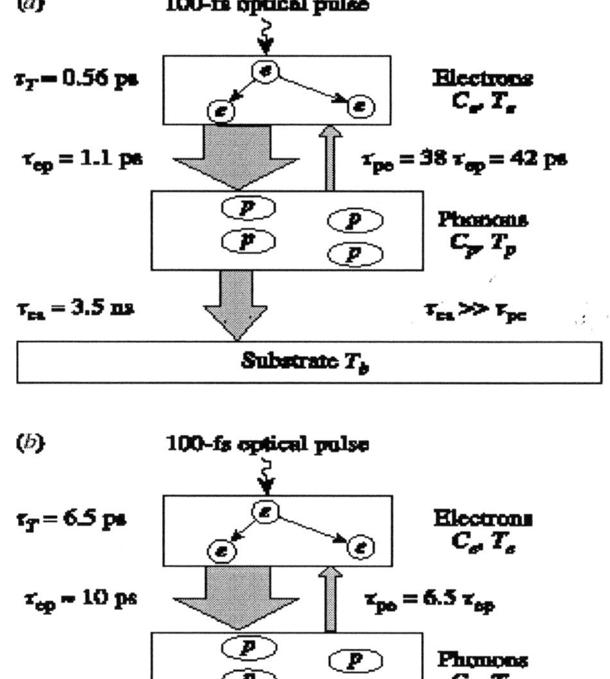

Figure 4. Hot-electron relaxation diagrams and characteristic time constants for (a) thin-film YBCO [9] and (b) ultrathin NbN film [8]. (Reproduced by permission).

Appendix A2, Courtesy, Ref Appl. Phys. Lett, 1999, 73, 3939

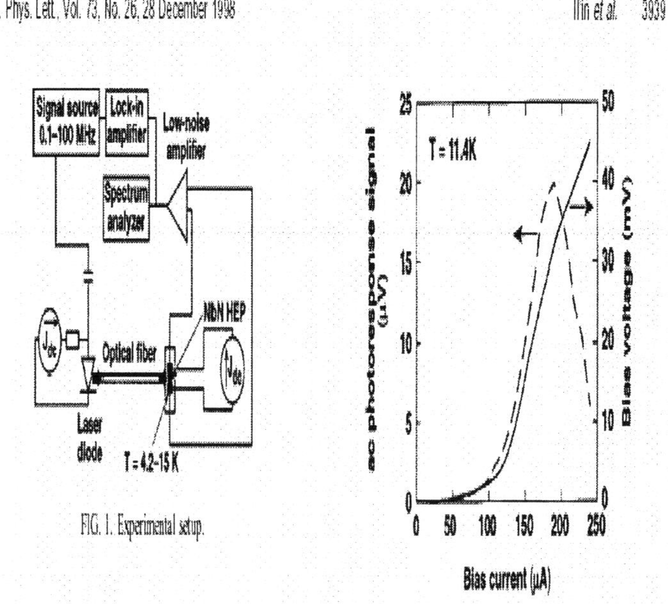

Appendix A3, Courtesy Ref J. Appl. Phys., 1994, 75, 3698

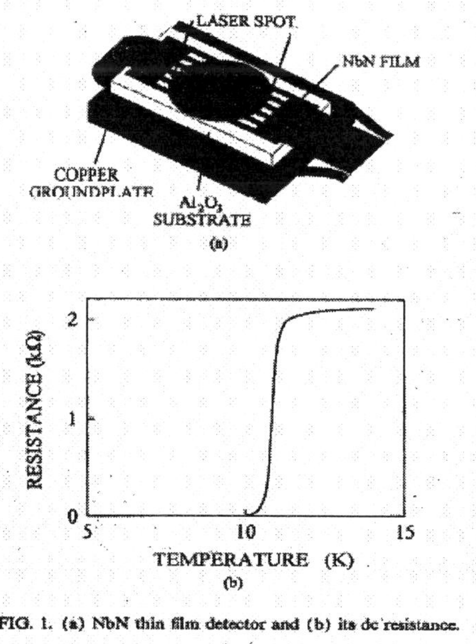

FIG. 1. (a) NbN thin film detector and (b) its dc resistance.

Appendix A4, Courtesy Ref[2] Physica C, 2002, 378-381, 372

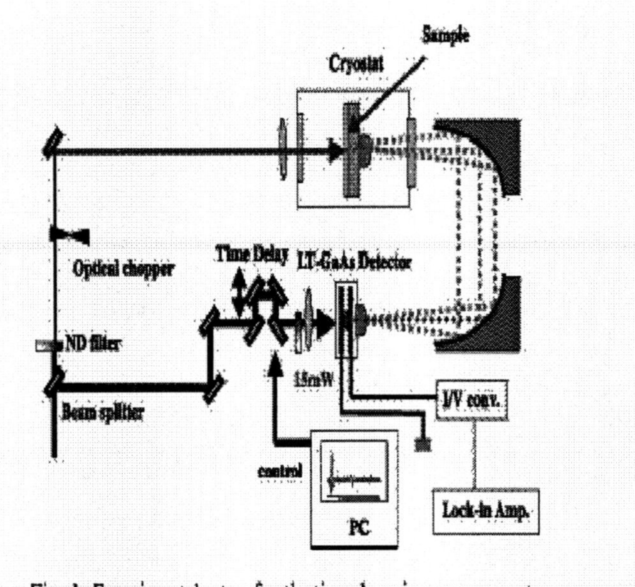

Fig. 1. Experimental setup for the time domain measurements.

Appendix A5, Courtesy Ref[3] Physica C, 2001, 362, 314

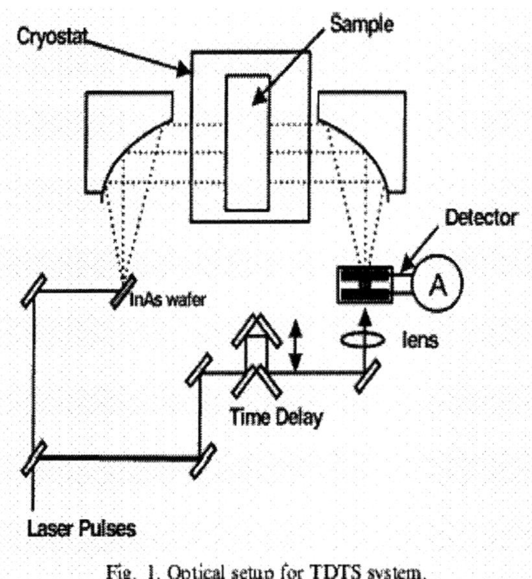

Fig. 1. Optical setup for TDTS system.

Physica C 362 (2001) 314-318

Apendix B1, Courtesy Ref[28] Rep. Prog. Phys., 1988, 51, 459

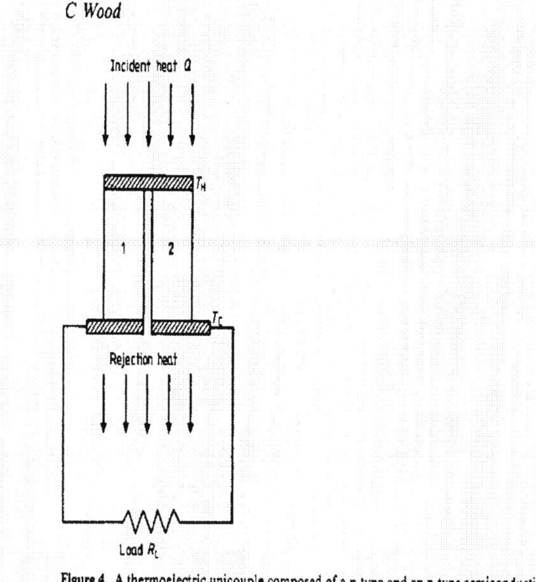

Figure 4. A thermoelectric unicouple composed of a p-type and an n-type semiconducting leg.

Appendix B2, Courtesy Ref[28] Rep. Prog. Phys., 1988, 51, 459

478   C Wood

$(ZT)_{max}$ and the depth of the Fermi level below the conduction band at optimum doping both increase with the value of a material parameter

$$\beta = 8.952 \times 10^{-6}(\mu/\Lambda_{ph})(T/300)^{5/2}(m^*)^{3/2}. \tag{3.16}$$

A plot, derived from their curves, showing the variation of $(ZT)_{max}$ with $\beta^*(\propto \beta)$ for various values of scattering parameter r is shown in figure 9.

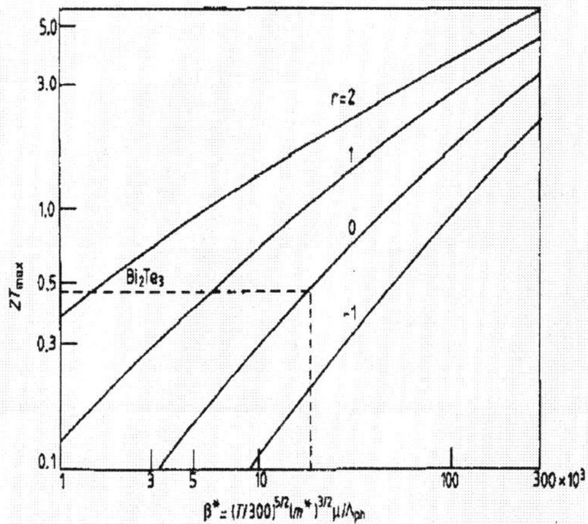

Figure 9. Variation of $(ZT)_{max}$ with the parameter β^* for various scattering mechanisms, $r = -1, 0, 1$ and 2.

Appendix B3, Courtesy Ref[28] Rep. Prog. Phys., 1988, 51, 459

Table 2. Physical properties of thermoelectric materials.

Compound	Structure	Melting point (K)	Energy gap (eV)	Coefficient of energy gap (10^{-4} eV K^{-1})	Mobility Electron (cm^2 V^{-1} s^{-1})	Mobility Hole (cm^2 V^{-1} s^{-1})
Bi_2Te_3	Rhombohedral	859	0.13	-1	1 200	500
Bi_2Se_3	Rhombohedral	979	0.28	-2	600	
Sb_2Te_3	Rhombohedral	895	0.3			275
Sb_2Se_3	Orthorhombic	890	1.2	-7	15	45
$Bi_2Te_{2.7}Se_{0.3}$	Rhombohedral					
$Bi_{0.5}Sb_{1.5}Te_3$	Rhombohedral					
$PbTe$	NaCl FCC	1197	0.33	$+4.5$	1 620	750
$(PbTe)_{0.997}Na_{0.003}$						
$(PbTe)_{0.99}Na_{0.01}$						
$(PbTe)_{0.9997}PbI_{0.0003}$						
$(PbTe)_{0.999}PbI_{0.001}$						
$(PbTe)_{0.75}(SnTe)_{0.25}$						
$GeTe$	NaCl FCC	997	~ 1.0			
$(GeTe)_{0.95}(Bi_2Te_3)_{0.05}$						
$AgSbTe_2$	NaCl FCC	843	~ 0.3			140
$(AgSbTe_2)_{0.75}(GeTe)_{0.25}$						230
$(AgSbTe_2)_{0.25}(GeTe)_{0.75}$						
$(AgSbTe_2)_{0.1}(GeTe)_{0.9}$						
$InSb$	Zinc blende	809	0.1		80 000	750
$InAs$		1215	0.45		33 000	460
$(InAs)_{0.6}(GaAs)_{0.4}$		1250	0.65			
$(InAs)_{0.9}(InP)_{0.1}$			0.5			
$(InSb)_{0.45}(GaSb)_{0.55}$			0.35			
$(InSb)_{0.35}(GaSb)_{0.65}$			0.45			
$(Insb)_{0.65}(GaSb)_{0.35}$			0.25			
$Cu_{1.97}Ag_{0.03}Se_{1.0045}$						
Ge	Diamond	1209	0.7	-5	4 500	3500
Si		1693	1.1	-5	1 300	500
$Si_{80}Ge_{20}$		1300	1.0		400	250
$Si_{80}Ge_{20}(GaP)_{0.01}$						

Appendix B4, Courtesy Ref[28] Rep. Prog. Phys., 1988, 51, 459

Table 2. (continued)

Density of states effective mass		Lattice thermal conductivity (mW cm⁻¹ K⁻¹)	Z_{max}		Temperature at Z_{max} (K)	Comments
Electron m_e^*	Hole m_h^*		n-type (10⁻³ K)	p-type (10⁻³ K)		
0.58	1.07	15*	2.6	2.2	120	Onset of intrinsic conduction precludes use above ~600 K
		15				
		12	3.0			
		11		3.3		
0.3		21	2.0	1.6	270	Loss of Te precludes use above ~800 K
				1.3	370	
				1.2	510	
			1.7		270	
			1.3		670	
		12	1.6		900	
	1.75 or 3.2	10→30		0.9	800	
				1.7	750	
		6		1.8	650	
				1.2	400	
					700	
				1.5	600	
0.014		150	0.5		650	Intrinsic at low temperature
0.025		290	0.45		1000	
		~50	0.95		700	
		~150	0.7		1000	
				0.5	700	
			0.93		600	Unstable due to ionic conduction
			1.15		630	
		~8		1.04	1000	
		600				
		1500				
		40	0.85	0.6	1200	
			1.1	0.7	1200	

13 456   LITTLETON, TRITT, KC

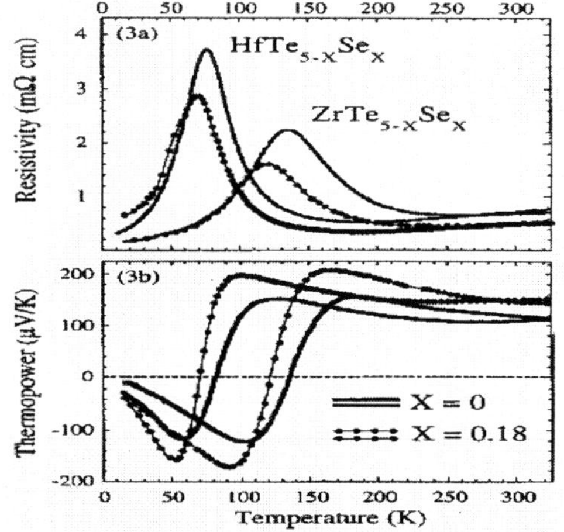

FIG. 3. (a) Resistivity ρ as a function on temperature for HfTe₅₋ₓSeₓ and ZrTe₅₋ₓSeₓ for $x=0$ and 3.6% up to $T=500$ K. (b) The absolute thermopower as a function on temperature for HfTe₅₋ₓSeₓ and ZrTe₅₋ₓSeₓ for $x=0$ and 3.6% up to $T=500$ K.

Appendix B5, Courtesy Ref[14] Phys. Rev. B, 1999, 13457

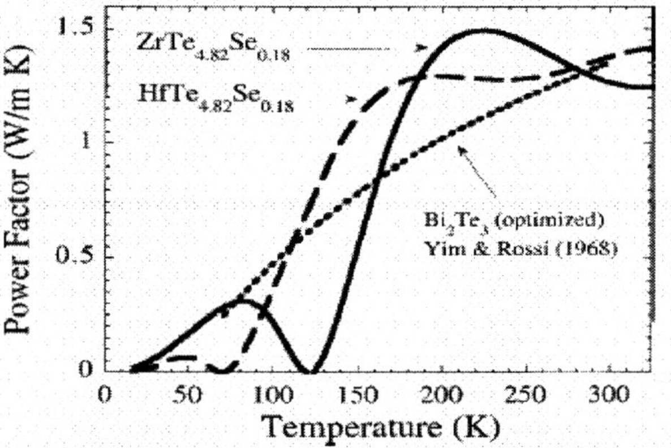

FIG. 4. Power factor, $\alpha^2 T/\rho$ (or $\alpha^2 \sigma T$), as a function of temperature for HfTe$_{4.82}$Se$_{0.18}$ and ZrTe$_{4.82}$Se$_{0.18}$. These are compared to the power factor for optimized Bi$_2$Te$_3$ materials which are also shown.

Appendix B6, Courtesy Ref[7] Phys. Rev. B, 2004, 69. 035306-1

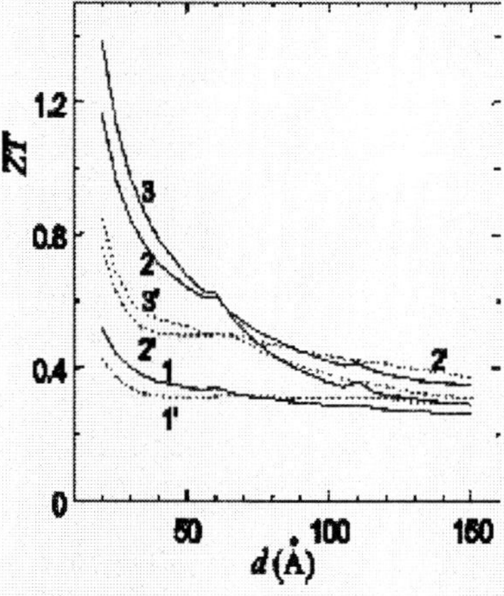

FIG. 4. Thermoelectric figure of merit ZT of PbTe/Pb$_{1-x}$Eu$_x$Te quantum well as function of well width d for (100) oriented QW (curves 1, 2, 3) and for (111) ones (curves 1′, 2′, 3′). Curves 1 and 1′ are for $n = 10^{18}$ cm^{-3}, 2 and 2′ for $n = 5 \times 10^{18}$ cm^{-3}, 3 and 3′ for $n = 10^{19}$ cm^{-3}.

Appendix B7, Courtesy Ref.[7] Phys. Rev. B, 2004, 69. 035306-1

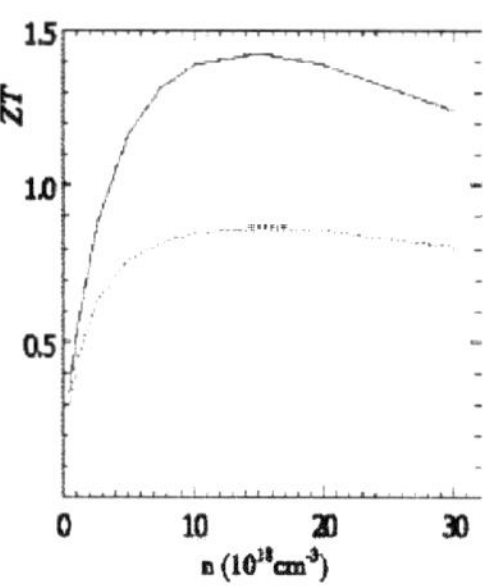

FIG. 5. Figure of merit ZT of PbTe/Pb$_{1-x}$Eu$_x$Te quantum well as a function of carrier density n for (100) (solid line) and (111) QW's (dotted line).

Appendix B8, Courtesy Ref[9] Phys. Rev. B, 1993, 47, 16631

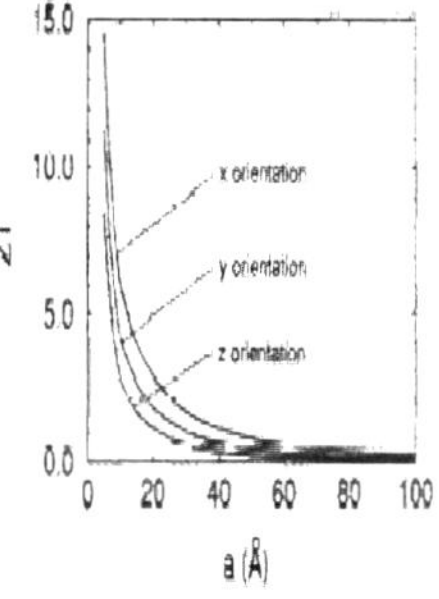

FIG. 1. Plot of $ZT(\eta^*)$ vs wire width a for 1D wires fabricated along the x, y, and z directions.

Appendix B9, Courtesy Ref[10] Phys. Rev. B, 2000, 62, 4610

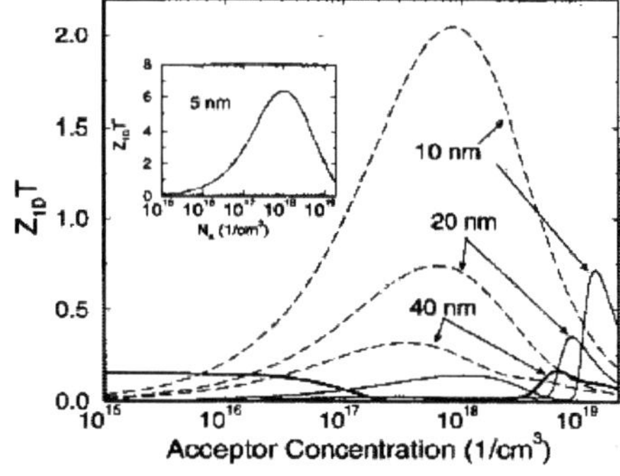

FIG. 9. Calculated $Z_{1D}T$ for p-type Bi nanowires of different diameters oriented along the trigonal axis at 77 K as a function of acceptor dopant concentration N_a. The dashed curves represent the results assuming there are no T-point holes. The inset shows $Z_{1D}T$ calculated as a function of N_a for 5 nm nanowires.

Appendix B10, Courtesy Ref[11] Appl. Physs. Lett. 2003, 82, 415

A. A. Balandin and O. L. Lazarenkova   417

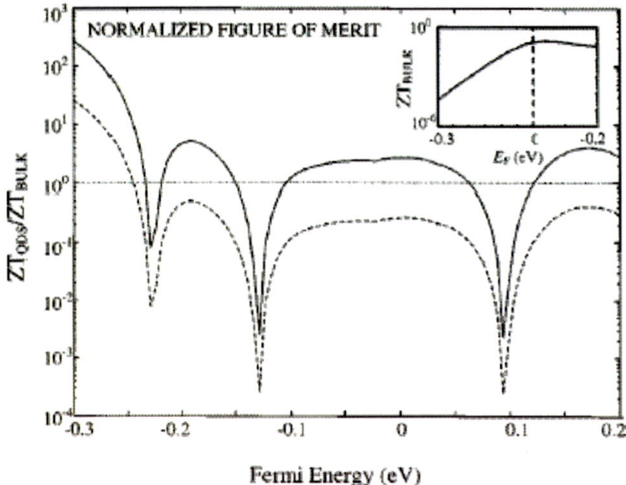

FIG. 3. Room temperature thermoelectric figure-of-merit ZT for p-type Ge/Si quantum dot superlattice normalized to the bulk Si value vs Fermi energy. Note an order of magnitude ZT increase in a wide range of QDS parameters over bulk silicon value. QDS parameters are the same as in Fig. 2. Solid line corresponds to $\kappa_L = 15$ W m^{-1} K^{-1}, while dashed line corresponds bulk Si value of $\kappa_L = 156$ W m^{-1} K^{-1}. The inset shows ZT for the bulk Si vs Fermi energy.

Appendix B11, Ref[25] Nature, 2001, 413, 597

NATURE | VOL 413 | 11 OCTOBER 2001 | www.nature.com

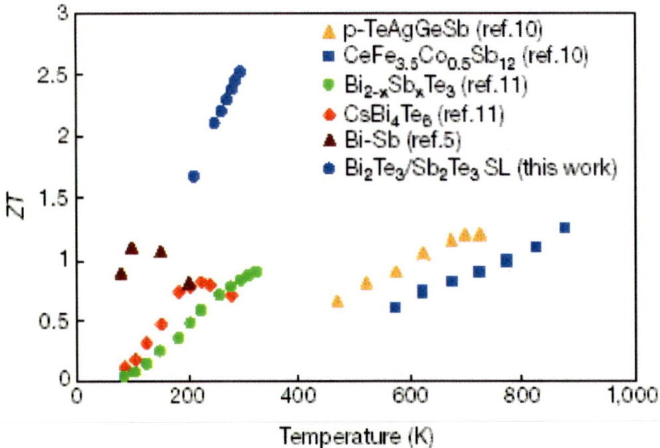

Figure 3 Temperature dependence of ZT of 10Å/50Å p-type Bi$_2$Te$_3$/Sb$_2$Te$_3$ superlattice compared to those of several recently reported materials.

Appendix B12, Courtesy Ref[6] Appl. Phys. Lett. , 1996, 68, 1996

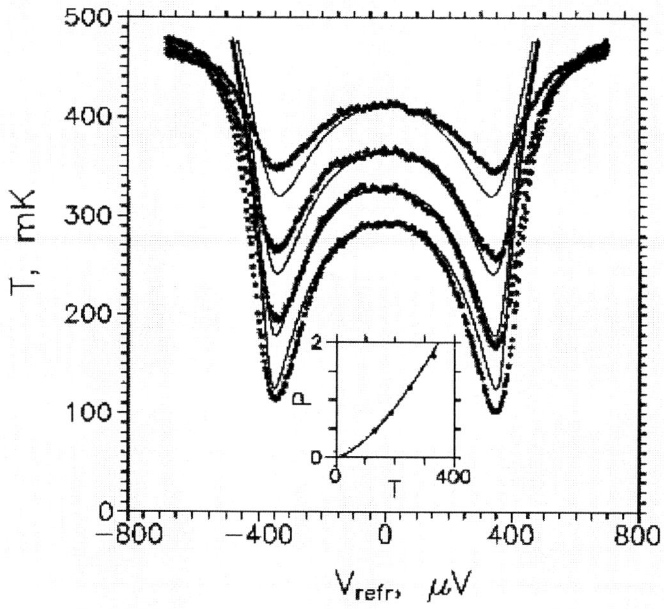

FIG. 3. SINIS refrigerator performance at various starting temperatures. Notations are the same as in Fig. 2. Dots are the experimental data, while the solid lines show the theoretical fit with one fitting parameter Σ for all curves. The inset shows the cooling power P [pW] from the fits (dots), together with the analytical result from Eq. (2).

Appendix B13, Courtesy Ref. Appl. Phys. Lett., 1994, 65, 3123

3124   Appl. Phys. Lett., Vol. 65, No. 24, 12 December 1994

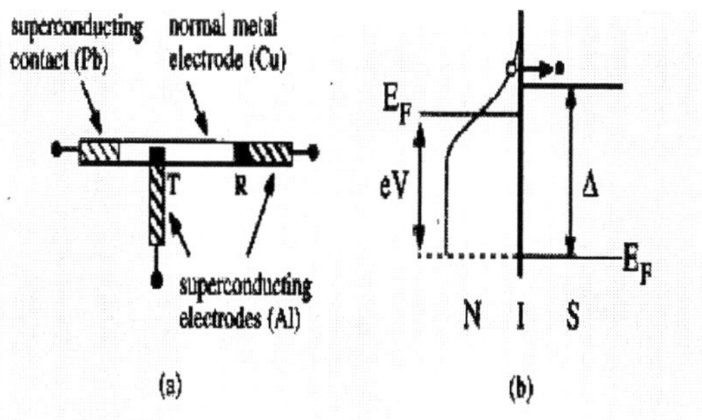

FIG. 1. (a) Schematic of the electron refrigerator. The thermometer and refrigerator tunnel junctions, labeled T and R are depicted as black squares. (b) Energy level diagram for the refrigerator junction. The junction is biased close to the superconducting gap Δ, so that only electrons whose energy is higher than E_F tunnel out of the normal electrode.

Appendix B14, Courtesy Ref[28] Rep. Prog. Phys., 1988, 51, 459

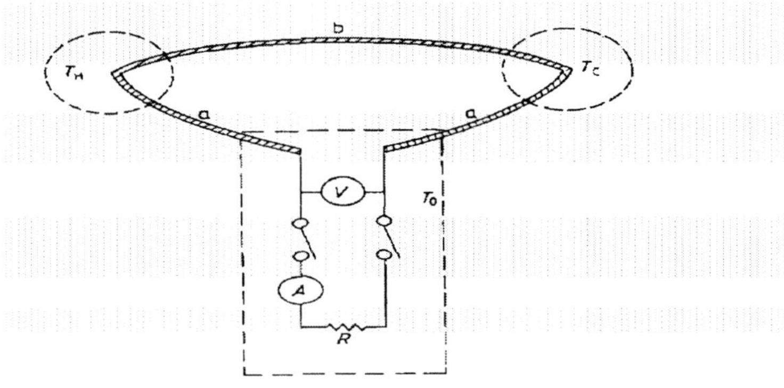

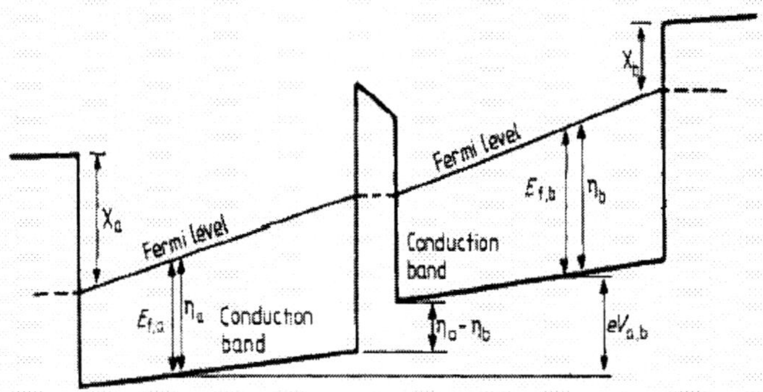

Figure 2. Potential energy diagram of the contact between two dissimilar metals in a temperature gradient.

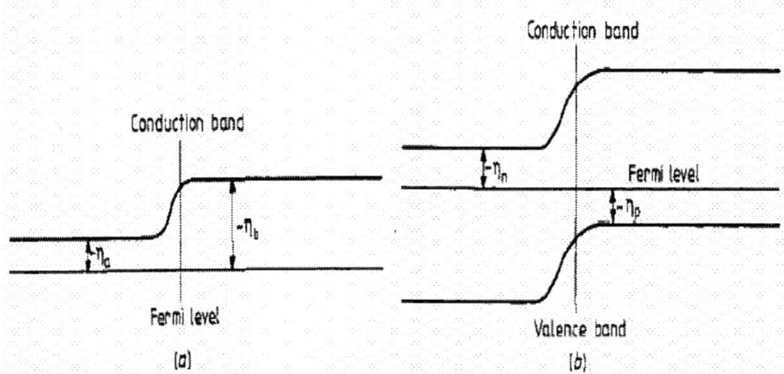

Figure 3. Potential energy diagram (a) for the contact between two n-type semiconductors and (b) a p-n junction.

Appendix C1 Courtesy Ref. J. Appl. Phys. 2000, 88, 6758

J. Appl. Phys. 82 (10), 15 November 1997

J. Appl. Phys., Vol. 88, No. 11, 1 December 2000

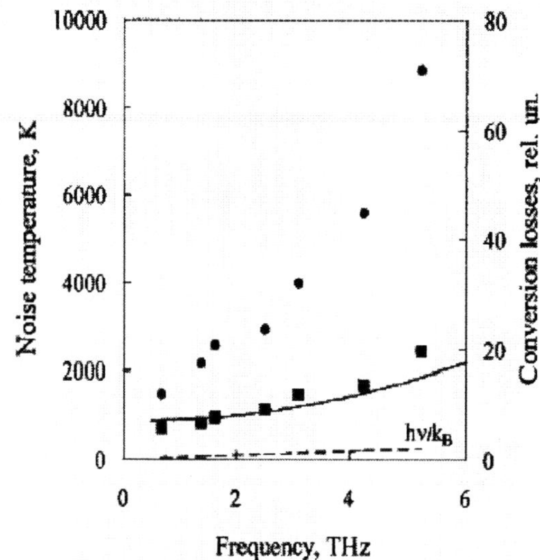

FIG. 3. DSB system noise temperature measured (circles) and corrected (squares) for losses in optical elements. Dashed line represents quantum limited noise temperature $h\nu/k_B$. Solid line shows calculated conversion losses. The scale of the right axis was adjusted in order to match calculated conversion losses and corrected noise temperature.

Appendix C2, Courtesy Ref[22], J. Appl. Phys, 1997, 88, 4719

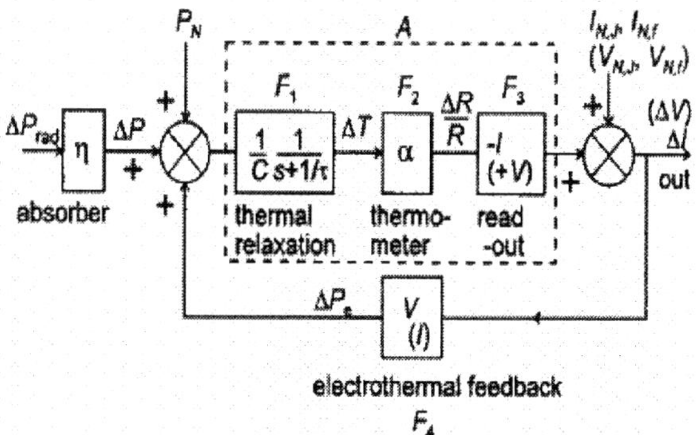

FIG. 1. Functional block diagram of a bolometer. The text between the brackets holds for current bias with voltage readout, the alternatives without brackets for voltage bias with current readout. The contributions to the signal from phonon noise (P_N) of the thermal conductance G, and from Johnson and $1/f$ noise $(V_{N,J}$ and $V_{N,f}$, respectively, $I_{N,J}$ and $I_{N,f})$ of the thermometer resistance are also indicated.

Appendix C3, Courtesy Ref.[35], J. Appl. Phys., 1997, 81,1581

1582   J. Appl. Phys., Vol. 81, No. 3, 1 February 1997

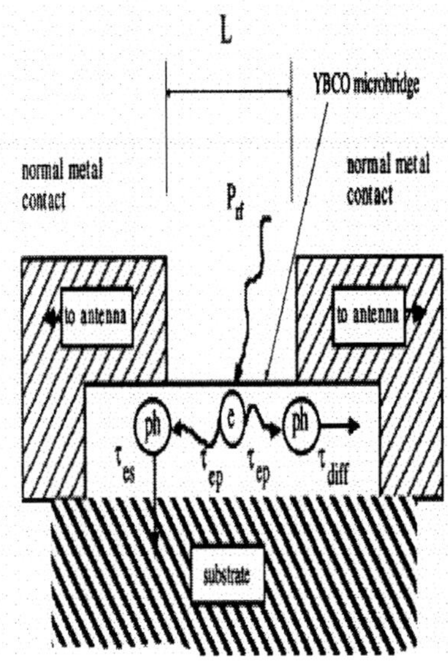

TABLE I. Physical parameters of YBCO at ~90 K.

Electron specific heat c_e $(J\,K^{-1}\,cm^{-3})$	0.025	Ref 19
Phonon specific heat c_p $(J\,K^{-1}\,cm^{-3})$	0.65	Ref. 20
Phonon thermal conductivity (in a–b plane) κ_p $(W\,K^{-1}\,cm^{-1})$	0.1	Ref. 25
Electron thermal conductivity (in a–b plane) κ_e $(W\,K^{-1}\,cm^{-1})$	0.01	Ref. 18
Thermal boundary resistance (MgO substrate) R_b $(K\,cm^2\,W^{-1})$	5.0×10^{-4}	Refs. 21, 22

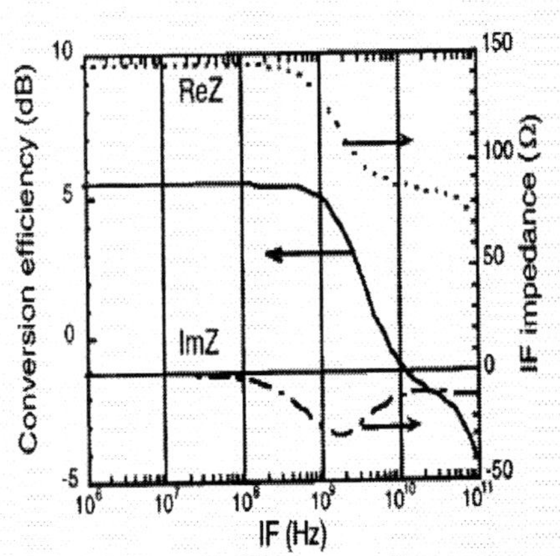

FIG. 6. Conversion efficiency vs IF and the IF impedance at the optimal operating point.

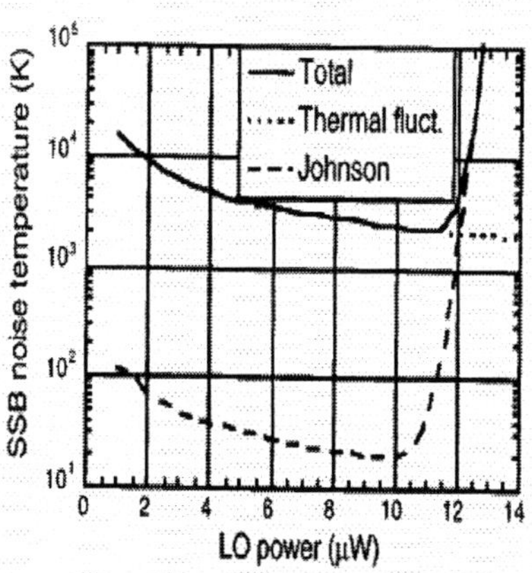

FIG. 8. Noise temperature as a function of the LO power in the vicinity of the optimum point. $T=66$ K, $d=10$ nm, $L=0.1$ μm.

Appendix C4, Courtesy Ref[Appl. Phys. Lett. 1996, 68, 853

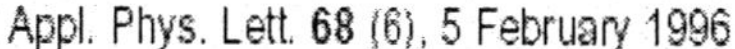

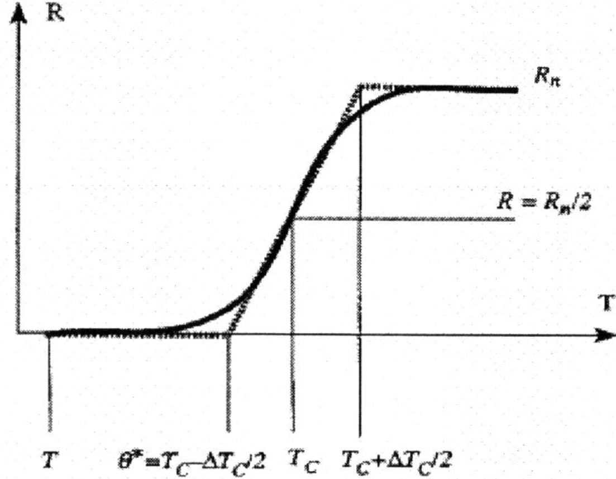

FIG. 1. A broken-line model of the superconducting transition.

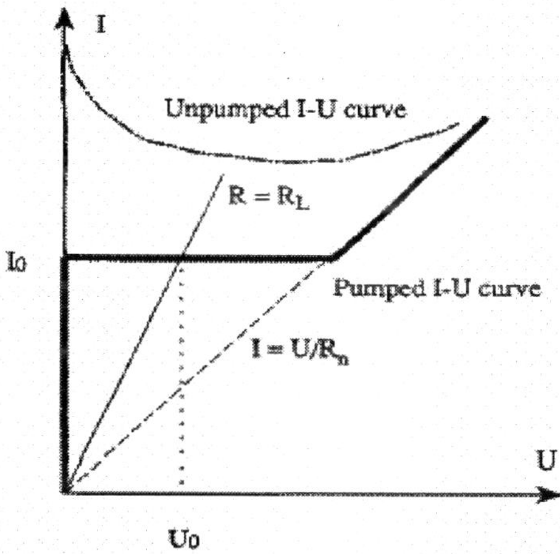

FIG. 2. Unpumped and pumped [$C=1$, $Z(0)=\infty$] IU characteristics.

References

[1]   D. M. Mittleman, J. Cunningham, M. C. Nuss, and M. Geva, *Appl. Phys. Lett.*, 1997, 71, 16-19.

[2]   H. Wald, P. Seidel, H. Schneidewind, Y. Tominari, H. Murakami, M. Tonoouchi, *Physics C*, 2002, 378-381, 372-376.

[3]   T. Kiwa, M. Tonouchi, and *Physics C*, 2001, 362, 314-318.

[4]   O. V. Misochko, K. Sakai, S. Nakashima, *Physics C*, 2000, 329, 12-16.

[5]   D. Semenov, G. N. Gol'tsman, and R. Sobolewski, *Supercond. Sci. Technol.*, 2002, 15, R1-R16.

[6]   M. M. Leivo, J. P. Pekola, and D. V. Averin, *Appl. Phys. Lett.*, 1996, 68, 1996-1998.

[7]   Sur, A. Casian, and A. Balandin, *Phys. Rev. B*, 2004, 69, 035306-1-7.

[8]   D. A. Broido, T. L. Reinecke, *Phys. Rev. B*, 2001, 64, 045324-1-10.

[9]   L. D. Hicks, and M. S. Dresselhaus, *Phys. Rev. B*, 1993, 47, 16631-16634.

[10] Yu-M. Lin, X. Sun, and M. S. Dresselhaus, *Phys. Rev. B*, 2000, 62, 4610-4623.

[11] A. Balandin, and O. L. Lazarenkova, *Applied Physics Lett.*, 2003, 82, 415-417.

[12] L. Pope, T. M. Tritt, M. A. Chernikov, and M. Feuerbacher, *Phys. Rev. B*, 1999, 75, 1854-1856.

[13] G. S. Nolas, D. T. Morelli, T. M. Tritt, *Annu. Rev. Mater. Sci*, 1999, 29, 89-116.

[14] R. T. Littleton IV, T. M. Tritt, J. W. Kolis, and D. R. Ketchum, *Phys, Rev. B*, 60, 1999 , 13453-13457.

[15] M. M. Kaila, J. *Superconductivity*, 2004, 17, 339-343.

[16] V. Klmel, A. Kirilyuk, A. Tsvetkov, R. V. Pisarev, and Th. Rasing, *Nature*, 2004, 429, 850-853.

[17] S. Ferreira, E. C. Marino, and M. A. Continentino, *Physics C*, 2004, 408-410, 169-170.

[18] X. Sun, Z. Y. Jia, Y. H. Huang, J. W. Harell, D. E. Nikles, K. Sun, and L. M. Wang, *J. Appl. Phys.*, 2004, 95, 6747-6749.

[19] S. S. Kang, D. E. Nikles, and J. W. Harell, *J. Appl. Phys.*, 2003, 93, 7178-7180.

[20] F. Khapikov, J. W. Harrell, H. Fujiwara, and C. Hou, *J. Appl. Phys.*, 2000, 87, 4954-4956.

[21] M. Lindgreen, M. Currie, C. Williams, T. Y. Hsiang, P. M. Fauchet, R. Sobolewski, S. H. Moffat, R. A. Hughes, J. S. Preston, and F. A. Hegman, *Appl. Phys. Lett.*, 1999, 74, 853-856.

[22] M. J. M. de Nivelle, M. P. Bruijn, etal, *J. Appl. Phys.* 1997, 82, 4719-4728

[23] M. M. Kaila, J. W. Cochrane, and G. J. Russell, *Supercond. Sci. Technol.*, 1997, 10, 763-765.

[24] M. M. Kaila, *J. Instn. Telecom. Engrs.*, 1969, 15, 671-675.

[25] Rama Venkatasubramanian, Edward Siivola, Thomas Colpitts and Brooks O'Quinn, *Nature*, 2001, 413, 597-602.

[26] D. M. Rowe and C. M. Bhandari, *Modern Thermoelectricas, Holt*, Rinehart and Winston Ltd., Eastbourne, East Sussex, U.K., 1983.

[27] H. J. Goldsmid, *Thermoelectric Refrigeration*, Plenum, New York, 1964.

[28] Wood, Materials for Thermoelectric Energy Conversion, *Rep. Prog. Phys.*, 1988, 51, 459-539.

[29] *Solid State Physics*, J. S. Blakemore, 1985, Cambridge University Press, New York, USA.

[30] M.Galffy, Ch. Hohn, and A. Freimuth, *Annalen der Physik*, 1994, 3, 215-224.

[31] M. Nahum, T. M. Eiles and John M. Martinis, *Appl. Phys. Lett.*, 1994, 65, 3123-3125.

[32] M. M. Kaila, *Physica C*, 2004, C 406, 205-209.

[33] M. M. Kaila, *Supercond. Sci. Technol.*, 2004, 17, 140-142.

[34] M. M. Kaila and G. J. Russell, *J. Phys.. D : Appll. Phys.*, 1998, 31, 1987-1990.

[35] S. Karasik, W. R. McGrath and M. C. Gaidis, *J. Appl. Phys.*, 1997, 81, 1581-1589.

[36] S. Karsik and A. I . Elantiev, 1996, 68, 853-855.

Chapter 2

USE OF ACOUSTIC EMISSION IN STUDYING HIGH-T_C SUPERCONDUCTING PHENOMENA

E. Dul'kin and M. Roth

Department of Applied Physics, The Hebrew University of Jerusalem,
Jerusalem 91904, Israel

Abstract

Acoustic emission is widely known as a nondestructive method for investigating the dislocation movement and accumulation accompanying plastic deformation as well as the generation and propagation of cracks in solid state materials subjected to mechanical stress. Other extensively studied sources of acoustic emission include martensitic phase transitions in metals and alloys under thermal ramping and martensitic-like structural phase transitions in ferroelectric and ferroelastic materials under both thermally and electric field induced stresses. During the last decade, the acoustic emission method has been successfully applied to studying the physical properties of high-T_c superconductors under variable temperature, electric current and external magnetic field conditions. The most important issues emphasized in the present review are: (i) superconducting and structural phase transitions in a wide temperature range, (ii) kinetics of superconducting ceramics sintering and oxygenation, (iii) dislocation mechanisms of mechanical work hardening during long term thermal cycling and (iv) magnetic flux penetration into the superconductor and flux lines pinning and interaction. Most of the results have been obtained with YBCO ($YBa_2Cu_3O_x$) ceramics, yet some properties of BISCCO ($Bi_2Sr_2CaCu_2O_x$) high-T_c superconducting composite tapes have been also addressed. We show that by monitoring the acoustic emission bursts it is possible to measure the temperature hysteresis of phase transitions and to reveal their order, to determine the temperature of maximal oxygen absorption (and calculate the absorption kinetic coefficient) as well as to measure the lower critical magnetic field H_{cl} and the full penetration field H^* under electrical current transport. The cumulative results demonstrate that acoustic emission method is an indispensable tool for studying the high-T_c superconducting phenomena.

Introduction

The class of phenomena whereby transient elastic waves (in the ultrasound range) are generated by the rapid release of energy from localized sources within a material is known as acoustic emission (AE), which is associated with structural reconstructions within the solid state under the influence of external forces [1]. AE is widely known as a nondestructive method for investigating the kinetics of defect production, such as movement and accumulation of dislocations accompanying plastic deformation and their annihilation, twinning and movement of twin walls and of phase boundaries (PB) as well as the generation and propagation of cracks in solid state materials subjected to mechanical stress [2]. Being a very sensitive, AE is able to predict the destruction of the material on the earl stage of the load.

Another extensively studied source of AE includes martensitic phase transitions (PT) in metals and alloys under thermal ramping [3-13]. A detailed analysis of the different mechanisms underlying this particular phenomenon reveals that the greatest contribution to the AE accompanying martensitic PT is made by processes associated with the generation (or annihilation) and movement of dislocations in the metal structure. The origin of dislocation production is the crystallographic mismatch between the original and the new phases at the PB. In case of coherent matching of the phases, or an ideal PB, the AE signal is very low or absent. Thus, the AE activity, $\dot{N}$ (sec^{-1}), measured by means of a piezoelectric transducer reveals the temperature of the martensitic PT, T_m, and is commensurate with the degree of crystallographic coherence of the phases during the PT. Moreover, AE reflects the dislocation density changes in course of durable thermal cycling through the PT region, also called phase work hardening (PWH). The appearance of the PWH is indicative of creation and accumulation of dislocations, their movement and interaction, as well as of interaction between dislocations and grain boundaries (GB). The result of the latter can be both annihilation and filling of GBs. Upon repeated thermal cycling, the number of incremented dislocations decreases from cycle to cycle, and the AE signal tends to decrease exponentially in process of metal hardening [11,13]. However, there are some cycles during the process of dislocation accumulation which involve annihilation of dislocations, and this annihilation is accompanied by an additional enhanced AE activity [13].

A martensitic-like AE response is observed also in course of the phase transitions in ferroelectric and ferroelastic crystals. By employing the AE method in studying the $BaTiO_3$ and $SrTiO_3$ ceramic materials it has been possible to detect all ferroelectric- ferroelectric-paraelectric PTs [14]. More elaborate investigations show that AE is an indispensible method for characterizing also the smeared PT in $BaTiO_3$ defect-containing crystals [15-19]. Specifically, the AE measurement allows distinguishing the Curie temperature (T_c) of the defected surface layer, which is shifted by a few degrees below the T_c of the pure crystal bulk. In $PbTiO_3$ crystals, for example, a minimum AE response is obtained in the case of domain twinning through the ferroelectric PT, in similarity with the martensitic PT that is due to formation of a coherent PB [20]. AE has also been applied to studying the dependence of the dislocation density in $PbTiO_3$ crystals on the PB orientation, in relation to the direction of the thermal field gradient; an AE maximum has been found when the PB is oriented at about 45° angle relatively to the direction of thermal field gradient [21]. In this particular case, cracks sometimes shout on the PB causing a very power AE output [22]. Similarly to the martensitic

PT, the PWH has been detected in PbTiO$_3$ crystals [23] as well as in (Na$_{1-x}$Li$_X$)NbO$_3$ binary solid solution ceramics [24] in course of prolonged thermal cycling. This is manifested by the appearance of an $\dot{N}$ maximum during the 6th cycle in PbTiO$_3$ (at the background of the exponentially decreasing AE activity) and $\dot{N}$ minimuma during the 3rd and 8th cycles in (Na$_{1-x}$Li$_X$)NbO$_3$, as it has been observed earlier in NiTi-based alloys [13]. This identity in the AE behavior in metals and ferroelectrics emphasizes the universality of the AE method in its use for studying structural PTs in the solid state matter. The appearance of PWH is also observed in relaxor ferroelectric Pb(Mg$_{1/3}$Nb$_{2/3}$)O$_3$ crystal through the diffusive PT [25]. In the (Na$_{1-x}$Li$_X$)NbO$_3$ ceramics, AE accompanies two high-temperature ferroelectric-ferroelectric and ferroelectric- paraelectric PT, which have not been detected by the traditional dielectric method [26]. A similar uniqueness of the AE method has been demonstrated in the case of Ba$_{0.85}$Sr$_{0.15}$TiO$_3$ posistor ceramics, where the PT cannot be detected by resistance measurements [27]. On the other hand, in PbZrO$_3$ and PbHfO$_3$ crystals AE is absent through the antiferroelectric-antiferroelectric PT because of the complete coherence of these phases, yet weak AE accompanies the antiferroelectric-ferroelectric PT and strong AE accompanies the ferroelectric-paraelectric PT due to the corresponding incoherent PBs [28]. A similar AE effect is observed in Pb(Fe$_{0.5}$Nb$_{0.5}$)O$_3$ crystals through the ferroelectric-ferroelectric-paraelectric PT [29].

The phenomena described above show clearly that AE is an indispensable method of studying many kinetic features of the structural PTs in solid state materials, such as determining the T$_c$ values and characterizing the ΔT$_c$ hysteresis, identifying the order of PTs, estimate the degree of inter-phase coherence at phase boundaries, defining the degree of hardening based on the PWH data [30], etc. It is well known, that high-T$_c$ superconductors undergo PTs as well, including the superconducting transition. Some of the high-T$_c$ superconductors have a perovskite-like crystallographic structure, like the ferroelectric crystals, and they undergo structural PTs similar to the martensitic-like transitions. This explains the extensive efforts that have been made during the last decade to apply the AE method to investigating the high-T$_c$ superconducting phenomena. Below, we review the main results recently obtained with the best characterized YBCO (YBa$_2$Cu$_3$O$_x$) and BISCCO (Bi$_2$Sr$_2$CaCu$_2$O$_x$) high-T$_c$ superconductors.

Samples and Experimental Procedure

Superconducting grade YBCO samples are usually prepared as follows. The batch of YBa$_2$Cu$_3$O$_X$ is synthesized by a solid state reaction of analytically pure powder materials, Y$_2$O$_3$, BaCO$_3$ and CuO. After accurate weighing and careful intermediate grinding, the initial mixture is sintered in air at 920°C for 12 h. The resulting mixture is powdered and then compacted in bulk rectangular blocks of 10×10×40 mm dimensions under 10 MPa pressure. These blocks are further sintered at 950°C in air for 24 h and subsequently cooled in oxygen at a rate of about 100°C/h. Smaller rectangular shaped samples of 5x5x20 mm dimensions are cut from these blocks. These samples typically exhibit the high superconducting transition temperature, T$_c$ = 92 K, and a narrow transition width, ΔT$_c$ = 0.2 K. The material density is about 5.4 g/cm^3, grain size is in the 6-9 μm range. The preparation of BISCCO superconducting composite tapes is described elsewhere [31].

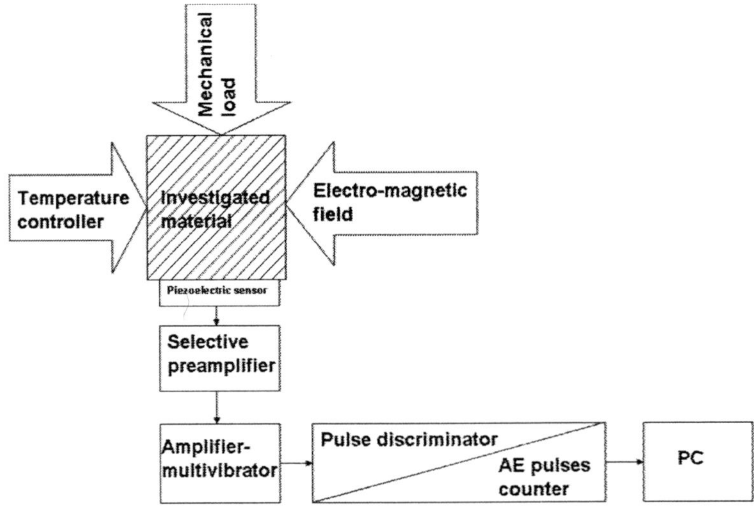

Fig. 1. Operational scheme for studying materials by AE method under mechanical, temperature or electromagnetic loading.

The common experimental procedure of AE measurements is simple, and the basic setup is presented schematically in Fig. 1. Due to an external force of mechanical, thermal or electromagnetic nature, the investigated material produces elastic (ultrasonic) waves, which are converted to electrical signals by direct coupling to a piezoelectric sensor. Then output of the piezoelectric sensor is amplified through a frequency-selective low-noise preamplifier, filtered and additionally amplified through an amplitude discriminating amplifier and converted to voltage pulses through an amplifier-multivibrator, which are counted and displayed in time units. Usually, three parameters of the AE are being measured: (i) total signal amplitude $\sum A$, (ii) total number of pulses $\sum N$ and (iii) activity $\Delta N/\Delta t = \dot{N}$ (s^{-1}). The latter parameter is most commonly determined.

It is noteworthy that both in the case of low- and high-temperature experiments it is undesirable to subject the AE sensor to nonambient temperatures. Therefore, a quartz glass waveguide is usually introduced as a buffer transmitting the ultrasonic waves from the studied material to the AE sensor [2]. There are three specific AE experimental setups allowing for convenient high- and low-temperature measurements, which are described in some detail below:

1. *High-temperature set up* (Fig.2). The sample is glued with a high-temperature epoxy resin to the polished end of the fused quartz acoustic waveguide. A piezoelectric PZT-19 ceramic sensor is glued to the opposite end of the waveguide and connected to a 500 kHz band-pass preamplifier. The sample comprising the top part of the waveguide are mounted in a resistance furnace. A Ch/Al thermocouple is attached to the waveguide near the sample. Two pinned rods connected to an external differential dilatometer are monitoring the sample size.

The AE activity $\dot{N}$ and thermal expansion ΔL are simultaneously measured during heating and cooling at a rate of about 1-2 K/min with or without an oxygen flow.

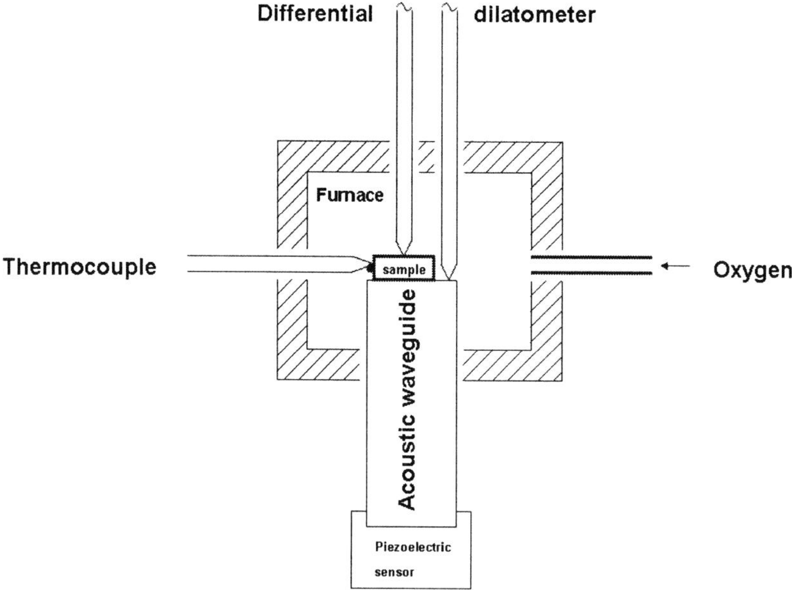

Fig. 2. High-temperature setup for the simultaneous measurements of thermal expansion and AE activity, $\dot{N}$, in the 0 - 1000°C temperature range.

2. *Low-temperature setup (a)* (Fig. 3). The configuration of the experiment is the same as above, with the addition of an induction coil and use of liquid nitrogen vapour for cooling the sample. The induction coil is used to measure the magnetic susceptibility χ at the frequency of 1 MHz. The AE activity $\dot{N}$, susceptibility χ and thermal expansion ΔL are simultaneously measured during heating and cooling at a rate of about 1-2 K/min.

3. *Low-temperature set up (b)* (Fig. 4). Hereby, the sample is glued to the bottom end of the acoustic waveguide, while the piezoelectric sensor is adhered to its top end. A similar 500 kHz band-pass preamplifier is used. The sample with the lower part of the waveguide is submerged into liquid nitrogen. The temperature is monitored by a Cu-K thermocouple attached to the waveguide near the sample. DC electric current is applied through two silver epoxy contacts on opposite side ends of the sample. The liquid nitrogen Dewar flask is mounted between the two poles of a DC magnet. AE activity $\dot{N}$ is measured at 77K in the presence of an electric current flow or magnetic field.

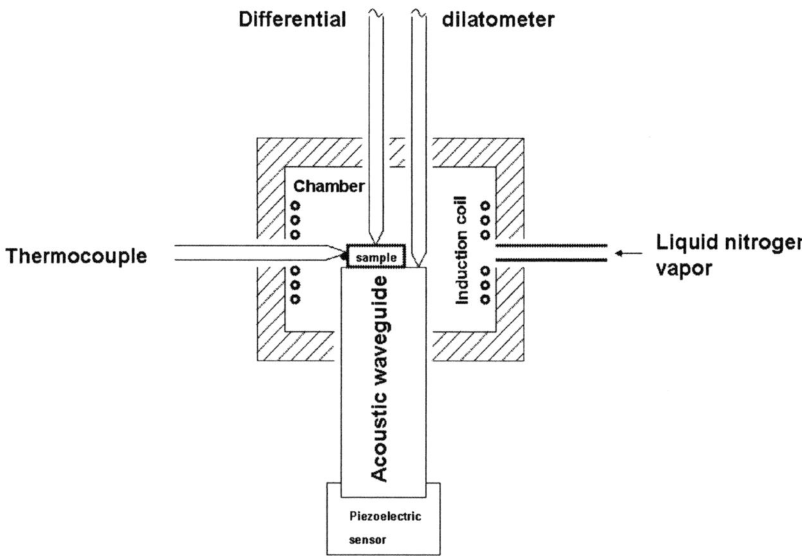

Fig. 3. Low-temperature setup (a) for simultaneous measurements of thermal expansion, AE activity, $\dot{N}$, and magnetic susceptibility in the 77-300K temperature range.

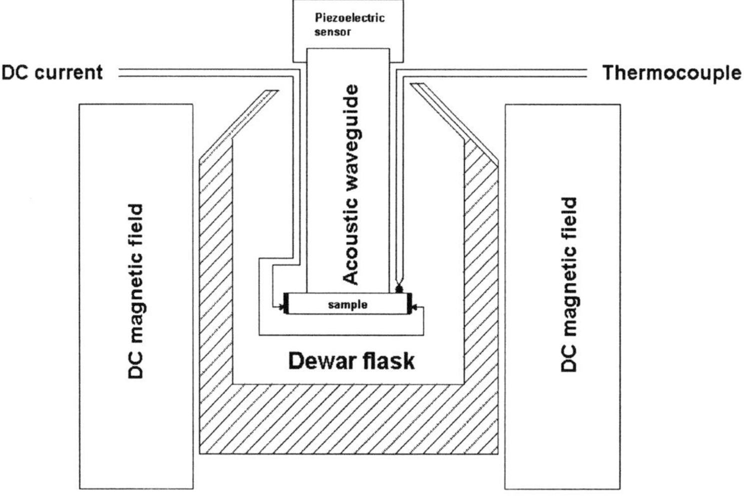

Fig. 4. Low-temperature setup (b) for measurements of AE under external magnetic field and transport current conditions.

Sintering and Oxygenation

Practical application of oxide superconductors requires bulk materials with greatly improved current-carrying capacities. Since it is well known that the critical current density of sintered YBCO ceramics appears to be limited by intergrain resistance, it is essential to control the grain growth. During the process of ceramics sintering, spontaneous grains growth is observed experimentally. In YBCO, the grains appear in the temperature range of about 800-900°C and they continue to grow in size with a rate proportional to the temperature gradient in the material volume [32]. The anisotropic expansion and contraction of the grains during thermal processing produce an AE signal, which carries information about the size of the grains formed.

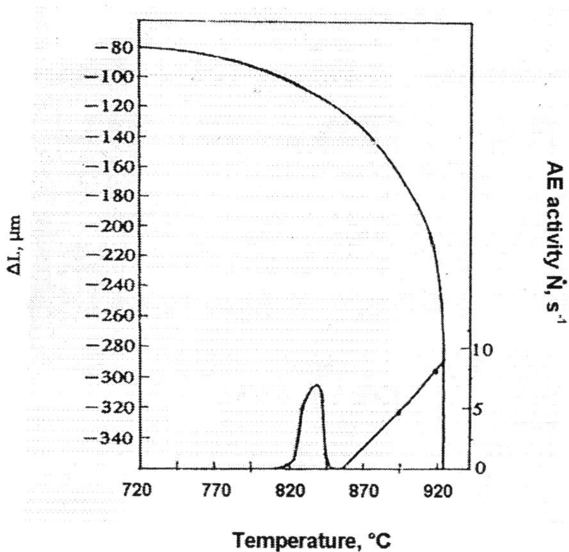

Fig. 5. AE activity and mechanical deformation as a function of temperature during sintering of YBa$_2$Cu$_3$O$_X$ compacted powder.

In order to study the grain growth kinetics in course of sintering, the prereacted YBCO material has been pressed into 10×10×40 mm rectangular blocks, and the latter have been introduced as samples into the high-temperature setup presented in Fig. 2. The results of both AE and dilatometric measurements during the ceramics sintering are shown on Fig. 5 (also see ref. [33]). Apparently, the sample shrinks as the sintering temperature increase, while two characteristic shrinkage stages can be observed. The first stage of gradual shrinking covers a broad temperature range from 720 to about 900°C. This is followed by a rapid shrinkage stage in a narrow temperature range of 900-930°C, but the dilatation   L remains constant above 930°C.

The AE activity of YBCO displays three characteristic stages. AE is initiated above 810°C, and a relatively sharp activity peak is observed in the 820-840°C temperature range. This is followed by a narrow second stage, where essentially no AE can be detected. However, above 850°C, the AE activity reemerges and increases nearly linearly with further temperature ramping. Just above this temperature, regular grain growth commences. Fig. 6 presents the polarized optical micrographs of YBCO sintered at three different temperatures, 890, 920 and 950°C. The average grain sizes are in the 0.5-0.8, 1.0-1.5 and 2.2-3.5   m respectively, clearly growing with the increase of sintering temperature. Thus, on-line monitoring of the AE activity describes usefully and accurately the sintering process, including the onset of sintering at 820°C and the initial temperature of grain formation at 850°C.

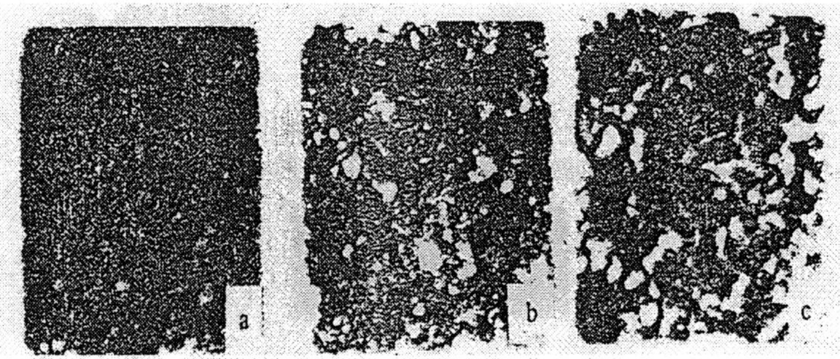

Fig. 6. Polarized optical microphotographs of grains grown during sintering of compacted $YBa_2Cu_3O_X$ powder. Grains with average sizes of 0.5-0.8 (a), 1.0-1.5 (b) and 2.2-3.5 mm (c) are obtained for sintering at 890, 920 and 950°C repectively.

The measured AE temperature dependence may be interpreted within the framework of a qualitative model [34]. The model suggests that sintering proceeds in three stages. During the first stage, the pellet of pressed powder shrinks, and the material density increases. The shrinkage is accompanied by considerable mechanical stresses in the powder generating AE. During the second stage, a fluid glass phase appears at a higher temperature. This glass phase acts as a "lubricant", facilitating the material's further shrinkage. Such "lubrication" decreases the friction in the system, which is confirmed by absence of AE. During the third stage, grain growth starts due to recrystallization. Unstrained crystallites take up material and grow into the neighboring strained (heavily plastically deformed) areas of the same phase, being gradually increased in size. This results in an increase in the area of the stressed intergranular boundaries, which involves climb or cross-slip of dislocations as they rearrange into the moving boundary. For larger grains, or larger intergranular area, the relief of plastic deformation strain is accompanied by an increased AE associated with the recrystallization process. Therefore, AE can be used efficiently as a nondestructive method for grain size technological monitoring during the sintering process.

Sintering is only the first important step of the YBCO preparation. The second necessary step is oxygenation, which is crucial for obtaining the material in its superconducting state. Oxygen content determines the actual superconducting PT temperature, or so-called critical temperature (T_c). On cooling, after the sintering, of the initially tetragonal $YB_2Cu_3O_X$ phase in oxygen atmosphere the supercondicting orthorhombic II-phase (O-II) appears at about 650°C. In this process, inflowing oxygen ions engage in occupying some of the vacant sites to form O-Cu-O chains. When the oxygen stoichiometric coefficient x ~ 6.5, most of the alternating O-Cu-O chains are filled with oxygen. The T_c of O-II is close to 60 K. The lattice strain associated with the incorporation of extra oxygen ions and the consequent T→O-II phase transition is accommodated by crystallographic twinning in the (110) plain. The 60K O-II phase nucleates and grows gradually in the tetragonal matrix. On further oxygenation, oxygen ions fill the vacant sites in the O-Cu-O chains completely as x reaches the value of 7. The lattice parameter a becomes half of that in the O-II phase implying that the O-II phase transforms to a new phase, O-I, with the T_c increasing to about 90K. Since the O-II→O-I phase transition is similar to the martensitic-like PT in FE materials, AE is expected to be a suitable method ofo studying the oxygenation processes in the $YB_2Cu_3O_X$ material.

An AE response has been indeed registered in YBCO on cooling in air after sintering in the 911-922°C temperature range [35]. A sharp and intense peak of AE activity has occurred in a relatively narrow temperature range between 800 and 600°C. The authors have attributed this peak to generation of cracks as the oxygen enters the crystallographic lattice and induces thermal expansion anisotropy. Indeed, small cracks have been observed at the surfaces of the ceramic samples after oxygenation by polarized light microscopy [36]. It is noteworthy that the pronounced AE signal jump occurs at 650°C, where the T→O-II PT takes place. The latter has been also confirmed by AE measurements on cooling in oxygen atmosphere [37]. The AE activity starts rising on cooling below 700°C and it reaches its maximum at 650°C. Then $\dot{N}$ decreases slowly, revealing an additional small maximum near 380°C, and becomes negligible below 300°C. Further studies have revealed that the AE has actually two peaks of $\dot{N}$ on cooling after sintering: at around 650°C and 605°C (Fig. 7) [38]. The first peak corresponds to the T→O-II PT, while the second is obviously attributed to the maximum adsorption of oxygen.

If, indeed, the 605°C peak is associated with oxygen adsorption, it is possible to study the kinetics of this adsorption as well. In general, the adsorption rate can be approximated by an exponential Avrami curve, namely ~ exp(-t/τ), where τ is the relaxation time only slightly dependent on temperature in YB$_2$Cu$_3$O$_X$ [39]. We have studied the $\dot{N}$ variation experimentally. An YBCO pellet pressed into a rectangular shape 10×10×40 mm sample has been sintered at 920°C during 3 hours and then cooled in the air down to 605°C. Then the temperature has been stabilized and an oxygen flow has been introduced to the furnace. Fig. 8 shows that at the beginning of oxygenation the AE signal increases abruptly and then decreases exponentially as $\dot{N}$ = 40exp(-t/1.7) [40], or an average value of τ = 1.7 sec in obtained.

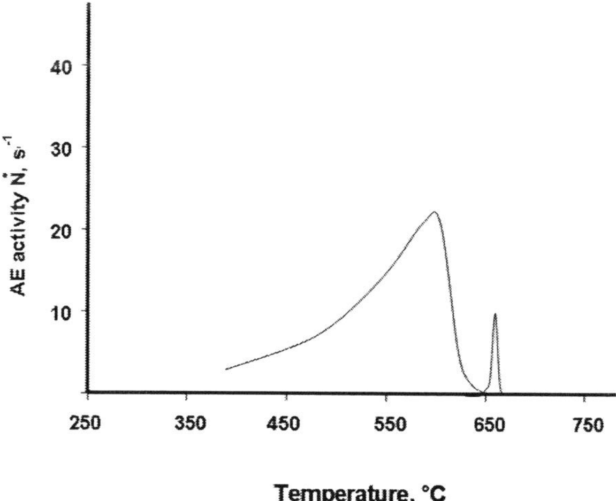

Fig. 7. Temperature dependence of AE activity during cooling of YBa$_2$Cu$_3$O$_X$ ceramics after sintering. First peak at 650°C corresponds to the T→O-II PT and the second peak at 605°C - to highest adsorption of oxygen.

It is interesting to compare the above kinetic results with the mass-spectroscopic data of Krishnan et al. [41] obtained during oxygen desorption at 550°C (insert of Fig. 8). The shape of the curve is similar to the behavior of AE during oxygenation at 605°C. Krishnan et al. have shown that the initial fast desorption depends on the average grain size of the ceramic sample porous matrix. The subsequent exponential decrease depends on oxygen kinetics within the sample, and the desorption (R) varies by the following law: $R \approx 3.5\exp(-t/508)$, where $\tau = 508$ sec. The discrepancy in the values of τ is explained by the large difference in the densities and their grain sizes of the various samples. The specific variation of the AE activity can be also compared with the change in electrical resistivity of the $YB_2Cu_3O_X$ material, which decreases exponentially ($\rho \sim \exp(-\delta/\delta_0)$, where $\delta = 7-x$) during the oxygenation process [42].

AE exhibits interesting features also during sintering of superconducting BISCCO tapes [43]. Tapes composed of the highest-T_c Bi-2223 phase ceramics enclosed in silver cladding have been studied most extensively. The specific feature of such tapes is that cracks arise between primary and secondary sintering, due to intermediate rolling. Therefore, the secondary sintering of Bi-2223/Ag tapes after rolling up to a strain of $\varepsilon \sim 18\%$ has been studied by both the AE and magnetic susceptibility methods for comparison. During the second (post-rolling) thermal treatment of the tape, a broad band of AE from is detected in the temperature range from 570 to 660°C. This broad band can be interpreted on the basis of other data obtained during *in situ* studies of crack generation and healing in Bi-2223/Ag tapes [41]. According to the BISCCO phase diagram [44], a liquid phase exists in the 400-660°C temperature range. Magnetic susceptibility (χ'') measurements have been carried out during heating through this temperature interval. The results show a clear narrowing of the χ'' peak, which is characteristic of enhanced electrical connectivity between the ceramic grains due to healing of the cracks [45]. Consequently, the process of liquid-phase healing of the rolling-induced cracks in the tapes can be regarded as the source of AE.

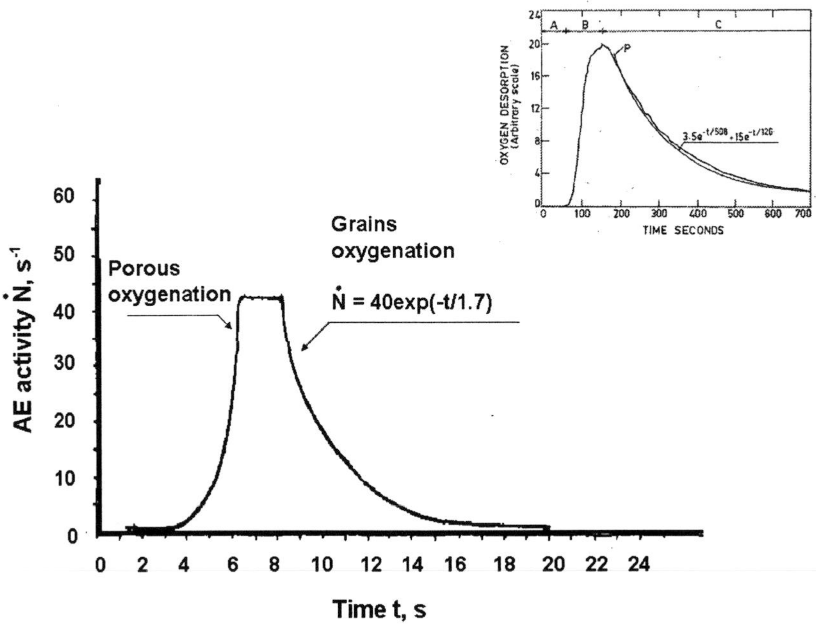

Fig. 8. AE activity associated with the kinetics of $YBa_2Cu_3O_X$ ceramic oxygenation after sintering (insert shows the mass analysis results during oxygen desorption, after [41]).

On cooling the Bi-2223/Ag tape after secondary sintering at 730°C, an AE signal is detected below 230°C as well [46] (see Fig. 9). The AE activity is weak near 230°C, but increases in intensity with decreasing temperature down to room temperature. In order to explain the appearance of AE signals we consider the relative deformation of the silver clad and the ceramic core of the Bi-2223/Ag composite tape. The thermal expansion coefficient (α) of silver and Bi-2223 ceramics are $\alpha_{Ag} = 20.5 \cdot 10^{-6}$ K^{-1} and $\alpha_c = 13.6 \cdot 10^{-6}$ K^{-1} respectively, which implies stronger contraction of silver on cooling. In view of the high plasticity of silver (yield stress 65 MPa) and sufficiently large strength of the sintered ceramic (yield stress 150 MPa), the silver envelope cannot contract the underlying ceramics effectively and experiences considerable tensile strain. Based on the difference in the thermal expansion coefficients, this strain is of the order of 0.35% for the tape cooled from 730°C to 230°C. This exceeds the average elastic limit (0.2%) for metals. The associated plastic deformation causes the formation and movement of dislocations and is accompanied by AE. The observed strong AE activity ($\dot{N}$ values reach several hundreds of s^{-1} at RT), is about an order of magnitude larger than that of martensitic phase transitions in metals.

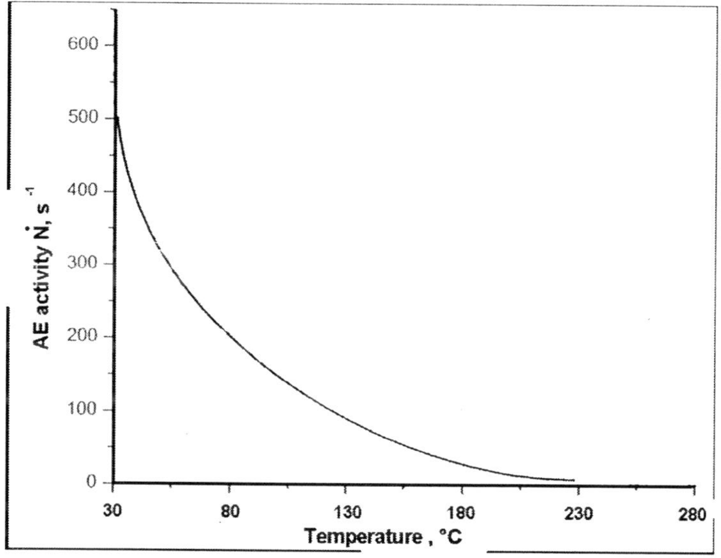

Fig. 9. AE activity of a Bi-2223/Ag tape recorded on cooling after secondary sintering at 730°C.

High Temperature PT

It has been well established that the oxygenated YB$_2$Cu$_3$O$_X$ material loses oxygen atoms during heating, linearly with temperature up to $\sim$ 360°C and superlinearly at higher temperatures. Unexpectedly, a small additional oxygen loss of the order of 0.1% of the YB$_2$Cu$_3$O$_X$ ceramic sample mass has been revealed in the 265-350°C temperature range [47]. This effect is accompanied by an appreciable variation in the a and b lattice constants of the orthorhombic structure due to oxygen loss. The rhombic distortion coefficient $(b - a)/b$ has been studied during heating of YB$_2$Cu$_3$O$_X$ crystals from 20 to 700°C [48]. As the temperature increases, the distortion coefficient decreases linearly and reaches a zero value at the OI→T

PT above 630°C. However, there is an anomaly, in a form of a small plateau, in the narrow 270-300°C range. It has been concluded that this effect is a consequence of oxygen redistribution in the basal plane at the expense of oxygen loss from the crystal. This conclusion is supported by the low temperature resistivity measurements of $YB_2Cu_3O_X$ crystals, namely the existence of a certain amount of the oxygen deficient O-II phase in the O-I phase bulk is detected [49]. Similar conclusions can be drawn from the measurements of the c lattice constant of $YB_2Cu_3O_X$ films during their heating, where an irreversible jump up of c at 250°C has been observed [50].

The established distortion of the crystallographic lattice in the 250-350°C range is the reason behind a number of macroscopic anomalies observed in the measurements of microhardness [49], internal friction [51], Young's modulus [52] and, of course, AE. The AE activity has been detected on heating of oxygenated YBCO samples from 200 to 400°C and explained in terms of a reversible (second order) O-II ↔ O-I PT [53]. Simultaneous AE and thermal expansion measurements reveal a reproducible AE which correlates well with the thermal expansion local minimum near 270°C on heating of an $YB_2Cu_3O_X$ ceramic sample oxygenated to saturation [54].

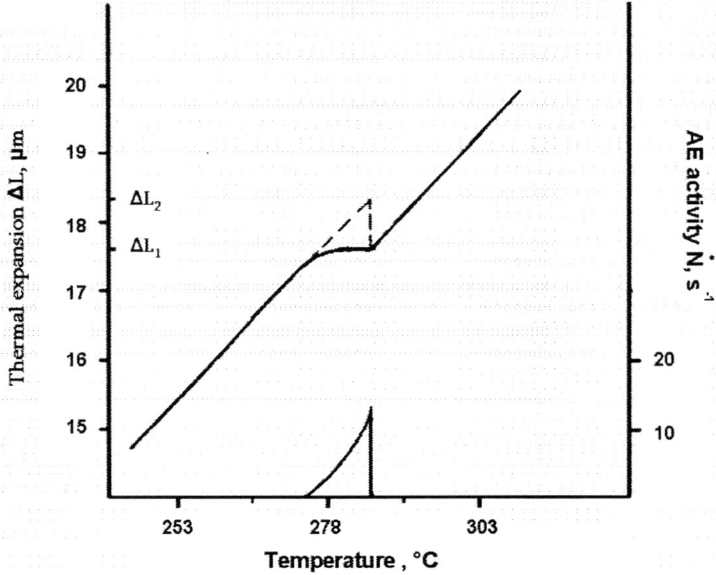

Fig. 10. Dilatometric anomaly at 550K, accompanied by AE, during heating of $YBa_2Cu_3O_X$ ceramics.

Other workers [55] have attempted to freeze in the oxygen atoms in samples exposed to cycled annealing at 260°C, by quenching them in liquid nitrogen. The T_c value of such samples can be as hugh as 97 K. However, subsequent annealing above 100 K reverses the T_c back to the nominal 92 K value on repeated cooling. This effect is explained by redistribution of oxygen vacancies between the different sublattices of the material at $T \geq 100K$. The suggested model may possibly explain also the thermodiffusion of oxygen atoms revealed by internal friction measurements near the T_c, but only in samples initially annealed in the 270-300°C temperature range [56]. These processes are usually accompanied by changes in the unit cell parameters.

Elaborate dilatometric measurements of YBCO show that the c lattice constant decreases on heating thorough the 240-300°C temperature interval [57]. The thermally induced deformation rises linearly below 240°C, but saturates on further heating to 285°C (Fig. 10). The saturation region is accompanied by AE activity. On heating above 285°C, the thermal expansion rises linearly again at the same rate as at lower temperatures. Since for normal thermal expansion the actual sample elongation, ΔL_1, at 285°C is smaller than the nominally expected elongation, ΔL_2, the observed thermal expansion anomaly is associated with a negative elongation ($-\Delta L$). By assuming that the orthorhombic disorder coefficient roughly equals 0.5 throughout the ceramic sample, Δc, can be expressed as :

$$0.5 \, \Delta c = - \, c \, (\Delta L/L), \qquad (1)$$

where the sample length L = 5 mm and c = 11.673Å at 298 K. using the experimental data of Fig. 10 and eq. (1), we obtain Δc = - 0.003Å. The 'elongation' is thus negative to an extent preventing the oxygen atoms from entering the sample lattice. We can further calculate the stoichiometric coefficient x by the linear regression equation connecting the x value and thw c lattice constant size [58] :

$$x = 91.9179 - 7.2757 \, (c - \Delta c) \qquad (2)$$

Together with the previous results, eq. (2) yields x = 7.01.

The large value of the experimentally obtained stoichiometric coefficient is in obvious contradiction with the previously quoted data, such as oxygen losses [47] in associated with an abrupt increase of the c lattice constant [50] and the T_c increase [55] correlated with the c constant decrease during heating through the 240-300°C temperature interval. (We note, however, that the abrupt increase in the c constant has been observed in thin films, where the oxygen diffusion kinetics can differ in comparison with the bulk material.) To explain this contradiction, the existence of different mechanisms of oxygen desorption induced by the O-II↔O-I PT, can be considered. For example, oxygen atoms may enter to so-called apical oxygen O4 sites during the desorption process. These sites are located between the CuO2 superconducting planes and the CuO1 electron reservoir chains, so the oxygen atom in an O4 site plays a role of an "electronic bridge". It can be deduced from the thermal expansion data of the $YB_2Cu_3O_X$ unit cell, based on X-ray diffraction measurements [59], that the Ba-CuO1 bond distance, including the O4 site, suffers a large negative dilatation during oxygenation. Therefore, oxygen atoms entering the apical sites are responsible for the c lattice constant decrease, and the associated increase of x up to 7.01 seems meaningful. Moreover, the increase in apical oxygen content in $YB_2Cu_3O_X$ raises the oxidation degree of cooper, and the T_c of such material may increase upon quenching to the liquid nitrogen temperature after annealing near 260°C, in accordance with the experimental data [55].

Since the apical oxygen plays an important role in the charge transfer between CuO2 superconducting planes and CuO1 electron reservoir chains [60], the hole density changes too. Consequently, heating of thee samples through the 240-300°C temperature interval may influence also the critical current (J_c) value. In order to examine the corresponding J_c behavior, a batch of fully oxygenated YBCO samples have been devided into two parts [61]. The first part has been heated up to 300°C and then quenched in air to RT. The second part has been also heated to 300°C, but then cooled slowly down to RT within the furnace. Each of these sample parts has been thermally cycled between RT and 300°C more than 10 times. During this thermal cycling the AE signal has been measured. At the beginning and end of each thermal cycle the *I-V* characteristics have been recorded at 77 K as well, and the J_c values have been determined using the 1 μV/cm criterion.

The two sets of *I-V* curves are presented in Fig. 11, and the corresponding J_c dependences on the thermal cycle number n are given in the insert. Apparently, the J_c values of quenched samples decreases more slowly than those of the cooled samples. Both dependences are exponential and can be fitted by $J_c = 8500\exp(-n/3.3)$ and $J_c = 8500\exp(-n/0.42)$ for the quenched and slowly cooled samples respectively. The moderate rate of the J_c drop in quenched samples is presumably due to oxygen incorporation into the apical sites of the sample lattice during heating over the 240-300°C temperature interval.

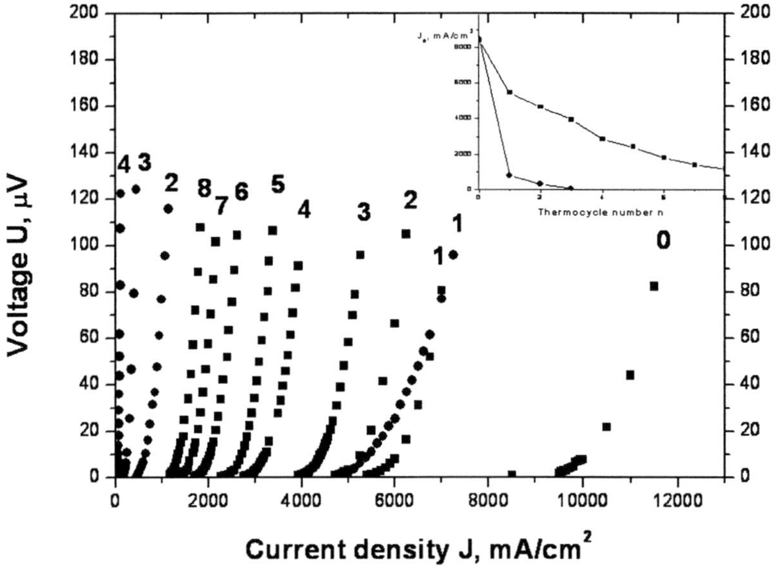

Fig. 11. *I-V* characterictics of YBa$_2$Cu$_3$O$_X$ ceramics during thermal cycling; insert shows evolution of I$_c$ as a function of thermal cycle number. (Squares and circles represent the first and second part of samples respectively).

The results of parallel AE measurements are presented in Fig. 12. The AE activity of slowly cooled samples initially high and reduces gradually with each thermal cycle. The quenched samples behave in a more complex way. Their AE activity is low initially, but rises abruptly through the 3rd and 6th cycles. These AE rising steps correlate well with the relatively faster drops of J_c after the 3rd and 6th cycles. Similar phenomena are observed in ferroelectrics, as it has been already described above, where $\dot{N}$ reaches a maximum value during the 6th thermal cycle in PbTiO$_3$ crystals [23] and drops to minimum values during the 3rd and 8th cycles in (Na$_{1-X}$Li$_X$)NbO$_3$ binary solid solution ceramics [24]. The latter phenomenon has been described earlier also in NiTi-based alloys [13]. By analogy with ferroelectrics and metal alloys, one can conclude that PWH takes place in YB$_2$Cu$_3$O$_X$ during thermal cycling over the 230-270°C range. To our knowledge, this is the first observation of the possible PWH phenomenon in high-T$_c$ superconducting materials. We suggest that the PWH in YBCO ceramics is caused by the O-II→O-I PT due to oxygen redistribution in the quenched material. The enhanced decrease of the J_c during the 3rd and 6th cycles in quenched YBCO can be ascribed to migration of dislocations, formed at the O-II→O-I PBs, to the GBs of the ceramic samples. This causes an increase in the electrical resistivity of the GBs and a corresponding reduction in the J_c value. The suggested mechanism implies that PWH influences the intergrain weak links in the ceramic samples of YB$_2$Cu$_3$O$_X$.

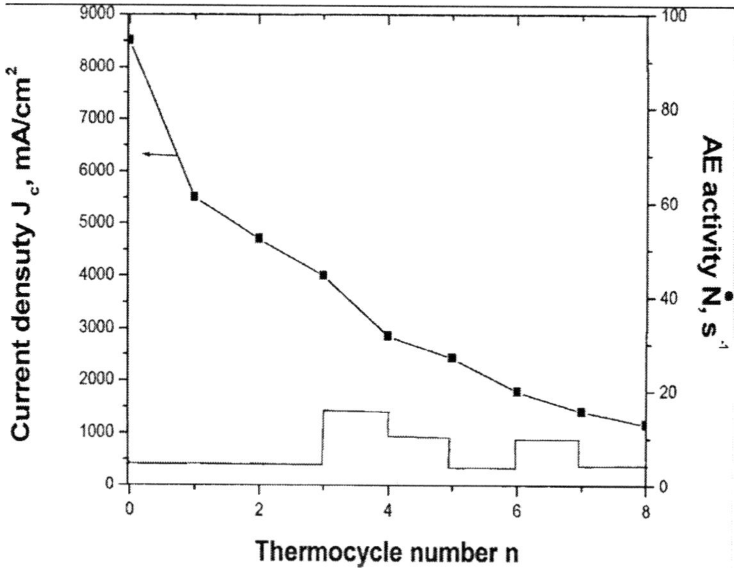

Fig. 12. AE activity and I_c evolution of YBa$_2$Cu$_3$O$_X$ ceramics during thermal cycling through 220-320°C.

The change in the properties of weak links as well as the oxygen losses inevitably affects the bulk electrical resistivity of the samples. The expected sharp resistivity increase has been indeed observed in YB$_2$Cu$_3$O$_X$ films during the first thermal cycle on heating to 300°C [62]. Subsequent heating has not reproduced the resistivity increase, which has been explained by severe oxygen loss from the film surface, accompanied also by the T$_c$ decrease. We have observed the simultaneous resistivity and AE activity peaks on heating bulk YB$_2$Cu$_3$O$_X$ ceramic samples (around 290°C), and the results are shown in Fig.13 [63]. Consequent (post-heating) X-ray diffraction measurements point at the existence of planar defects in the ceramics structure [64]. These planar defects resemble thin regions (15 - 20 layers) of highly ordered alternation of two phases, one with a structure containing single chain layers (as in YBa$_2$Cu$_3$O$_7$) and the other with a structure of double chain layers (as in YBa$_2$Cu$_4$O$_8$). It is noteworthy that thin films with a defective YBa$_2$Cu$_4$O$_8$ structure had been reported to exhibit T$_c$ of about 96 K [65]. This is in good agreement with the fact of the T$_c$ increase to 97 K in the samples quenched in liquid nitrogen after annealing at 260°C [55]. Thus the formation of double phase alternating structures on heating within the 250-300°C becomes an alternative explanation of the T$_c$ increase in YB$_2$Cu$_3$O$_X$ samples, in contrast to the hypothesis of oxygen vacancies' redistribution at T $\geq$ 100K described above.

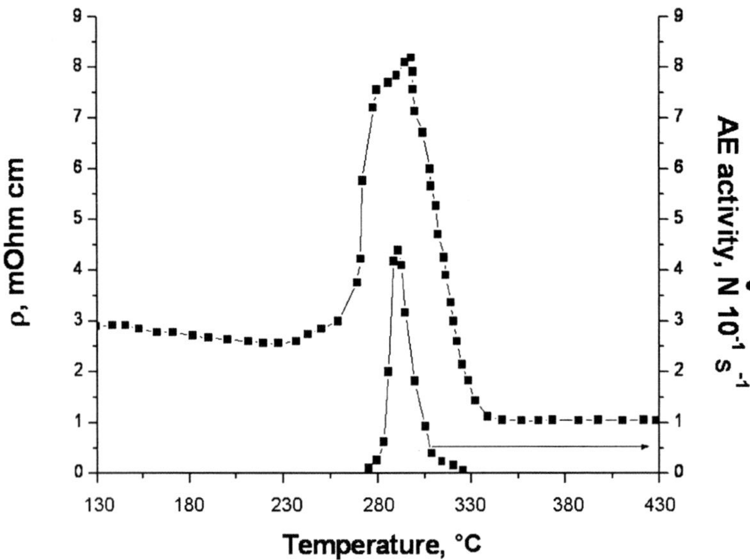

Fig. 13. Simultaneous measurements of electrical resistivity and AE of $YBa_2Cu_3O_X$ ceramics during heating through 220-320°C.

Low Temperature PT

The elastic and anelastic properties of the $YBa_2Cu_4O_X$ superconducting material have been intensively studied by various methods also below RT. Some anomalies (abrupt changes) in the lattice parameters have been revealed by X-Ray diffraction measurements at 250K and 125K [66] without any change in the symmetry of the crystallographic structure. In contrast, Raman spectra measurements do reveal structural anomalies at low temperatures [67]. These anomalies are explained in terms of oxygen atoms redistribution in the basal plane, inducing some distortion of the lattice, which is supported by internal friction measurements in the 237-244K temperature range revealing a 7K temperature hysteresis [68]. However, this internal friction hysteresis is absent after oxygen desorption. Acoustic measurements have allowed detecting the hysteresis of Young's modulus near 250K, and the existence of a martensite-like PT has been suggested as a possible explanation [56]. Ultrasonic attenuation data of polycrystalline YBCO at 10 MGz [69] show a plateau between 220 and 250K, which confirmes the results of Ref. [56]. The data on direct dilatometric measurements are in a rather poor mutual agreement. For example, the thermal expansion coefficient, α, is claimed to have a sharp maximum at 240K on heating, but none on cooling [70]. Other authors report on a smeared minimum of α at 220K [71] or on complete absence of dilatometric anomalies in the reviewed temperature range.

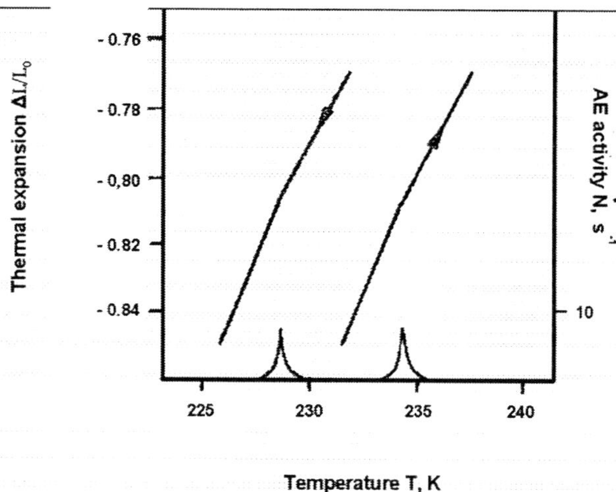

Fig. 14. Simultaneous measurements of thermal expansion and AE during heating and cooling of YBa$_2$Cu$_3$O$_X$ ceramics on heating through 200-300 K.

There is no consensus on the PT mechanisms as well. The existence of a 1st-order PT at 240K into another orthorhombic phase has been suggested based on the measurements of internal friction and vibrational frequency (elastic constants) hysteresis [73]. This transition is attributed to nonelastic accommodation of a 2nd-phase misfitting domains generated by thermal cycling. A similar PT has been revealed at 220K, and its existence is reported to be a sensitive to sintering conditions, namely to oxygen stoichiometry [74]. However, the frequency hysteresis [73] is very narrow. Other authors [75] ascribe to structural modifications related to the oxygen redistribution in the Cu1-O4 basal plane of the orthorhombic YBa$_2$Cu$_3$O$_X$ that is not accompanied by changes in the crystallographic lattice symmetry. Thus a possibility of an isostructural 2nd-order PT from one orthorhombic phase into another due to oxygen redistribution in the basal plane is alternatively suggested.

Indeed, theoretical calculations of the T – x phase diagram show that oxygen redistribution along the Cu1-O4 chains near 240 K can be expected [76]. It is predicted that the orthorhombic structure of YBa$_2$Cu$_3$O$_X$ allows for four stable phases to exist, which differ with respect of the type of oxygen ordering along the chains in the basal plane. Therefore, both 1st- and 2nd-order PTs are possible in the low-temperature 200-250K range.

We have used the low-temperature experimental setup (a) described above for simultaneous AE and dilatometric measurements in the 220-250K temperature range through thermal cycling [77]. The main results are presented in Fig. 14. An AE activity peak appears at 234K on heating, but at a lower temperature of 228K on reverse cooling. The AE temperature hysteresis is thus as small as 6 K, in correspondence with other reports [68]. The maximum values of $\dot{N}$ in both peaks is too small for a 1st-order PT implying the occurance of a 2nd-order transition. This interpretation is supported also by the dilatometric measurement presented in the sam figure. The dilatation curves (both on heating and cooling) are monotonic, without discontinuities, but rather change of slopes, at the PT temperatures. The dilatation data are used to determine the relevant thermal expansion coefficients, α. The apparent hysteresis in the α values is shown in Fig. 15. The obtained data have been also used to calculate the molar heat capacity, c_m, from the Grüneisen's formula:

$$c_m = \frac{3\alpha V_m}{\gamma \chi},$$

$$(3)$$

where the molar volume $V_m = 11 \cdot 10^{-6}$ m^3, the isothermal compressibility $\chi = 6 \cdot 10^{-12}$ m$^2 \cdot$N^{-1} and the Grüneisen constant $\gamma = 3.4$ [71]. The calculated heat capacity values before and after the PT are approximately 25.9 and 19.4 J$\cdot$K$^{-1}\cdot$mole^{-1} respectively. We recall that the 2nd-order PT under consideration is associated with oxygen redistribution. The calculated jump in heat capacity values through the PT, $\Delta c_m = 6.5$ J$\cdot$K$^{-1}\cdot$mole^{-1}, can be used to evaluate activation energy for oxygen migration [77], namely:

$$\Delta E = \frac{T V_i \Delta c_m}{1.6 V_m} \cdot 10^{19},$$

$$(4)$$

where $V_i = 170$ Å^3 is the unit cell volume at 250 K. From eq. (4) we obtain $\Delta E = 0.15$ eV, which is in good agreement with the oxygen thermal activation energies of 0.16-0.19 eV deduced from the elastic-energy-dissipation measurements in YBa$_2$Cu$_3$O$_X$ [78]. Summarizing the above experimental and theoretical results, we conclude that an isostructural oxygen displacement leading to a 2nd-order order PT between two different orthorhombic phases takes place in YBa$_2$Cu$_3$O$_X$ near 240K.

Superconducting PT

The superconducting PT in YBCO has been extensively studied by electrical, magnetic and acoustic methods since the discovery of the high-T_c superconductivity. The accumulated results show obvious acoustic anomalies just below the critical temperatures of ~ 90-92 K. For example, the anomaly of Young's modulus is registered at 87 K, and it is attributed to the formation of crystallographic twins within the YBCO ceramics grains [79]. The absorption of ultrasound near 88K is explained by variation of the piezoelectric properties of the material in the framework of the Drude theory [80]. Small local minima in the ultrasonic wave velocity are observed at around 84K, which are claimed to be cuased by 2nd-order structural PTs [81]. Simultaneous measurements of the ultrasonic wave velocity and magnetic susceptibility have been performed as well [82]. A discontinuity in the sound velocity has been observed at 87K, whereas the magnetic susceptibility measurements yield T_c ~ 90 K. This discrepancy is explained in terms of the changes in the electron conducting properties in the superconducting state (creation of Cooper's pairs) and their contribution to the ultrasound velocity. The results of combined measurements of the ultrasound velocity and thermal expansion are reported in Ref. [83]. The ultrasound velocity behavior is similar to [82], but no thermal expansion anomalies have been registered. In contrast, some dilatation anomalies below the T_c have been observed by other workers [84,85]. Unfortunately, these anomalies are poorly developed and nearly smeared out in the 75-90 K temperature range.

We address now the results of AE measurements employed for studying the superconducting PT in YBa$_2$Cu$_3$O$_X$. Unfortunately, there is a poor agreement between the published results and, sometimes, lack of clarity. For example, a broad 30K wide AE activity peak has been observed with the peak maximum coinciding with the T$_c$ value, yet concurrent magnetic measurements reveal a superconducting PT with a width of 2 K. In contrast, the results of AE measurements under DC current conditions show that two $\dot{N}$ peaks exist, one below the T$_c$ at 86-87K and the other in the 95-98K range, and the temperature interval between the peaks depends on the current magnitude [87]. However, the followed measurements shown that AE has the peaks higher than T$_c$ at 110 – 115 K without any current in YBa$_2$Cu$_3$O$_{7-X}$ ceramics with 6.28<(7-X)<6.8, but due to high speed of heating ~ 0.5 K/s from 77 K [88]. A continuous AE spectrum lacking any features has been also observed in YBCO in the 60-110K temperature range [89].

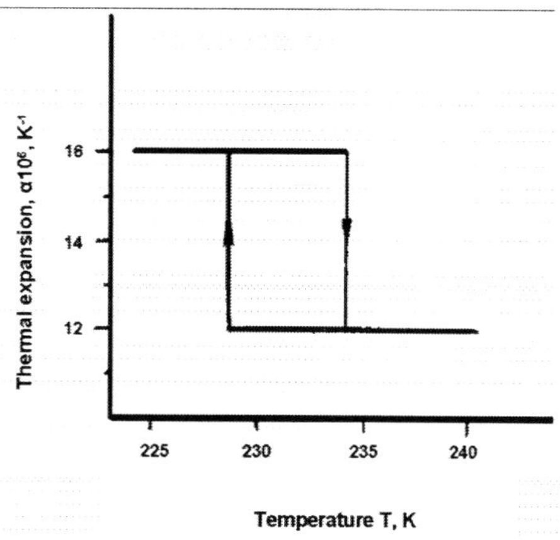

Fig. 15. Hysteresis of the thermal expansion coefficient.

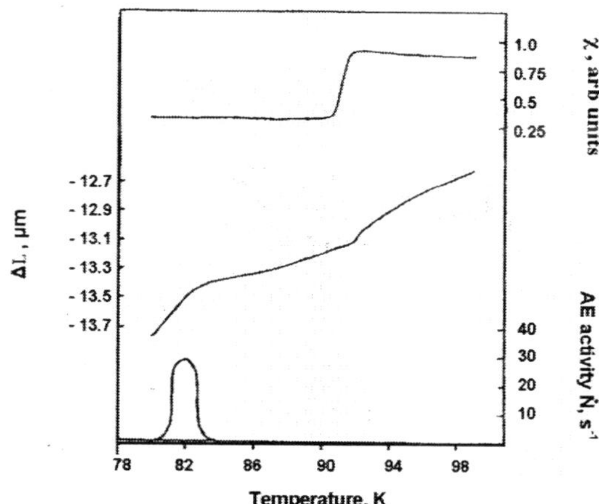

Fig.16. Simultaneous measurements of AE, thermal expansion and magnetic susceptibility in the superconducting PT range of YBa$_2$Cu$_3$O$_X$ ceramics.

Over a decade ago, one of us has conducted a series of combined measurements comprising AE, thermal expansion (ΔL) and magnetic susceptibility (χ) using apparatus similar to the low-temperature set up (a) described above [90]. The results of these measurements are shown in Fig. 16. The dilatometric curve has two inflection points, at 83 and 92K, exhibiting no discontinuity. The first inflection is accompanied by AE, the second -

by a 'jump' of χ. The relative change in χ is considerably large, about 50%, and the superconducting PT width is about 1K. The magnitude of $\dot{N}$ in the AE spectrum is too small to represent a 1st-order PT. These results imply, therefore, that a 2nd-order structural PT take place around 823 K, below the T_c, in agreement with the results of other workers quoted above [80]. This structural PT can be understood in terms of the interaction between ultrasonic phonons and conduction electrons, which produces attenuation in the normal state and decay below T_c as the electrons become paired [91]. The symmetry of the subsystem associated with paired electrons may coincide or not with the symmetry of the crystallographic lattice [92]. Therefore, the superconducting transition is able to induce certain lattice instability and, as a consequence, lead to a structural PT below T_c [93]. A fit of the BCS theory to experimental results of elastic energy dissipation due to electron-phonon interaction gives an average value of $T_c \approx 86.5K$, while the electrical resistivity measurements performed on the same YBCO ceramic sample yield $T_c = 94K$. This provides an additional explanation to our AE results indicating that a 2nd-order PT occurs at 82K, below the T_c, in the $YBa_2Cu_3O_X$ material.

Flux Penetration and Mixed State

In his classic paper, Abrikosov has demonstrated [94] that the magnetic field (MF) flux lines (FL) start penetrating the type-II superconductor when the applied MF reaches the lower critical value, H_{c1}. Above this field, each FL is accompanied by a vortex of persistent current (supercurrent) surrounding a normal resistance core. These FLs thread the material's volume along the applied MF direction. Upon entering the material, the FLs are trapped by the pinning centers (defects, including grain boundaries, twins, etc.), initially at the surface and then in the volume of the sample. Under the transport current (I) condition, the repulsive Lorentz force ($I{\times}\mu H$) acts on the FLs in such a way as to drive them deeper into the material if they can overcome the opposing pinning forces, and the mixed state begins to form [95]. Since the defects are connected along each FL, the mechanical parameters of the material ought to change due to pinning. This fact has been exploited in the internal friction measurements [96] aimed at studying the magnetic field penetration and inward movement into the material. It has been found that the MF begins to penetrate the YBCO material at H_{c1} of about 90 Oe, in good agreement with the results obtained by traditional magnetic measurements.

On the other hand, trapping at pinning centers forces the FLs to transfer their kinetic energy to the defects, and the latter begin to vibrate inducing the acoustic waves in the material as a whole. Excitation of AE may be expected therefore as the MF penetrates the material. Indeed, AE is observed in Nb-Ti wire under electrical current transport [97]. AE starts to increase when the current exceeds some threshold value, and no other sources are present except the flux motion. More detailed studes shows that intense acoustic signals are rather generated at FL jump events in Nb-Ti wires [98]. This conclusion has been successfully confirmed in the case of Nb-Ti wires and multifilamentary composites [99].

Recently, the AE method has been applied also to studying the FL penetration process in high-T_c superconductors, such as $YBa_2Cu_3O_X$. The low-temperature experimental setup set up (b) (Fig. 4) has been used for this study. The self-induced MF enhances as the DC current I

flowing through the sample wire increases (external MF is absent). Fig. 17 shows that two AE activity peaks appear, at 0.5 and 2.7A values of the DC transport current. The first peak is less intense, and it is assumed to mark the onset of MF penetration into the sample [100]. In order to substantiate the model, we expand the magnetic field $H(r)$ just outside the wire, induced by the transport current, into the Taylor's series around the radius r_0 at the surface of the sample, taking into account only the first two terms:

$$H(r) = H(r_0) + \frac{dH}{dr}\bigg|_{r=r_0} (r - r_0).$$

(5)

The first term on the right-hand side of eq. (5) equals zero, because the MF is absent in the Meissner state within the sample. Then for r $\cong$ r$_0$, outside the sample surface, we present the field differential as

$$dH = H(r)\frac{dr}{(r - r_0)}.$$

(6)

By integrating eq. (6) we obtain, for $r \geq r_0$:

$$H = \frac{2I}{r_0}\ln|r - r_0|.$$

(7)

This result implies that the MF tends to infinity, yet the flux can penetrate the sample as deep as the coherence length, $\xi = 1.2 \cdot 10^{-9}$m [101]. Substituting into eq. 7 the current value 0.5A, determined by AE (Fig. 17) and taking $(r-r_0) \approx \xi$, we obtain $H \approx 100$ Oe. This value of the magnetic field is in a good agreement with $H_{c1} = 90$ Oe deduced from the internal friction measurements [96]. Thus, AE is a very convenient and useful tool for determining the H_{c1} values in superconductors [102].

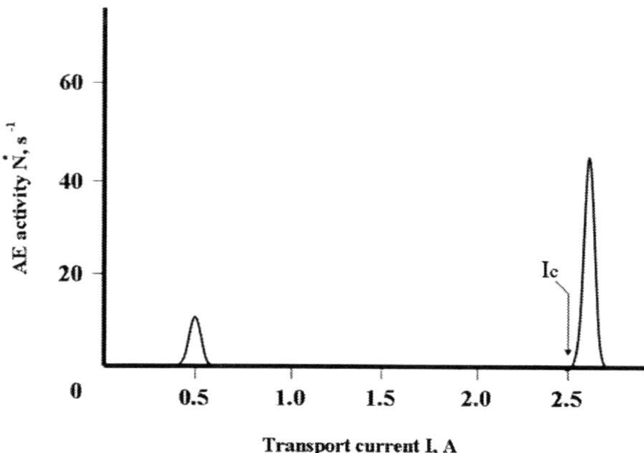

Fig. 17. AE activity as a function of electrical transport current flowing through YBa$_2$Cu$_3$O$_x$ ceramics at 77 K.

The second AE activity peak in Fig. 17 is about four times more intense than the first peak, and it appears slightly above the critical current (I_c). Obviously, the difference in the $\dot{N}$ magnitude is related to the difference in the number of FLs. The first peak reflects only FL pinning in the near-surface region, while FLs forming the mixed state interact with numerous pinning sites at a deeper volume. As the transport current reaches the I_c value, the Lorentz force overcomes the pinning forces, and FLs start moving through the sample [95]. The FL movement is accompanied by AE due to atomic movement, as in the case of FL jumps in Nb-Ti wires described above. Moreover, the movement of FLs is perpendicular to I and, therefore, active energy losses sharply increase due to the evolution of Joule heat within the normal core of the moving FLs. Such sharp increase of power losses has been registered during electrical current transport through Bi-2223/Ag at I ~ (1.1-1.3)·I_c values [103-105]. Thus, AE can be also successfully applied to detecting the current overload in high-power superconducting circuits.

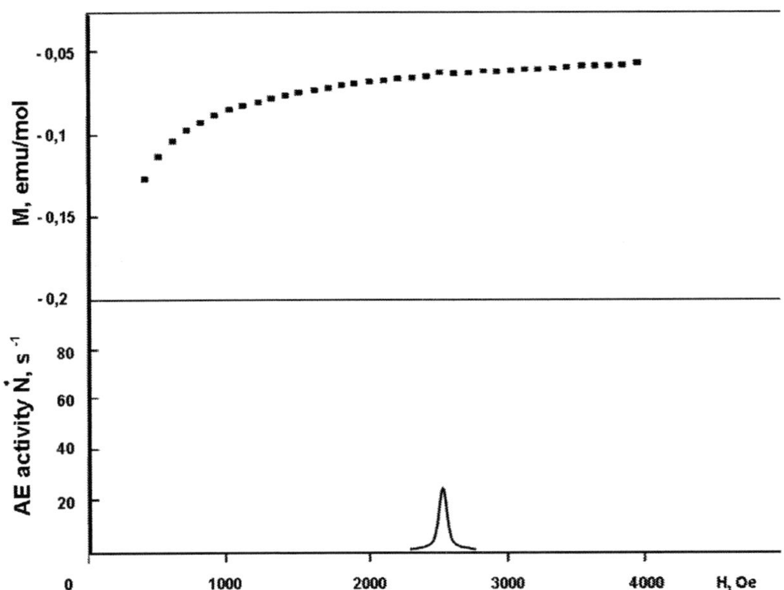

Fig. 18. Magnetic susceptibility anomaly at 2500 Oe accompanied by AE on the external magnetic field increase in $YBa_2Cu_3O_X$ ceramics at 77 K.

In order to reach the mixed state in the entire sample volume, an external MF must be applied. As the external MF increases, FL penetrate deeper into the material until they reach the center of the sample. To explain the penetration mechanism, Bean [106] has assumed that I_c is independent of the MF value inside a sample. The actual MF value, at which two FL fronts meet in the middle of a sample, is known as the full penetration MF (H*). As the FL fronts meet, the common FL structure begins to form. When the common FL structure is established, a mixed state exists at a field H_s [107]:

$$H_s = H_{c1} + 2H^*$$

(8)

The mixed state field defined above is not a fundamental constant, but rather a parameter depending on the individual sample properties, such as grain size, processing route, density, etc. The H_s value in $YBa_2Cu_3O_X$ bulk ceramic samples is routinely determined by measuring the inflection of the slope of I_c measured as a function of an externally applied MF [107]. However, this slope changes gradually and smoothly, which does not allow to determine the H_s value accurately enough. In contrast, AE shows a sharp activity maximum at H_s (Fig. 18), which correlates well with the magnetic anomaly near 2500 Oe [108]. Both the magnetic anomaly and the AE peak have been detected on the MF increase only, but none on the MF decrease.

In order to understand the origin of AE in course of MF penetration into the sample under current transport, we revisit the sources of PT-induced AE in ferroelectrics. As already mentioned above, certain PTs occur during movement of two or more PBs through the ferroelectric crystal [30]. These PBs moving from the crystal periphery to the center may have differently oriented ferroelectric domain structures. When the PBs meet, a common domain structure is formed through reorientation of the domains, which is. accompanied by excitation of strong AE. Similarly vigorous atomic movement takes place when the FLs are traversing the superconducting sample, from the periphery to the center, and strong AE is generated as a result.

Finally, we calculate the H_s and H^* values and compare the results with other data, such as obtained by the internal friction measurements [96]. In order to account for the sample dimensions (5 mm thickness in our case versus 2 mm thickness elsewhere [96]), we recalculate the the H_s and H^* values as 1000 and 450 Oe respectively (see eq. (8)). These values are in a reasonable agreement with H* ~ 350 Oe and H$_s$ ≥ 600 Oe as determined from the internal friction studies.

Conclusions

The results reviewed above demonstrate that AE is an efficient and inexpensive method suitable for studying many aspects of high-T$_c$ superconductors, from the materials' processing (sintering and oxygenation) to fundamental investigations of their superconducting and structural phase transitions and performance in devices. We have described comprehensively for the first time how AE can be applied to investigation of phase work hardening as well as to determination of the lower critical magnetic field H_{cl} and of the mixed state magnetic field H_s. The generation of strong AE under current transport in superconducting wires emphasizes one of the more important industrial applications of the method, namely an early current overload alarm in high-power superconducting circuits. Thus, AE is becoming a universal method of studying the entire range of phenomena associated with high-T$_c$ superconductivity, and it steadily becomes a routine characterization tool of a growing number of materials both in research laboratories and in industry.

References

[1]   Kaiser *J. Archiv Eisenhuttenwessn* 1953, Bd 24, 42 – 45.

[2]   Boyko V.S.; Garber R.I.; Kosevich A.M. *Reversible Crystal Plasticity* AIP NY 1994.

[3]   Speich G.R.; Fisher R.M. *ASTM STP* 1972, 505, 140 – 151.

[4]   Speich G.R.; Scswoeble A.J. *ASTM STP* 1975, 571, 40 – 58.

[5]   Ievlev I.Yu.; Melechin V.P.; Minz R.I.; Segal' V.I. *Phys. Solid. State* 1974, 15, 1761 - 1763.

[6]   Kulemin A.V.; Solov'ev V.A.; Zambrzhitskii V.N.; Zhiryakov V.V. *Phys. Metal. Metallogr.* 1979, 48, 135 – 140.

[7]   Baram J.; Gefen Y; Rosen M. *Scripta Metall.* 1981, 15, 835 - 838.

[8]   Picornell C.; Segni C.; Torra V.; Lovey F.C.; Rapacioli R. *Thermochimica Acta* 1987, 113, 171 - 183.

[9]   Plotnikov V.A.; Monasevich L.A.; Pascal' Yu. I. *Phys. Solid. State* 1985, 27, 3117.

[10]   Parfenov O.E.; Chernyshov A.A. *Phys. Solid State* 1988, 30, 1611 - 1613.

[11]   Vives E.; Ortín J.; Mañosa L.; Ràfols I.; Pérez–Magrané R.; Planes A. *Phys. Rev. Lett.* 1994, 72, 1694 – 1697.

[12]   Andreev V.N.; Pikulin V.A.; Frolov D.I. *Phys. Solid State* 2000, 42, 330 – 333.

[13]   Plotnikov V.A.; Kokhanenko D.V. *Tech. Phys. Lett.* 2002, 28, 833 - 835.

[14]   Kalitenko V.A.; Perga V.M.; Salivonov I.N. *Phys. Solid State* 1980, 22, 1067 - 1068.

[15]   Dul'kin E.A.; Gavrilyachenko V.G.; Semenchev A.F. *Phys. Solid State* 1992, 34, 863 – 864.

[16]   Dul'kin E.A.; Gavrilyachenko V.G.; Semenchev Λ.F. *Phys. Solid State* 1993, 35, 1016 – 1018.

[17]   Dul'kin E.A. *Ferroelectrics Letters*, 1996, 20, 157 – 162.

[18]   Dul'kin E.A.; Raevskii I.P. *Ferroelectrics* 2000, 239, 381 – 387.

[19]   Dul'kin E.A.; Raevskii I.P. *Ferroelectrics* 2000, 242, 53 – 58.

[20]   Dul'kin E.A. *Crystallogr. Rep.* 1995, 40, 340 – 341.

[21]   Dul'kin E.A.; Gavrilyachenko V.G. *Tech. Phys. Lett.* 1997, 23, 835 – 836.

[22]   Dul'kin E.A. *Crystallogr. Rep.* 1994, 39, 669 – 670.

[23]   Dul'kin E.A.; Gavrilyachenko V.G.; Semenchev A.F. *Phys. Solid State* 1995, 37, 668 – 669.

[24]   Dul'kin E.A.; Grebenkina L.V.; Pozdnyakova I.V.; Reznichenko L.A.; Gavrilyachenko V.G. *Tech. Phys. Lett.* 1999, 25, 70 – 71.

[25]   Dul'kin E.A.; Raevskii I.P.; Emel'yanov S.M. *Phys. Solid State* 2003, 45, 158 – 162.

[26]   26.Dul'kin E.A.; Grebenkina L.V.; Gavrilyachenko V.G. *Tech. Phys. Lett.* 2001, 27, 873 – 874.

[27]   Dul'kin E.A.; Raevskii I.P. *Tech. Phys. Lett.* 1996, 22, 201- 202.

[28]   Dul'kin E.A.; Gavrilyachenko V.G.; Fesenko O.E. *Phys. Solid State* 1997, 39, 654 – 655.

[29]   Dul'kin E.A.; Raevskii I.P.; Emel'yanov S.M. *Phys. Solid State* 1997, 39, 316 – 317.

[30]   Dul'kin E.A. *Mat. Res. Innovat.* 1999, 2, 338 – 340.

[31]   Reissner M.; Steiner W.; Stroh R.; Hörhager S.; Schmid W.; Wruss W. *Physica C* 1990, 167, 495-502.

[32]   Dul'kin E.A. *Superconductivity* 1994, 7, 104-107.

[33]   Ikegami T. *Acta Metall.* 1987, 35, 667- 675.

[34]   Staines M.P.; Flower N.E. *Supercond. Sci. Technol.* 1991, 4, 5232-234.

[35]   Diko P.; Crabbes G. *Supercond. Sci. Technol.* 2003, 16, 90–93.

[36]   Richrdson T.J.; De Jonghe L.C. *J. Mater. Res.* 1990, 5, 2066 – 2074.

[37]   Dul'kin E.A. *Tech. Phys. Lett.* 1995, 21, 533 – 534.

[38] Baikov Yu. M.; Shalkova E.K.; Ushakova T.A. *Superconductivity* 1993, 6 449 – 453.

[39] Dul'kin E.A. J. *Superconductivity* 1998, 11, 281 – 282.

[40] Krishan K.; Natarajan D.V.; Purniax B. Pramana – *J. Phys.* 1988, 31, L327 – L335.

[41] Zhang H.; Ye H., Du1 K.; Huang X.Y.; Wang Z.H. *Supercond. Sci. Technol.* 2002, 15, 1268 – 1274.

[42] Dul'kin E.; Beilin V.; Yashchin E.; Roth M.; Grebenkina L.V. *tech. Phys. Lett.* 2001, 27, 387 – 388.

[43] Frello T.; Roulsen H.E.; Gottschalk A.L.; Andersen N.H.; Bentzon M.D.; Schmidberger J. *Supercond. Sci. Technol.* 1999, 12, 293 – 300.

[44] Polyanskii A.; Beilin V.; Goldgirsh A.; Yashchin E.; Roth M.; Larbalestier D. The 2000 International Workshop on Superconductivity, June 19 – 22, 2000, Matsue-shi, Shimane, Japan, 253 – 256.

[45] Dul'kin E.; Beilin V.; Yashchin E.; Roth M.; Grebenkina L.V. *Tech. Phys. Lett.* 2002, 28, 489 – 490.

[46] Doronina G.A.; Fotiev V.A. *Superconductivity* 1989, 2, 40 – 46.

[47] Vlasko-Vlasov V.K.; Dorosinskii L.A.; Indenbom M.V.; Osip'yan Yu.A. *Superconductivity* 1990, 3, 60 – 66.

[48] Bobrov V.S.; Vlasko-Vlasov V.K.; Emel'chenko G.A.; Indenbom M.V.; Lebedkin M.A.; Osip'yan Yu. A.; Tatarchenko V.A.; Farber B.Ya. *Phys. Solid. State* 1989, 31, 597 – 601.

[49] Stenger M.; Ockenfuss G.; Reese M.; Konigs T.; Wordenweber R.; Barre C.; Prieur J.-Y.; Joffrin J. *Solid State Commun.* 1996, 98, 777 – 780.

[50] Xie X.M.; Chen T.G. *Supercond. Sci. Technol.* 1992, 5, 290 – 294.

[51] Ivanov O.N.; Dibova O.V.; Kudryash V.I. *Superconductivity* 1992, 5, 1936 – 1938.

[52] Subbarao E.C.; Srikanth V. *Physics C* 1990, 171, 449 – 453.

[53] Dul'kin E.A. *Superconductivity* 1992, 5, 182 – 183.

[54] Mamalui' A.A.; Palashnik L.S.; Bilya P.V. *Phys. Sol. State* 1990, 32, 1703 – 1705.

[55] Gridnev S.A.; Ivanov O.N. *Superconductivity* 1992, 5, 1139 – 1164.

[56] Dul'kin E.A. J. *Superconductivity* 2001, 14, 497 – 499.

[57] Ono A. *Jpn. J. Appl. Phys.* 1987, 26, L1223 – L1225.

[58] Varela M.; Arias D.; Sefriouri Z.; Leon C.; Ballesteros C.; Pennycook S.J.; Santamaria J. *Phys. Rev. B* 2002, 66, 134517-1 - 5.

[59] Ha D.H.; Byon S.; Lee K. *Physica C* 2000, 340, 243 - 250.

[60] Dul'kin E. J. *Superconductivity* 2003, 16, 953 – 956.

[61] Kittelberger S.; Bolz U.; Huebener R.P.; Holzapfel B.; Mex L.; Schwarzer R.A. *Physica C* 1999, 312, 7 – 20.

[62] Dul'kin E. *Supercond. Sci. Technol.* 2004, 17, 954 – 956.

[63] Kameneva M. Yu.; Romanenko A.I.; Anikeeva O.B.; Kuratieva N.V. *Physica C* 2004, 408 – 410, 816 – 817.

[64] Kapitulnik A. *Physica C* 1988, 153 – 155, 520 - 526.

[65] Yin H.-G.; Gao Y.-M.; Du J.-J.; Chen B.-P.; Jiang J.-Y.; Wang X. *Appl. Phys. Lett.* 1988, 52, 1899 – 1900.

[66] Liu J.-G.; Wang W.-K.; Chang D.-W.; Huang X.-M.; Wang Y.-Y.; Chen G.-H.; Zou K.-P. *Inter. J. Modern Phys B* 1987, 1, 513 - 515.

[67] Cannelli G.; Cantelli R.; Cordero F. *Phys. Rev. B* 1988, 38, 7200 – 7202.

[68] Gridvev S.A.; Ivanov O.N. *Superconductivity* 1992, 5, 1139 – 1164.

[69]   K.J. Sun; M. Levy; B.K. Sarma; P.H. Hor; R.L. Meng; Y.Q. Wang C.W. Chu *Phys. Lett. A* 1988, 131, 541 – 544.

[70]   Popov V.P.; Morgun V.N.; Voronov A.P. *Superconductivity* 1990, 3, 147 – 150.

[71]   Vorob'ev G.P.; Kodomtseva A.M.; Kozei Z.A.; Krynetskii I.B.; Levitin R.Z.; Nikishin S.B.; Snegirev V.V.; Sokolov V.I. *Superconductivity* 1989, 2, 57 – 64.

[72]   Podgornikh S.M.; *Superconductivity* 1989, 2, 106 – 113.

[73]   Cannelli G.; Cantelli R.; Cordero F.; Costa G.A.; Ferritti M.; Olcess G.L. *Europhys. Lett.* 1988, 6, 271 - 276.

[74]   Mi Y.; Schaller R.; Berger H.; Benoit W.; Sathish S. *Physica C* 1991, 172, 407 - 412.

[75]   Mamsurova L.G.; Pigal'skiy K.S.; Sakun V.P.; Sushin A.I.; Shcherbakova G. *Physica C* 1990, 167, 11 - 19.

[76]   Mamsurova L.G.; Pigal'skiy K.S.; Sakun V.P.; Sushin A.I.; Shcherbakova G. *JETP* 1990, 71, 544 – 549.

[77]   Dul'kin E. J. *Superconductivity* 2002, 15, 527 – 529.

[78]   Cannelli G.; Cantelli R.; Cordero F.; Ferretti M.; Verdini L. *Phys. Rev. B* 1990, 42, 7925–7930.

[79]   Khaidarov T.; Makhmudov A. Sh.; Khidirov I. *Superconductivity* 1990, 3, S255 – S257.

[80]   Kim T.J.; Mohler E.; Grill W. *Alloys and Compounds* 1994, 211, 318 – 320.

[81]   Abd-Shukor R. *Jpn. J. Appl. Phys.* 1992, 31, L1034 – 1036.

[82]   Bishop D.J.; Ramirez A.P.; Gammel P.L.; Batlogg B.; Rietman E.A.; Cava R.J.;

[83]   Millis A.J. *Phys. Rev. B.* 1987, 36, 2408 - 2410.

[84]   Jericho M.N.; Simpson A.M. *Solid State Communn.* 1988, 65, 989 – 990.

[85]   Popov V.P; Morgun V.N.; Voronov A.P. *Superconductivity* 1990, 3, 147 – 150.

[86]   Podgornikh S.M. *Superconductivity* 1989, 2, 106 – 113.

[87]   Serdobolskaya O. Yu.; Morozova G.P. *Phys. Solid State* 1989, 31, 1439 – 1440.

[88]   Wozny L.; Mazurek B. *Physica B* 1991, 173, 309 - 312.

[89]   Gaiduk A.L.; Zherlitsyn S.V.; Fil' V.D. *Low Temper. Phys.* 1990, 16, 217 – 218.

[90]   Dul'kin E. *Superconductivity* 1993, 6, 246 – 248.

[91]   Almond D.P. J. Phys. C : *Solid State Phys.* 1988, 21, L1137 – L1139.

[92]   Poluektov Yu.M. *Low Temp. Phys.* 1994, 20, 161 – 167.

[93]   Cannelli G.; Cantelli R.; Cordero F.; Costa G.A.; Ferretti M.; Olcese G.L. *Phys. Rev. B* 1987, 36, 8907 – 8909.

[94]   Abrikosov A.A. *JETP* 1957, 5, 1174 - 1182.

[95]   Nikolo M.; *Supercond. Sci. Technol.* 1993, 6, 618 - 623.

[96]   Arzhavitin V.M.; Efimova N.N.; Ustimenkova M.V.; Finkel' V.A. *Phys. Sol. State* 2000, 42, 1398 – 1402.

[97]   Pasztor G.; Schmidt C. J. *Applied Physics* 1982, 53, 3918 - 3920.

[98]   Tsukamoto O.; Iwasa Y. J. *Applied Physics* 1983, 54, 997 – 1007.

[99]   Zheng-Kuan J.; Zhuo-Min C.; Ren D.L.; En-Yue W.; Zian W.; Zhi-Oin Zh. *Acta Physica Sinica* 1984, 33, 530 – 537.

[100] Dul'kin E. *Tech. Phys. Lett.* 2004, 30, 200 – 201.

[101] Turchinsraya M.; Smith D.T.; Roytburd A.L.; Kaiser D.L. *J. Appl. Phys.* 2000, 88, 1541 - 1546.

[102] Dul'kin E.; Beilin V.; Yashchin E.; Roth M. *Supercond. Sci. Technol.* 2002, 15, 1081 – 1082.

[103] Savvides N.; Herrmann J,; Reilly D. *Physics C* 1998, 306, 129 - 135.

[104] Flesher S,; Cronic L.T.; Conway G.E. *Appl. Phys. Lett.* 1995, 67, 3189 - 3191.

[105] Fukunaga T,; Abe T,; Oota A. *Appl. Phys. Lett.* 1995, 66, 2128 - 2130.

[106] Bean C.P. *Rev. Mod. Phys.* 1964, 36, 31 – 39.

[107] Altshuler E.; Garcia S.; barrosso J. *Physica C* 1991, 177, 61 - 66.

[108] Dul'kin E.; Roth M. *Supercond. Sci. Technol.* 2003, 16, 361 – 363.

In: New Topics in Superconductivity Research
Editor: Barry P. Martins, pp 73-106

ISBN: 1-59454-985-0
© 2006 Nova Science Publishers, Inc.

Chapter 3

VAN HOVE SCENARIO FOR HIGH TC SUPERCONDUCTORS

J. Bok[1,2] and J. Bouvier[2]

[1]Laboratoire Matériaux et Phénomènes Quantiques UMR 7162 - PARIS 7 -
[2]Solid State Physics Laboratory – ESPCI -10, rue Vauquelin.75231
PARIS Cedex 05, FRANCE

Abstract

We give a general description of our approach which explains many physical properties in the superconducting and normal states of almost all 2D high T_c superconductors (HTSC). This 2D character leads to the existence of Van Hove singularities (VHs) or saddle points in the band structure of these compounds. The presence of VHs near the Fermi level in HTSC is now well established. We review some physical properties of these materials which can be explained by this scenario, in particular: the critical temperature T_c, the anomalous isotope effect, the superconducting gap and its anisotropy, and thermodynamic and transport properties (eg: Hall effect). The effects of doping and temperature are also studied, and they are directly dependent of the position of the Fermi level relative to the VHs position. We show that these compounds present a topological transition for a critical hole doping $p \approx 0.21$ hole per CuO_2 plane. Most of these compounds are disordered metals in the normal state, we think that the Coulomb repulsion is responsible for the loss of electronic states at the Fermi level, leading to a dip, or the so-called "pseudo-gap".

Introduction

Twenty years after the discovery of the high temperature superconductivity in cuprates compounds [1], the exact mechanism of superconductivity is still not yet understood. All these compounds are strongly anisotropic and almost two dimensional, due to their CuO_2 planes, where superconductivity mainly occurs. It is well known that in 2 dimensions, electrons in a periodic potential show a logarithmic density of states (DOS), named Van Hove singularity (VHs) (Van Hove (1953) [2]). The Van Hove scenario is based on the assumption that, in high critical temperature superconductors cuprates (HTSC), the Fermi

Level (FL) lies close to such a singularity (Labbé-Bok (1987)) [3]. This hypothesis has been confirmed by many experiments, in particular by Angular Resolved Photoemission Spectroscopy (ARPES) [4-9] in different compounds as:

$La_{2-x}Sr_xCuO_{4-\delta}$ (LSCO), $Bi_2Sr_2CuO_6$ (Bi 2201), $Bi_2Sr_2CaCu_2O_8$ (Bi 2212), $YBa_2Cu_3O_{7-\delta}$ (Y123), $YBa_2Cu_4O_8$ (Y124) and $Nd_{2-x}Ce_xCuO_{4+\delta}$ (NCCO). These experiments establish a general feature: in very high T_c superconductors cuprates ($T_c \sim 90$ K) Van Hove singularities are present close to the Fermi level. This is probably not purely accidental and we think that any theoretical model must take into account these experimental facts.

The origin of high T_c in the cuprates is still controversial and the role of these singularities in the mechanism of high T_c superconductivity is not yet established, but we want to stress that the model of 2D itinerant electrons in presence of VHs in the band structure has already explained a certain number of experimental facts.

In this review paper:
We recall what are the VHs physics in 1, 2, 3 dimensions.

We give a rapid description of the band structure of the CuO_2 planes.

We compute the critical temperature T_c [3,10,11], the anisotropic superconducting gap [10]. We show the importance of screening and Coulomb repulsion [10,12]. We explain the anomalous isotope effect [13], the very small values of the coherence length [14,15].

We compute the DOS in these compounds and apply this result to the calculations of various physical parameters: the conductance of tunnelling junctions, the specific heat [11], the magnetic susceptibility [16]. The variation of all these properties with hole doping (from underdoped UD to overdoped OD samples) and temperature are obtained and compared with the experiments. The agreement is very satisfactory. The variation with the doping is linked to the distance of the FL from the singularity level ($\xi_F - \xi_S$), so does the variation with the temperature due to the Fermi-Dirac distribution.

Transport properties in the normal state are described. We show that $\xi_F - \xi_S$ is critical for these properties, leading to Fermi liquid or marginal Fermi liquid [17].

We compute the Hall coefficient and its variation with doping and temperature [18]. We show that the experimental results may be explained by the topology of the Fermi surface (FS) which goes from hole-like to electron-like as the hole doping is increased. The critical doping, for which a topological transition is observed and calculated is $p = 0.21$ hole per CuO_2 plane.

A so-called "pseudo-gap" is observed in the normal state of cuprates. These compounds are disordered metals if we refer to their coefficient of diffusion, which is very low. The Coulomb interaction between electrons must be taken into account as shown by Altshuler and Aronov (1985) [19]. The main effect is to open a dip in the DOS at the FL. We show that this explains the observed features of the "pseudo-gap", value, anisotropy and variation with doping [20].

In conclusion we show that VHs play an important role in HTSC, and that by taking them into account, we may explain most of their normal and superconducting properties.

VHS Physics in 1, 2, 3 Dimensions

Van Hove singularities [2] are general features of periodic systems. They are topological properties of the electronic band structure (BS) and do not depend on the particular form of the BS. In one dimension (1D), they give a divergence of the DOS varying as $1/\sqrt{\xi-\xi_s}$, where ξ_s is the energy of the singularity level. In two dimensions (2D), near VHs the variation of the DOS is logarithmic varying as:

$$\ln\left[\frac{D}{\xi-\xi_s}\right]$$

(Figure 1) where D/2 is the width of the singularity. In three dimensions (3D) the divergence is removed and we have a truncated 2D DOS.

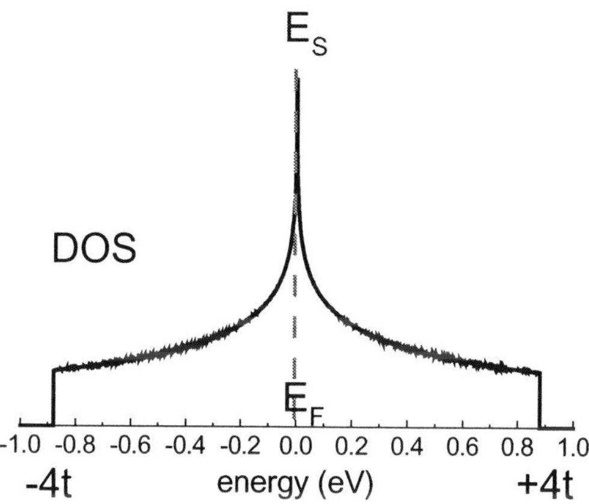

Figure 1: Density of States of a 2D system. ($E_F \equiv \xi_F$)

Calculation of T_c

Labbé-Bok [3] have computed the band structure for the bidimensional CuO_2 planes of the cuprates, considered as a square lattice (quadratic phase). The simplest band structure we can take for a square lattice is :

$$\xi_k = -2t\left[\cos k_x a + \cos k_y a\right] \tag{1}$$

where t is an interaction with nearest neighbours. This gives a square Fermi surface with saddle points, or VHs., at $[\,0\,,|\pi|\,]$ positions of the Brillouin zone (B.Z.), and a logarithmic D.O.S. with a singularity : $n(\xi) = n_1 \ln|D/(\xi-\xi_s)|$. The VHs. corresponds to half filling in this first calculation. We know that is not a good representation of the high T_c cuprates because for half filling (one electron per copper site) they are antiferromagnetic insulators. We think that the Fermi level is at VHs. for a doping level corresponding to 21 % of holes in each CuO_2 plane or 0.395 filling of the first B.Z., this is confirmed by the observations of Ino *et al* [7]. This can be achieved by taking into account the repulsive interaction between second nearest neighbours (s.n.n.) and the effect of the rhomboedric distorsion. For the repulsive interaction with s.n.n. the band structure becomes:

$$\xi_k = -2t\left[\cos k_x a + \cos k_y a\right] + 4\alpha t \cos k_x a \cos k_y a \tag{2}$$

where $\alpha t \equiv t'$ is an integral representing the interaction with s.n.n.. The singularity occurs for $\xi = -4\alpha t$, there is a shift towards lower energy. The Fermi surface at the VHs. is no longer a square but is rather diamond-shaped. More detailed calculations can be obtained in reference [14], taking also into account the rhomboedric distorsion.

The Labbé-Bok [3] formula was obtained using the following assumptions :

1- the Fermi level lies at the van Hove singularity
2- the B.C.S. approach :

- The electron-phonon interaction is isotropic and so is the superconducting gap Δ.
- The attractive interaction V_p between electrons is non zero only in an interval of energy $\pm\hbar\omega_0$ around the Fermi level where it is constant. When this attraction is mediated by emission and absorption of phonons, ω_0 is a typical phonon frequency.

In that case, the critical temperature is given by

$$k_B T_c = 1.13 D \exp\left[-\left(\frac{1}{\lambda} + \ln^2\left(\frac{\hbar\omega_0}{D}\right) - 1.3\right)^{1/2}\right] \tag{3}$$

where $\lambda = (1/2)\, n_1 V_p$ is equivalent to the coupling constant.

A simplified version of formula (1), when $\hbar\omega_0$ is not too small compared to D, is :

$$k_B T_c = 1.13 D \exp(-1/\sqrt{\lambda})$$

The two main effects enhancing T_c are

1- the prefactor in formula [3] which is an electronic energy much larger than a typical phonon energy $\hbar\omega_0$.

2- λ is replaced by $\sqrt{\lambda}$ in formula (3) in comparaison with the BCS formula, so that in the weak coupling limit when $\lambda < 1$, the critical temperature is increased. In fact it gives too high values of T_c, we shall see later that this is due to the fact that we have neglected Coulomb repulsion between electrons. Taking this repulsion into account we shall obtain values for T_c which are very close to the observed one.

As it is however, this approach already explains many of the properties of the high T_c cuprates near optimum doping.
   - The variation of T_c with doping
The highest T_c is obtained when the Fermi level is exactly at the VHs in this first calculation (formula (1)). For lower or higher doping the critical temperature decreases. That is what is observed experimentally [11].
   - The isotope effect
Labbé and Bok [3] showed using formula (3), that the isotope effect is strongly reduced for high T_c cuprates. Tsuei *et al* [21] have calculated the variation of the isotope effect with doping and shown that it explains the experimental observations.
   - Marginal Fermi liquid behaviour
In a classical Fermi liquid, the lifetime broadening $1/\tau$ of an excited quasiparticle goes as ε^2. The marginal Fermi liquid situation is the case where $1/\tau$ goes as ε. Theoretically marginal behaviour has been established in two situations (a) the half-filled nearest-neighbour coupled Hubbard model on a square lattice and (b) the Fermi level lies at a VHs (17,21). Experimental evidence of marginal Fermi liquid behaviour has been seen in angle resolved photoemission [22], infrared data [23] and temperature dependence of electrical resistivity [17]. Marginal Fermi liquid theory, in the frame work of VHs predicts a resistivity linear with temperature T. This was observed by Kubo et al [24]. They also observe that the dependence of resistivity goes from T for high T_c material to T^2 as the system is doped away from the maximum T_c, which is consistent with our picture; in lower T_c material the Fermi level is pushed away from the singularity.

Influence of the Coulomb Repulsion

It was also been shown that the singularity is in the middle of a wide band and that in these circumstances, the Coulomb repulsion μ is renormalized and μ is replaced by a smaller number μ^* [25], the effective electron-phonon coupling is $\lambda_{eff} = \lambda - \mu^*$ and remains positive [15]. We think that this fact explains the very low T_c observed in Sr_2RuO_4, where a very narrow band has been determined by ARPES [26].
   Cohen and Anderson [27] have shown that the electron-electron repulsion plays a central role in superconductivity. Assuming a constant repulsive potential $V_{kk'} = V_c$ from 0 to ξ_F they find that T_c is given by:

$$T_c \cong T_0 \exp\left[\frac{-1}{\lambda - \mu^*}\right] \tag{4}$$

With $\qquad \mu = N_0 V_c$ and $\qquad \mu^* = \dfrac{\mu}{1 + \mu \ln \xi_F / \omega_0}$

Cohen and Anderson [27] assumed that for stability reasons μ is always greater than λ. Ginzburg [28] gave arguments that in some special circumstances μ can be smaller than λ. Nevertheless if we take $\mu \geq \lambda$, superconductivity only exists because μ^* is of the order of $\mu/3$ to $\mu/5$ for a Fermi energy ξ_F of the order of 100 $\hbar\omega_0$. It is useless to reduce the width of the band W, because λ and μ vary simultaneously and μ^* becomes greater if ξ_F is reduced, thus giving a lower T_c. Superconductivity can even disappear in a very narrow band if $\lambda - \mu^*$ becomes negative.

We have shown [15] that high T_c can be achieved in a metal containing almost free electrons (Fermi liquid) in a broad band, with a peak in the D.O.S. near the middle of the band.

Taking a D.O.S., which is a constant n_0 between energies [- W/2, - D] and [+ W/2, + D] (the zero of energy is at the Fermi level) and is $n(\xi) = n_1 \ln|D / \xi| + n_0$ between - D and + D we find for T_c, the following formula:

$$k_B T_c = \frac{D}{2} \exp\left[0.819 + \frac{n_0}{n_1} - \sqrt{F} \right]$$

(5)

where

$$F = \left(\frac{n_0}{n_1} + 0.819 \right)^2 + \left(\ln \frac{\hbar\omega_0}{D} \right)^2 - 2 - \frac{2}{n_1} \left(n_0 \ln \frac{2.28\hbar\omega_0}{D} - \frac{1}{V_p - V_c^*} \right)$$

$$V_c^* = \frac{V_c}{1 + V_c \left[\frac{n_1}{2} \left(\ln \frac{D}{\hbar\omega_0} \right)^2 + n_0 \ln \frac{W}{2\hbar\omega_0} \right]}$$

We can have a few limiting cases for this formula : $n_1 = 0$: no singularity. We find the Anderson-Morel formula. $V_c = 0$ and $n_0 = 0$: this gives the Labbé-Bok (L.B.) formula.

There are many effects enhancing T_c

$\lambda - \mu^*$ is reduced by the square root, down to $\sqrt{\lambda_1 - \mu_1^*}$ when n_1 is large enough.

As $\lambda - \mu^* < 1$ the critical temperature is strongly increased because this factor appears in an exponential. The prefactor before the exponential is D, the singularity width instead of $\hbar\omega_0$. We expect $D > \hbar\omega_0$. For instance D may be of the order of 0.5 eV and $\hbar\omega_0$ about a few 10 meV (D/$\hbar\omega_0$ of the order of 5 to 10).

We have made some numerical calculations using formula (5) to illustrate the effect of Coulomb repulsion. We used two values of D : D = 0.9 eV corresponding to t = 0.25 eV and a much more smaller value D = 0.3 eV. These calculations show that the Coulomb repulsion does not kill superconductivity in the framework of the L.B. model. The general rule for high T_c in this model is to have a peak in the density of states near the middle of a broad band to renormalize the effective repulsion μ. For a narrow band, W, or D, is small, T_c decreases very rapidly as seen in Figure 2 . A recent case has been observed in Sr_2RuO_4 with a narrow band and T_c is small [26].

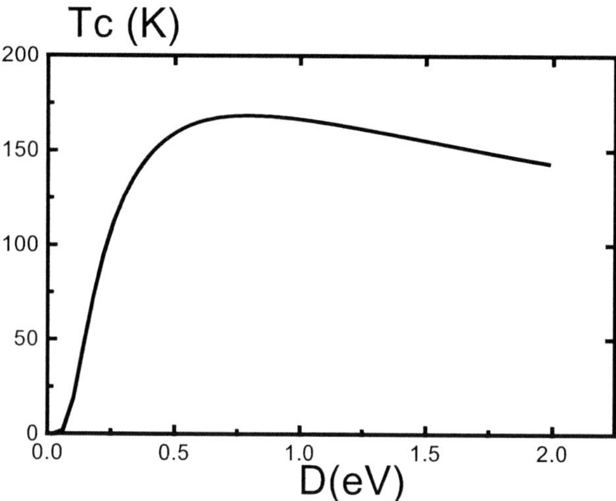

Figure 2: Effect of the width of the singularity D on T_c. n_0 and the total number of electrons per unit cell are maintained constant with this set of parameters.

Then $W = 2$ eV, $n_0 = 0.3$ eV/states/Cu, $n_1 = 0.2/D$.

In all these cases the calculations are made so that the total number of states of the band is one by Cu atom. Then $n_0 W + 2 n_1 D = 1$, and $\lambda = (n_0 + n_1) Vp$. In all these cases $\hbar\omega_0 = 0.05$ eV and $\lambda = 0.5$.

Gap Anisotropy

Bouvier and Bok [10] have shown that using a weakly screening electron-phonon interaction, and the band structure of the CuO_2 planes four saddle points: an anisotropic superconducting gap is found.

1   Model and Basic Equations

We use the rigid band model, the doping is represented by a shift $D_e = \xi_F - \xi_s$ of the Fermi level. This band structure is

$$\xi_k = -2t\left[\cos k_x a + \cos k_y a\right] - D_e \qquad (6)$$

The Fermi level is taken at $\xi_k = 0$.

We use a weakly screened attractive electron-phonon interaction potential :

$$V_{kk'} = \frac{-\left|g_q\right|^2}{q^2 + q_0^2} < 0$$

where g(q) is the electron phonon interaction matrix element for $\vec{q} = \vec{k'} - \vec{k}$ and q_0 is the inverse of the screening length.

We use reduced units: $X = k_x a$, $Y = k_y a$, $Q = qa$, $u = \dfrac{\xi}{2t}$, $\delta = \dfrac{D_e}{2t}$

We use the B.C.S. equation for an anisotropic gap :

$$\Delta_{\vec{k}} = \sum_{k'} \frac{V_{kk'} \Delta_{k'}}{\sqrt{\xi_{k'}^2 + \Delta_{k'}^2}} \tag{7}$$

We compute $\Delta_{\vec{k}}$ for two values of $\vec{k}$: Δ_A for $k_x a = \pi$, $k_y a = 0$ $\tag{8}$

$$\Delta_B \text{ for } k_x a = k_y a = \frac{\pi}{2}$$

We solve equation (7) by iteration. We know from group theory considerations, that $V_{kk'}$ having a four-fold symmetry, the solution Δ_k has the same symmetry. We then may use the angle Φ between the 0 axis and the $\vec{k}$ vector as a variable and expand $\Delta(\Phi)$ in Fourrier series:

$$\Delta(\Phi) = \Delta_0 + \Delta_1 \cos(4\Phi + \varphi_1) + \Delta_2 \cos(8\Phi + \varphi_2) + ... \tag{9}$$

We know that $\varphi_1 = 0$, because the maximum gap is in the directions of the saddle points. We use the first two terms. The first step in the iteration is obtained by replacing Δ_k by $\Delta_{av} = \Delta_0$ in the integral of equation (9). We thus obtain, for the two computed values : $\Delta_A = \Delta_{Max} = \Delta_0 + \Delta_1$ and $\Delta_B = \Delta_{min} = \Delta_0 - \Delta_1$, the following expression :

$$\Delta_{A,B}(T) = \lambda_{eff} \int_{u_{min}}^{u_{max}} \frac{\Delta_{av}(T)}{\sqrt{u^2 + u_{av}^2(T)}} I_{(A,B)}(u) \tanh\left(\frac{\sqrt{u^2 + u_{av}^2(T)}}{k_B T / t}\right) du \tag{10}$$

with
$$I_{A,B}(u) = \int_0^{x_o'} \frac{dx'}{\left[1 - \left[(\delta - u) - \cos x'\right]^2\right]^{1/2}} \frac{(q_o a)^2}{Q_{A,B}^2 + (q_o a)^2} \tag{11}$$

where $u_{min} = -\dfrac{\hbar \omega_c}{2t}$, $u_{Ma} = +\dfrac{\hbar \omega_c}{2t}$, $u_{av}(T) = \dfrac{\Delta_{av}(T)}{2t}$, $x_o' = a\cos\left(\dfrac{\delta - u}{2}\right)$

ω_c is the cut off frequency. In the following part of this work we will keep the value of $\hbar\omega_c = 60$ meV for the Bi2212 compound, a characteristic experimental phonon energy. This choice respects our approximation for $V_{kk'}$.

- For the choice of t, the transfer integral comes from the photoemission experiments and is t = 0.2 eV as explained in Reference [10].

- q_0a is adjusted, it is the Thomas Fermi approximation for small q's,

- λ_{eff} is adjusted so as to find the experimental value of Δ_{Max} and Δ_{min} and we find a reasonable value of about 0.5. λ_{eff} is the equivalent of λ-μ^* in the isotropic 3D, BCS model.

In fact the values of q_0a and λ_{eff} must depend of the hole doping level linked to D_e.

Here $q_0a = 0.12$ and $\lambda_{eff} = 0.665$.

2   Results

In Figure 3, we present the result of the iterative calculation (formula 7-11).

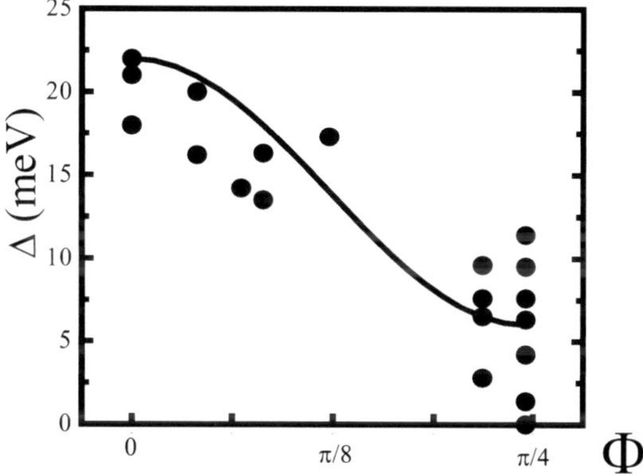

Figure 3: Anisotropic superconducting gap. Exact calculation for $\Phi = 0$ and $\pi/4$ This represents a s-wave anisotropic superconducting gap with no nodes in $\Phi = \pi/4$

In Figure 4, we present the variation of the various gaps Δ_{Max}, Δ_{min} and Δ_{av} with temperature at optimum doping, i.e. for a density of holes of the order of 0.20 per CuO_2 plane. We take in that case $D_e = 0$ and we find $T_c = 91$ K and an anisotropy ratio $\alpha = \Delta_{Max}/\Delta_{min} = 4.2$ and for the ratios of $2\Delta/k_BT_c$ the following values :

$$\frac{2\Delta_{Max}}{k_BT_c} = 6. \, , \quad \frac{2\Delta_{av}}{k_BT_c} = 3.7 \, , \quad \frac{2\Delta_{min}}{k_BT_c} = 1.4$$

This may explain the various values of $2\Delta/k_BT_c$ observed in experiments. Tunneling spectroscopy gives the maximum ratio and thermodynamic properties such as $\lambda(T)$ (penetration depth) gives the minimum gap.

In Figure 5 we present the same results, Δ_{Max}, Δ_{min}, Δ_{av} as a function of $D_e = \xi_F - \xi_s$ linked to the variation of doping.

In Figure 6 we plot the variation of the anisotropy ratio $\alpha = \Delta_{Max}/\Delta_{min}$ versus D_e. In Figure 7 the critical temperature T_c versus dx (variation of hole in the CuO_2 plane) from the optimal doping 0.20 hole per CuO_2 plane at dx=0 , dx is linked in our calculation to the variation of D_e .

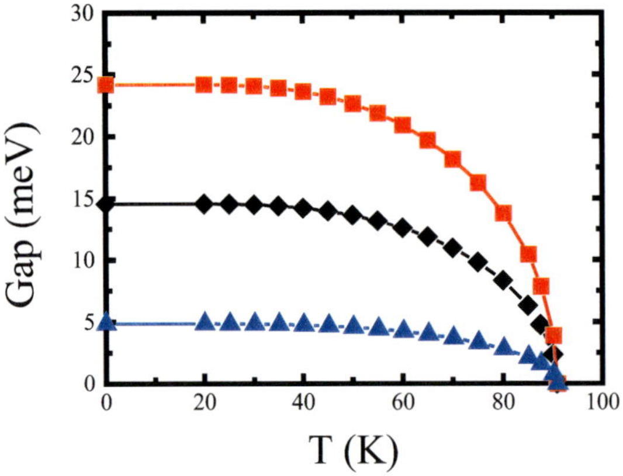

Figure 4: Variation of the various gaps Δ_{Max}, Δ_{min} and Δ_{av} versus temperature, at the optimum doping, i.e $D_e = \xi_F - \xi_s = 0$ in our model. With the following parameters, $t = 0.2$ eV, $\hbar\omega_c = 60$ meV, $q_0a = 0.12$, $\lambda_{eff} = 0.665$. The critical temperature found is $T_c = 90.75$ K

red square symbol $= \Delta_{Max}$, black diamond symbol $= \Delta_{av}$, blue up triangle symbol $= \Delta_{min}$

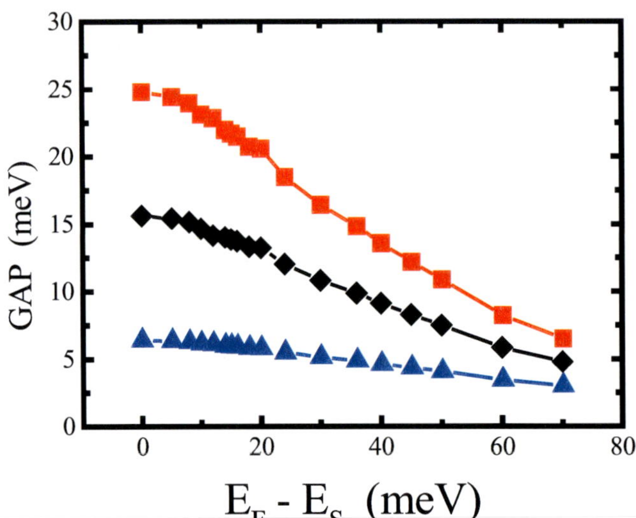

Figure 5: Variation of the various gaps Δ_{Max}, Δ_{min}, Δ_{av} versus $D_e = \xi_F - \xi_S \equiv E_F - E_S$, at $T = 0K$

red square symbol $= \Delta_{Max}$, black diamond symbol $= \Delta_{av}$, blue up triangle symbol $= \Delta_{min}$

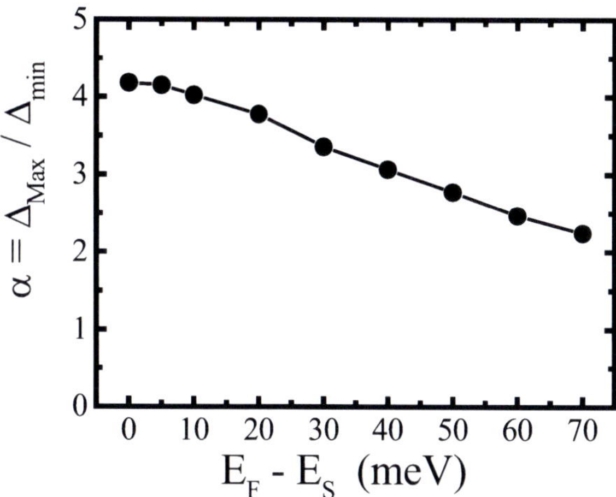

Figure 6: Variation of the anisotropy ratio $\alpha = \Delta_{Max}/\Delta_{min}$, versus $D_e = \xi_F - \xi_S \equiv E_F - E_S$.

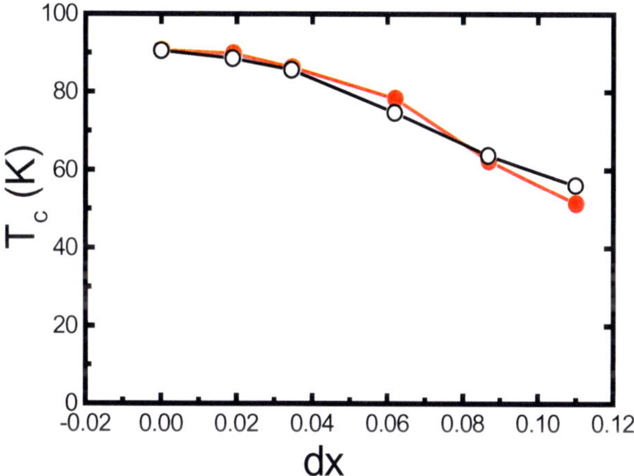

Figure 7: Comparison of the variation of T_c versus the variation of doping dx calculated in our model (red filled circles) and the experimental results of Koïke et al ref [29] (black open circles).

We observe of course that T_c and the gaps decrease with D_e or dx. The agreement with experiment [29] is very good Figure 7. We obtain a new and interesting result which is the decrease of the anisotropy ratio α with doping. This is confirmed by results on photoemission [30,31] where a maximum gap ratio $2\Delta_{Max}/k_B T_c = 5$ to 7 is observed at optimum doping with $T_c = 83$ K and $2\Delta_{Max}/k_B T_c = 3$ for an overdoped sample with $T_c = 56$ K, with a small gap $\Delta_{min} = 0\text{-}2$ meV for the both T_c, for a Bi2212 compound.

Density of States and Tunneling Spectroscopy

We have calculated the density of states of quasiparticle excitations in the superconducting state of high T_c [11,32] cuprates using the model of anisotropic gap that we have developed [10,32].

Here the D.O.S. is computed using the formula :

$$n(\varepsilon) = \frac{1}{2\pi^2} \frac{\partial A}{\partial \varepsilon} \tag{12}$$

where A is the area in k space between two curves of constant energy of the quasiparticle excitation ε_k given by :

$$\varepsilon_k^2 = \xi_k^2 + \Delta_k^2 \tag{13}$$

where ξ_k is the band structure (formula (6)). We use the same procedure and the same expression of Δ_k as before.

Figure 8 represents the variation of the D.O.S. as a function of ε for $T = 0$ K. This is similar to the experimental conductance (dI/dV versus the voltage V) of a N-I-S junction here we show the measurement made by Renner and Fisher [33] on a BSCCO sample. Δ_{Max} is located at the maximum peak and Δ_{min} at the first shoulder after the zero bias voltage, Figure 9. But for different values of $\xi_F - \xi_s$, we see a new maximum emerging, which is a signature of the van Hove singularity and a dip between this maximum and the peak at Δ_{Max}. This dip is seen experimentally in the STM tunneling experiments of Renner et al [33]. Figure 10 show the behaviour in fonction of the temperature, the value of the superconducting gap dependant in temperature are done by our calculations with formula (7-11).

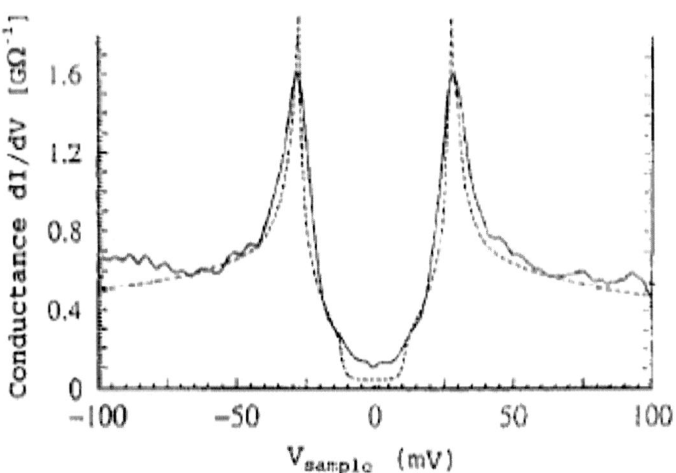

Figure 8: The best fit of the conductance measured by tunneling spectroscopy on BSCCO, N-I-S junction, by Renner and Fischer (Figure (10) of Reference [33]). solid line: fitted curve with $\Delta_{Max} = 27$ meV, $\Delta_{min} = 11$ meV, $t = 0.18$ eV, $\Gamma = 0.5$ meV at $T = 5$ K, dashed line : experimental curve.

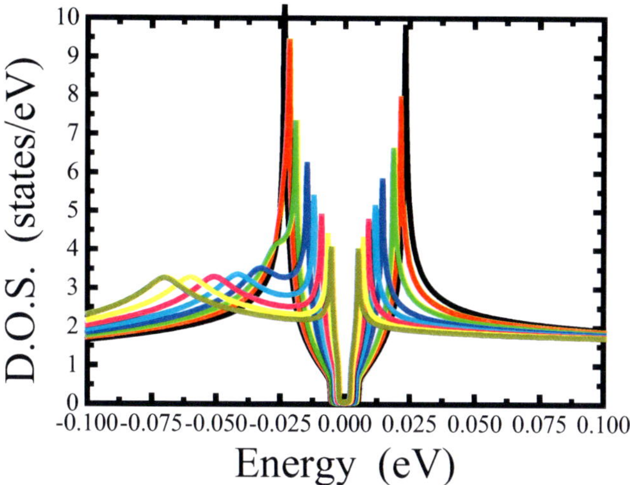

Figure 9: Variation of the D.O.S. versus the energy ε, for T = 0 K, that is similar at a NIS junction, for different values of the doping D = ξ_F - ξ_s, i.e. 0, 10, 20, 30, 40, 60 and 70 meV with Γ = 0.1 meV and Γ' = 5 meV in the model of Reference [32].

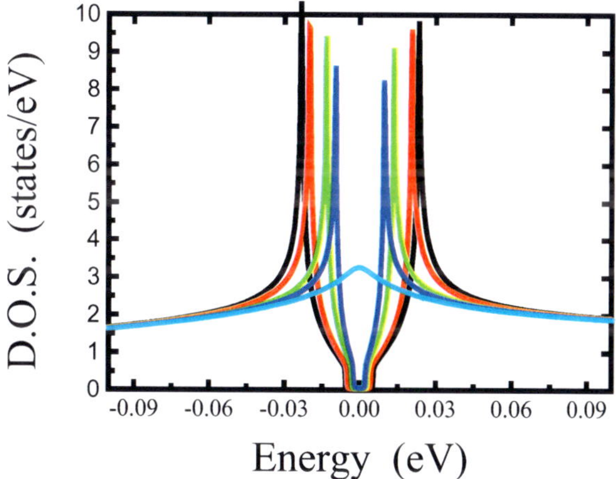

Figure 10: Variation of the in fonction of temperature T =0, 60, 80, 85, 91, Tc= 90.8 K, for the case ξ_F = ξ_s =0.

For the calculation of the conductance, we use the following formula

$$\frac{dI}{dV} = CN_0 \int_{-\infty}^{+\infty} N_S(\varepsilon) \left[-\frac{\partial f_{FD}}{\partial V}(\varepsilon - V) \right] d\varepsilon \qquad (14)$$

where f_{FD} is the usual Fermi-Dirac function; I and V are the current and voltage, C a constant proportional to $|T|^2$, the square of the barrier transmission, N_0 the D.O.S. of the normal metal that we assume constant, and $N_s(\varepsilon)$ the previously calculated D.O.S. in the anisotropic superconductor. We introduce a damping parameter Γ in order to take into account the effect of a low 3D interaction and of the surface impurities.

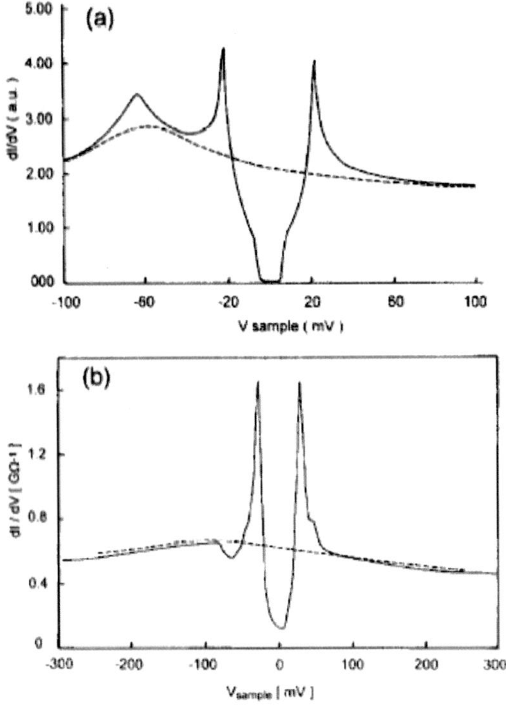

Figure 11: (a) Curves of the conductance calculated for a N-I-S junction. Solid line: in the superconducting state at T = 5 K with Δ_{Max} = 22 meV, Δ_{min} = 6 meV, Γ = 0.1 meV, t = 0.2 eV and D_e = -60 meV, Γ' = 5 meV. Dashed line : in the normal state at T = 100 K with Δ_{Max} = Δ_{min} = 0 meV, Γ = 0.1 meV, t = 0.2 eV and D_e = -60 meV, Γ' = 5 meV. (b) For comparison we show Figure (7) of Reference [33]. The maximum of the normal state conductance (or D.O.S.) at negative sample bias is well reproduced.

Specific Heat

1   Theoretical Calculation

The purpose of this chapter is to evaluate the influence of the VHs and the anisotropy of the gap on the specific heat calculated in the mean field B.C.S. approximation, i.e. we do not take into account the fluctuations near the critical temperature T_c. There are a great number of experiments measuring C_s. To compare our calculations to experiments, we must subtract the part due to fluctuations. These kind of adjustment have been made by various authors by using the fact that thermodynamic fluctuations are symmetric about T_c and can be easily evaluated above T_c [34,35]. Also we do not take into account the magnetic fluctuations in low temperature, nor the pair-breaking which may exist in overdoped sample. By the usual way, we obtain for C_s :

$$C_s(T) = \frac{2}{k_B T^2} \sum_k \frac{\exp(\varepsilon_k / k_B T)}{\left(1 + \exp(\varepsilon_k / k_B T)\right)^2} \varepsilon_k^2 - \frac{1}{k_B T} \sum_k \frac{\exp(\varepsilon_k / k_B T)}{\left(1 + \exp(\varepsilon_k / k_B T)\right)^2} \frac{\partial \Delta_k^2(T)}{\partial T} \quad (15)$$

We use the values of ε_k and Δ_k ($\Delta_{Max}(T, D_e)$ and $\Delta_{min}(T, D_e)$) given by formula (7-11) to evaluate the two integrals of formula (15) numerically. Near T_c we have a very good agreement between the calculated values and the following analytical formula :

$$\Delta_{Max,min} = \Delta_{Max,min}(T = 0) \, 1.7 \left(1 - (T / T_c)\right)^{1/2}$$

We see that the slopes $\partial \Delta^2 / \partial T$ do not depend on doping which simplifies the calculation of the second integral of formula (15). The results are presented in Figures 12 and 13 where we plot C_s versus T and $\Delta C/C|_{Tc}$ for various doping levels D_e.

We can make the following observations :

1- The jump in specific heat varies with doping $\Delta C/C|_{Tc}$ is 3.2 for $D_e = 0$ and 1.48 for $D_e = 60$ meV compared to 1.41, the B.C.S. value for a isotropic superconductor, with a constant D.O.S., N_0 in the normal state. The high value of $\Delta C/C|_{Tc}$ is essentially due to the VHs when it coincides with the Fermi level and the highest value of the gap Δ_k. With doping, the VHs moves away from ξ_F and $\Delta C/C|_{Tc}$ decreases toward its B.C.S. value.

2 - There is also a difference in the specific heat C_N in the normal state. For a usual metal with a constant DOS N_0, $\gamma_N = C_N / T$ is constant and proportional to N_0. Here we find $\gamma_N = a \ln(1 / T) + b$ for $0 \le D \le 30$ meV where a and b are constant. For $D_e = 0$ this behaviour has already been predicted by Bok and Labbé in 1987 [36]. The specific heat $C_N(T)$ explores a domain of width $k_B T$ around the Fermi level ξ_F. So for $D_e << k_B T_c$, the variation of γ_N above T_c is logarithmic. For $D_e > 30$ meV, at high temperature $T - T_c > D_e$, the B. L. law is observed, but for lower temperatures γ_N increases with T and passes through a maximum at T^*, following the law : T^* (meV) = 0.25 D_e (meV) or T^* (K) = 2.9 D_e (meV).

2   Comparison with Experiments

Because of the difficulty to extract exactly C_s from the experimental data, we will compare only the general features to our calculation. We see that the doping has a strong influence on T_c and all the superconducting properties, so we assume that its role is to increase the density of holes in the CuO_2 planes. To compare our results on the effect of doping on C_s with experiments, we have chosen the family of the $Tl_2Ba_2CuO_{6+\delta}$, studied by Loram et al [37], Figure. 9 of Reference [37], because they are overdoped samples, with only one CuO_2 plane. The family $YBa_2Cu_3O_{6+x}$ is underdoped for $x < 0.92$ and for $x > 0.92$ the chains become metallic and play an important role. However, recent results by Loram et al, Figure 2a of the Reference [38] on Calcium doped YBCO, $Y_{0.8}Ca_{0.2}Ba_2Cu_3O_{7-\delta}$, which are overdoped two dimensionnal systems, show a very good agreement with our results. We notice the displacement and the decrease of the jump in specific heat C_s with doping. The jump $\Delta C/C|_{Tc}$ = $\Delta\gamma/\gamma|_{Tc}$ = 1.67 [37], and 1.60 [38] greater than the B.C.S. value 1.41 for a metal with a constant DOS. We find theoretically this increase in our model due to the logarithmic VHs. The symmetrical shape of the peak of C_s, at low doping level, is due to the critical fluctuations. A subtraction of these fluctuations [34,35] gives an asymmetrical shape. For high doping levels the classical B.C.S. shape is found.

For $D_e = 0$, we find that γ_N is not constant but given by the logarithmic law [36]: $\gamma_N = a\ln(1/T) + b$. When D_e increases, the law changes, γ_N passes through a maximum for a value of T, T^*. This behaviour is clearly seen in the $YBCuO_{6+x}$ family [37]. We explain the high value $\Delta C/C|_{Tc} = 2.5$ for $x = 0.92$ in the YBCO family, and we find also the predicted variation of T*.

Our model, neglecting magnetic fluctuations gives an Arrhenius law for C_s at low temperature with a caracteristic energy which is Δ_{min}. We see that such a law is observed in $YBaCuO_{6.92}$ and for $Tl_2Ba_2CuO_6$ at optimum doping.

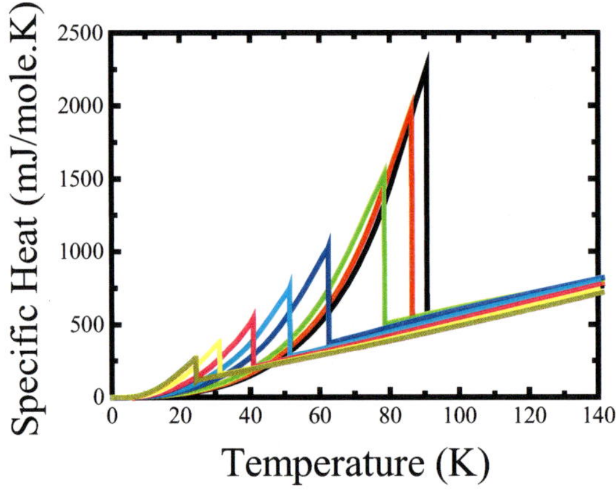

Figure 12: The calculated specific heat versus the temperature for the different value of $D_e = \xi_F - \xi_s = 0$, 10, 20, 30, 40, 50, 60 and 70 meV, linked to the variation of doping.

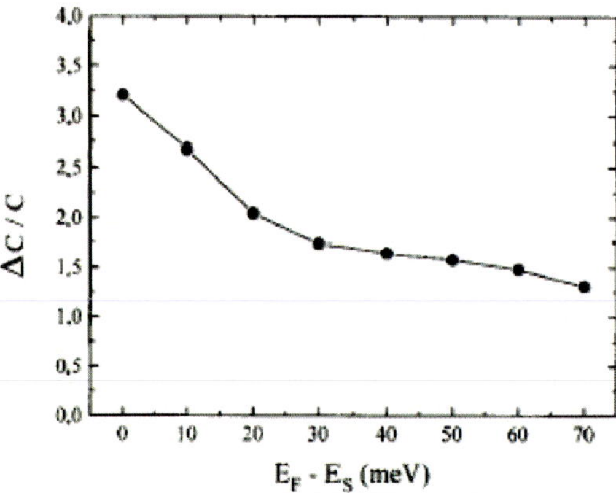

Figure 13: Variation of the jump in the specific heat, $\Delta C/C|_{Tc}$ versus D_e.

Van Hove Singularity and Characteristic Temperature T°

Several experiments on photoemission, NMR and specific heat have been analyzed using a normal state pseudo-gap [39]. In fact, all what is needed to interpret these data is a density of state showing a peak above the Fermi energy. To obtain the desired D.O.S. several authors [39] introduce a pseudogap in the normal state. This seems to us rather artificial, the above authors themselves write that the physical origin of this pseudogap is not understood.

We have shown that by using a band structure of the form formula (6), we may interpret the results obtained in the normal metallic state. We have computed the Pauli spin susceptibility using the following formula :

$$\chi_P = \frac{\mu_0 \mu_B}{B} \int_{-\infty}^{+\infty} n(\xi) \left(f_{FD}\left(\xi + \mu_B B\right) - f_{FD}\left(\xi - \mu_B B\right) \right) d\xi \tag{16}$$

The results fit well the experiments. We find a characteristic temperature T° where the variation of χ_p versus T goes through a maximum. We may express $D_e = \xi_F - \xi_S$ as a variation of doping $\delta p = p - p_0$, p_0 being the doping for which $\xi_F - \xi_S$, $p_0 = 0.20$ hole/copper atom in the CuO_2 plane. Figure 14 the Pauli susceptibility in the normal state for the different value of the doping, from metallic system to the metal-insulator transition. Figure 15 represents the various experimental points taken from Figure 5 of Reference [39] where the authors plot $E_g/k_B T_{cMax}$ versus p. We see that what the authors call pseudogap is exactly our $D_e = \xi_F - \xi_S$, the distance from the Fermi level to the peak in the D.O.S..

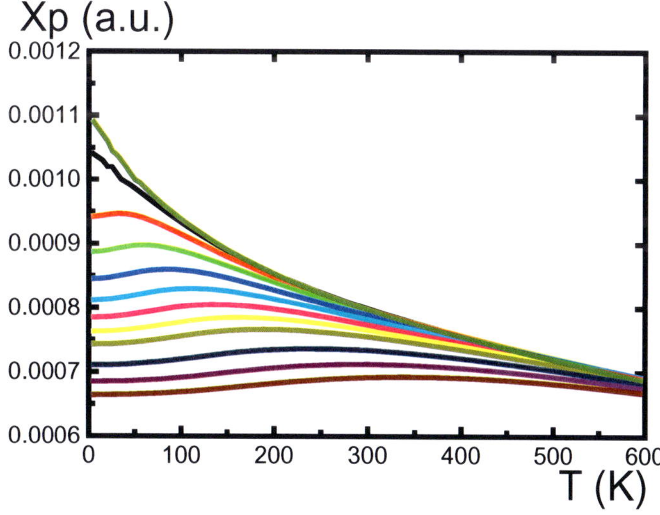

Figure 14: Pauli susceptibility in the normal state for the different value of the doping, decreasing from the top to the bottom. $\delta p = 0$, $p_0 = 0.20$ corresponding to the top curve, then δp varies from 0 to -0.15, from metallic system to the metal-insulator transition.

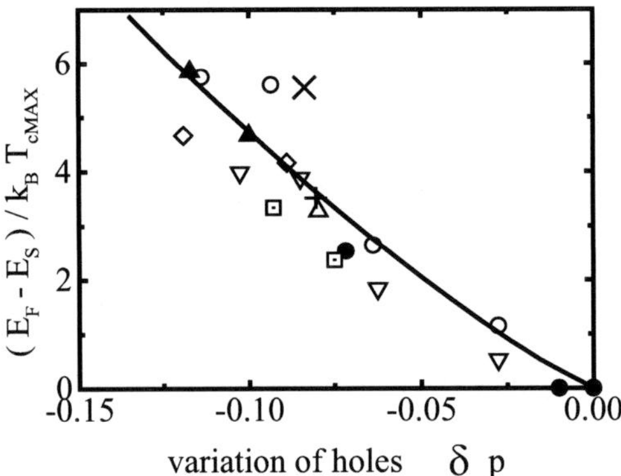

Figure 15: $D_e = \xi_F - \xi_S \equiv E_F - E_S$ divided by $k_B T_{cMAX}$ ($T_{cMAX} = 110$ K) versus the variation of the density of hole: solid line. The different symbols are the same as in the Figure 5 of the Reference [39], they represent the values of the so-called normal pseudogap divided by $k_B T_{cMAX}$ ($E_g / k_B T_{cMAX}$) obtained from NMR on different compounds. Our calculations are made with a transfer integral $t = 0.25$ eV, δp is taken as zero for $p = 0.20$.

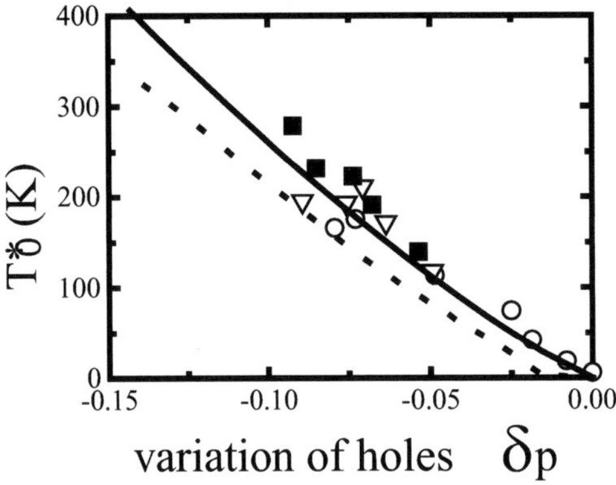

Figure 16: The temperature, $T°$, where the calculated χ_p (dashed line) and the specific heat (solid line) go through a maximum, versus δp. For comparaison we show the results presented in Figure 27 of Reference [40], the symbols are the same. (solid squares : from thermoelectric power, circles : from specific heat, triangles: from NMR Knight shift data).

We have also computed the electronic specific heat C_s [11] in the normal state using the same DOS. We find that $\gamma = C_s/T$ goes through a maximum with temperature T, at a value T* as found experimentally by Cooper and Loram [40]. In Figure 16 we compare our computed $T°$ with the experimental one (Reference [40]), the agreement is excellent. So we are able to interpret the NMR and specific heat data in the normal metallic state without invoking a pseudogap, but simply by taking into account the logarithmic singularity in the DOS.

We also explain the shift between the observed experimental optimum T_c, where $p = 0.16$ instead of 0.20, and the expected optimum T_c from our theory, i.e. where $D_e = 0$, by the fact that in first time in our gaps calculations we have not taking into account the variation of the 3D screening parameter q_0a in function of D_e. These calculations (see the following chapter) show the competition between the effect of the position of the V.H.s. and the value of q_0a for getting the optimum T_c, this competition depends on the compound. When the overdoping increases, i.e. the density of free carriers increases, then q_0a increases too, and in our model this leads to a decrease in T_c. It is why for $D_e = 0$, or $p = 0.20$, we have not the optimum T_c, and why the logarithmic law for χ_p is found in the overdoped range [11]. In the underdoped range in respect of the observed optimum T_c, (i.e. density of free carriers decrease), q_0a decreases too, but the Fermi level goes too far away from the singularity to obtain high T_c. Our results agree completely with the experimental observations.

Effect of Screening on the Gap Anisotropy and T_c

In the preceding part ("Gap Anisotropy" chapter) we have taken $q_0a = 0.12$ and the effective coupling constant $\lambda_{eff} = 0.665$ in order to fit the experimental values of the gap observed by ARPES and tunneling spectroscopy. We also have stressed the importance of q_0a in the value of the anisotropy ratio $\alpha = \Delta_{Max}/\Delta_{min}$. We shall now study in more details the influence of q_0a on α , T_c.

For α and T_c the calculations use equations 7-11 where q_0a is included in $V_{kk'}$.

For α, we adjusted our values of λ_{eff} to obtain a constant critical temperature of 90.75 K and an average gap of $\Delta_{av} = 14.50 \pm 0.15$ meV. This approximation is valid in the limit of weak screening ($q_0a < 0.2$).The results are presented in Figure 17. We see that increasing q_0a , or in other word going towards more metallic system or 3D, that the anisotropy of the gap decreases. There are no direct experiments to measure α as a function of q_0a. The photoemission experiments measure the anisotropy as a function of doping, so q_0a and $\xi_F - \xi_s$ vary simultaneously. But there is a decrease in α when the doping is varying [30,31].

For T_c, we keep the parameter $D_e = 0 = \xi_F - \xi_S$ and we resolve the self-consistant equation (7) varying q_0a, and adjusting Δ_{Max}, Δ_{min} and Δ_{av} .The results are presented in Figure 18. The effect of increasing the screening strength is to decrease T_c. An increase of the screening can be due to the proximity of ξ_F to ξ_S where the DOS is high which leads to a strong screening, and in the other side the hide DOS increase T_c . It is why we have to take into account these two effects to explain the phase diagram (see the previous chapter).

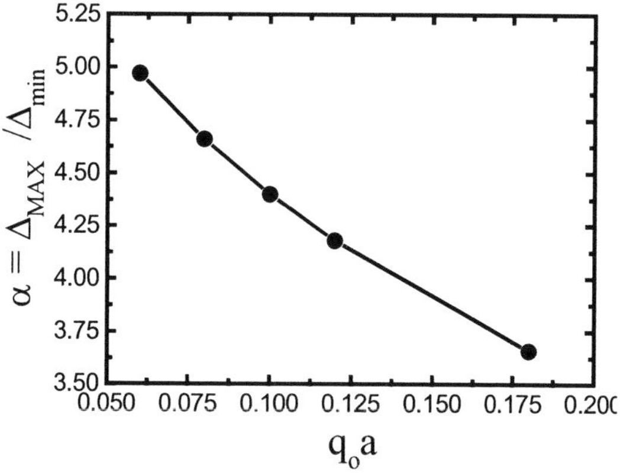

Figure 17: The anisotropy ratio $\alpha = \Delta_{Max}/\Delta_{min}$ versus the screening parameter q_0a.

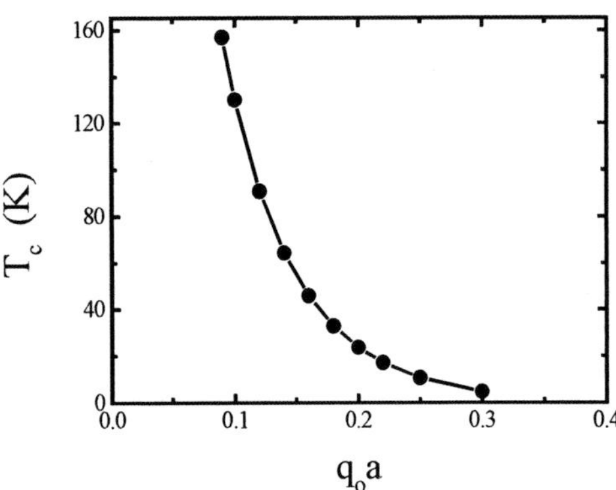

Figure 18: T_c versus the screening parameter q_0a.

We have shown that the effect of increasing q_0a is to transform the system in a metallic or more isotropic one.

Influence of Disorder in Superconductor Compounds

We can consider underdoped or overdoped cuprates as disordered conductors because the diffusion coefficient D can be as low as 10^{-5} $m^2.s^{-1}$. Under these conditions Coulomb interaction between electrons must be taken into account. The main effect is to open a dip in the DOS near the Fermi level. We show that this model explains most of the observed features of the so-called "pseudogap" in the normal state including its value, anisotropy and variation with doping.

1   Introduction

Many experiments made in the normal state of high Tc superconductor have revealed a so-called pseudogap. This pseudogap was observed in transport, magnetic, specific heat measurements and in scanning tunneling and ARPES measurements [41]. The pseudogap observed in the normal state seems to be a partial gap. It is related to a crossover temperature, named T*, below which its observation is possible. Many authors relate T* with magnetic phenomena. We propose another explanation for the pseudogap related to T*. It is mainly observed in underdoped samples, which are disordered and in which the mean free path and thus the diffusion coefficient is very low. Under these conditions, the diffusion length (L_D) becomes of the order of magnitude or smaller than the electron wavelength $1/k_F$. The materials are thus disordered conductors and the Coulomb repulsion becomes important (for a review see Altshuler and Aronov [19]).

2   Description of the Model Used

Altshuler and Aronov [19] have developed a theory to study the effect of the electron-electron interaction on the properties of disordered conductors. The conditions for its application

$k_F L_D \ll 1$ is also satisfied for underdoped cuprates. The theory has also shown that the interaction effects are most clearly pronounced in low-dimensionality systems. We compute the one particle DOS taking into account the Coulomb interaction in the self-energy term. We show that particle repulsion produces a dip in the DOS at the Fermi energy. In the cuprates, where the Fermi surface is very anisotropic, we find that the pseudogap appears first and is more pronounced in the directions of the saddle points (1,0) and equivalent of the CuO_2 planes, where the Fermi velocity is smaller. This is clearly seen in the ARPES experiments.

We take an anisotropic dispersion relation for the one electron energy ξ_k in the CuO_2 planes (bidimensional):

$$\xi_k = -2t(\cos X + \cos Y) + 4t' \cos X \cos Y + D_e \tag{17}$$

where $D_e = \xi_F - \xi_S + 4 t'$

The self-energy is computed using the following formula:

$$\Sigma_m = \Sigma_m^{ex} + \Sigma_m^H \tag{18}$$

where Σ_m^{ex} is the exchange part and Σ_m^H the Hartree part of the self-energy.

The exchange energy is given by:

$$\Sigma_{m,\xi}^{ex} = \frac{1}{2\pi v} \int_\xi^\infty d\omega \int \frac{d^3q}{(2\pi)^3} U\!\left(\vec{q}\right) \frac{Dq^2}{\omega^2 + \left(Dq^2\right)^2} \tag{19}$$

with $\vec{q} = \vec{k} - \vec{k}'$, D the diffusion coefficient. $U(\vec{q})$ is the Fourier transform of the long range Coulomb interaction and the term in Dq^2 the Fourier transform of the electron-electron correlation function. For $U(\vec{q})$ we take a screened Coulomb potential (the screening is tridimensional):

$$U(\vec{q}) = C / (q^2 + q_0^2) \tag{20}$$

where q_0^{-1} is the screening length. We then compute the DOS within a small angle $d\theta$, in the two directions (1,0) and (1,1), using a selfconsistent procedure.

3   Variation of the Coefficient of Diffusion D with Doping and k Space Direction

In a simple Fermi liquid, the diffusion coefficient is given by $D = (1/3)v_F l$, v_F is the Fermi velocity and l is the mean free path. For a given sample, with doping and disordered fixed, l is constant and v_F varies with direction, it is much smaller near the saddle point A $(0,\pm\pi)$ than at point B $(\pm\pi/2,\pm\pi/2)$. In underdoped samples there are disorder in the oxygen vacancies and crystalline defects. We assume that l is strongly reduced as the doping decreases until we reach a region where the crystalline order is restored in the insulating antiferromagnetic state.

$\xi_F - \xi_S$ varies slightly and v_F at point A is reduced, v_F at point B remains almost unchanged, so the anisotropy remains.

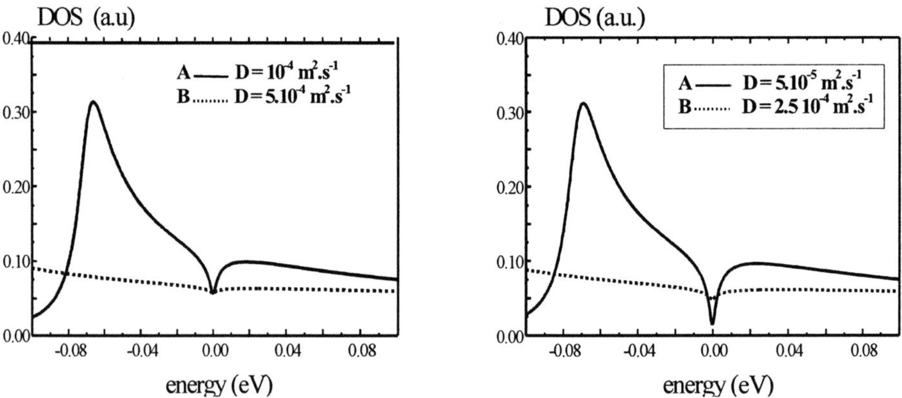

Figure 19: Calculated DOS with Coulomb interaction with different sets of values of D A : in the (1,0) direction, and equivalent directions - B : in the (1,1) direction

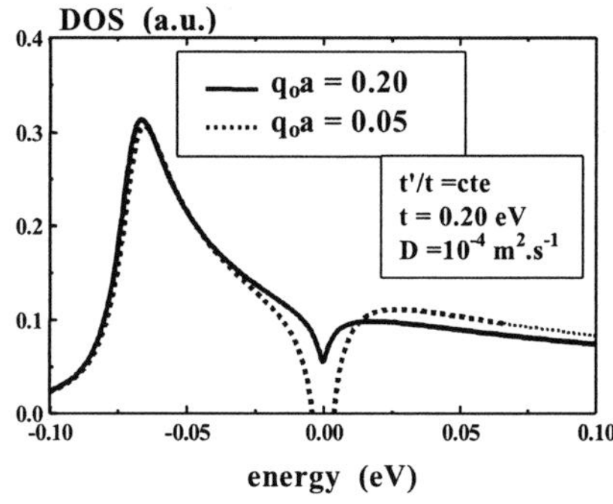

Figure 20: Effect of the screening on the DOS calculated with the Coulomb interaction term.

4   Effect of the the Coefficient of Diffusion, the Screening and the Bandwith

Our results are presented in Figure 19. We can see that our model explains why the pseudogap opens in the (1,0) direction and not in the (1,1) direction as seen in ARPES [41]. We have studied the effect of screening by varying q_oa, in the A direction, the result is shown on Figure 20. The decrease of q_oa increases the number of states in the wings and deepens the dip. The effect of varying the transfer integrals, t and t', i.e. the bandwidth, is less important.

5   Pauli Susceptibility and Disorder

We are able to calculate the total DOS including the exchange term in the self-energy, the dip created at FL by the disorder effect can be more or less deep or broad, depending for a given bandwidth of the coefficient of diffusion or the screening strength values. We study some cases and calculate the corresponding Pauli susceptibility, as already made but without disorder effect (see the previous chapter "VHs and $T°$ " and Reference [42], where the maxima in the Xp(T) curves were related to the high DOS at ξ_S. In Figure 21 we present our theoretical results for two different pseudogaps. In Figure 21b it is deeper and broader, at T^* , where these pseudogaps open, Xp(T) (dashed line) begins to be lower than Xp(T) (full line) without pseudogap. In Figure 21a this opening occurs after the temperature $T°$, where the high DOS at E_S begins to be filled, but in Figure 21b this opening occurs before this event. The consequence of these mixing effects give an effective $T°_1$ in our initial theory [42] and $T^* > T°$. This is a theoretical result to be discussed as all experimental results seem to give $T° > T^*$. Such bigger pseudogap probably occurs for lower doping, leading to $T^* < T°$.

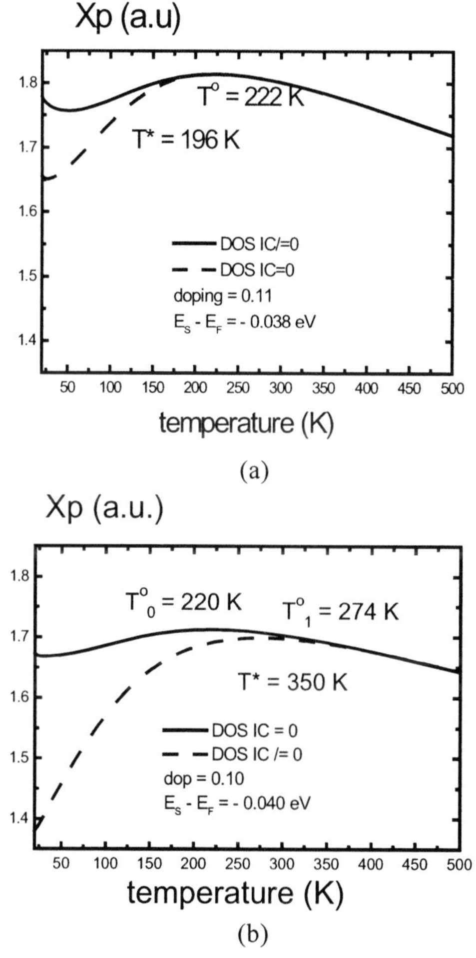

Figure 21: Calculated Pauli Susceptibilities

Full line = without disorder effect - Dashed line = with disorder effect. Figure 21a: for 0.11 hole doping. Figure 21b: for 0.10 hole, lower doping, with a deeper and broader dip, leading to a smaller Xp(T), to a more pronounced decrease, and to $T^* > T°$.

6   Discussion

Experiments reported by I. Vobornik *et al* [43] show the possibility of having disorder-induced pseudogaps comparable to those existing in underdoped Bi2212 samples. The pseudogap can also be observed in overdoped samples [43]. As we can see in our figures, the dip is less pronounced if either the screening or the diffusion coefficient is higher. These higher values exist in the overdoped regime, leading to a lower value of T*. Then it seems to be below T_c, so it cannot be observed in the normal state [44]. Therefore in varying the physical parameters in our model (screening, doping (i.e.: ξ_F - ξ_S), diffusion coefficient, bandwidth), we have a good explanation for the evolution of the pseudogap in the phase diagram. The pseudogap decreases from underdoped to overdoped region in agreement with these parameters. The pseudogap was observed in a non-superconducting region in scanning tunneling spectroscopy measurements made by T. Cren *et al* [45]. This shows that the pseudogap is not inevitably related to superconductivity, but is an intrinsic property of the material. The existence of the "Coulomb dip" in the HTSC and the Si doped metals [20,46], where we know it is due to disorder, confirms that disorder can be at the origin of the pseudogap.

Hall Effect in the Normal State of HTSC

1   Introduction

Many measurements of the Hall coefficient R_H in various high Tc cuprates have been published [47-51]. The main results are the following :

(i)   at low temperature T, $R_H \approx 1/p_{h0}e$, where p_{h0} is the hole doping, when T increases R_H decreases, and for highly overdoped samples becomes even negative [47].

(ii)   these authors are also able to define a temperature T_0 , where R_H changes its temperature behaviour, and they found that $R_H(T)/R_H(T_0)$ versus T/T_0 is a universal curve for a large doping domain (from $p_{h0} = 0.10$ to $p_{h0} = 0.27$).

We can explain these results by using the band structure for carriers in the CuO_2 planes. In particular, the existence of hole-like and electron-like constant energy curves, which give contributions of opposite sign to R_H. The transport properties explore a range of energy k_BT around the Fermi level, when T is increased more and more carriers are on the electron like orbits, resulting in a decrease of R_H.

2   Calculation of the Hall Coefficient

The constant energy surfaces of carriers in the CuO_2 planes are well describe by formula 17. It is very clearly seen [7] that the Fermi level crosses the saddle points (or VHs), at ξ_S , for a hole doping of $p_{h0} \approx 0.22$. For energies $E > \xi_S$ the orbits are hole-like, and for $E < \xi_S$ they are electron-like.

To compute the Hall coefficient we use the formula obtained by solving the Boltzmann equation. In the limit of low magnetic fields B, perpendicular to the CuO_2 plane, $\mu B \ll 1$, where μ is an average mobility of the carriers, R_H is given by:

$$R_H = \frac{\sigma_{xy}}{\sigma_{xx}^2} \frac{1}{B} \tag{21}$$

where σ_{xy} and σ_{xx} are the components of the conductivity tensor. We follow the approach given by N. P. Ong [52]:

$$\sigma_{xy} = \int_{E_{min}}^{E_{max}} \left(-\frac{\partial f_0}{\partial E} \right) \sigma_{xy}(E)\, dE \tag{22}$$

where f_0 is the Fermi Dirac distribution function, E_{min} and E_{max} are the bottom and the top of the band, and $\sigma_{xy}(E)$ is σ_{xy} computed on a constant energy surface.

For metals, where $k_B T \ll \xi_F$, σ_{xy} is usually chosen as $\sigma_{xy} = \sigma_{xy}(\xi_F)$, computed on the Fermi surface only, this is done by N. P. Ong [52]. In our case, $k_B T$ is not small compared to $\xi_F - \xi_S$, so when T increases the electron-like orbits as well as the hole-like orbits are populated. The electron-like orbits give a negative contribution to R_H, so that R_H decreases with temperature. This is our original approach to the problem. To compute R_H, we use the following method: We compute first $\sigma_{xy}(E)$ using the Ong approach. The idea is to draw the $\vec{l}$ curve swept by the vector $\vec{l} = \vec{v}_k\, \tau_k$ as $\vec{k}$ moves around the constant energy curve (C.E.C.). Then σ_{xy} reduces to:

$$\sigma_{xy} = \frac{2 e^3}{\hbar^2}\, A_l\, B \tag{23}$$

where A_l is the area enclosed by C.E.C., in the (l_x, l_y) plane. There may be secondary loops in the $\vec{l}$ curve. When the C.E.C. is non-convex, the $\vec{l}$ curve presents several parts where the circulation are opposite (see Reference [52] Figure 2). Then the effective density of carriers that must be taken in computing σ_{xy} is $n_e' = \Gamma\, n_e$ for the electron-like orbits, with $\Gamma < 1$, and $p_h' = p_h$ for the hole-like orbits, because for the hole-like orbits we can see that the C.E.C. have no non-convex parts. Finally we obtain for the Hall coefficient:

$$R_H = \frac{V}{e}\, \frac{p_h - b^2\, n_e'}{\left(p_h + b\left(p_{h0} - p_h\right)\right)^2} \tag{24}$$

where b is the ratio of the average mobilities of the carriers on the electron and hole like orbits. That is the mean value of $\langle \tau/m \rangle$, where τ is the relaxation time and m the effective mass. V is the volume of the unit cell. We adjust the Fermi level so that the total number of carriers p_{h0} remains constant. To compute Γ, we must know the scattering mechanisms and evaluate Γ. Γ was computed by Ong [52] assuming a constant $\vec{l}$, but this is not valid in our case because $\vec{l}$ is very small near the saddle points (hot spot), both v_k and mainly the relaxation time τ_k are small at this point. So we estimated a much smaller value of Γ, around $\Gamma = 0.2$ for E near ξ_S and going to $\Gamma = 1$ when E approaches E_{min}. We choose a function $\Gamma(E)$, varying from $\Gamma(\xi_S)$ to 1 for $\Gamma(E_{min})$.

$$\Gamma(E) = 1 - \alpha \left(E - E_{min}\right)^n$$

$$\alpha = \frac{1 - \Gamma(\xi_S)}{\left(\xi_S - E_{min}\right)^n} \quad , \quad \text{with} \quad n = 1$$

$n_e^{'}$, p_h are given by the following formulae:

$$n_e^{'} = \int_{E\,min}^{Es} A_e(E)\ \Gamma(E) \left(-\frac{\partial f_0}{\partial E}\right) dE \tag{25}$$

$$p_h = \int_{Es}^{E\,max} A_e(E)\ \Gamma(E) \left(-\frac{\partial f_0}{\partial E}\right) dE \tag{26}$$

$$p_h = \int_{Es}^{E\,max} A_h(e)\ \Gamma(E) \left(-\frac{\partial f_0}{\partial E}\right) dE \tag{27}$$

A_e is the area enclosed by the electron-like surfaces for $E < \xi_S$, A_h is the area enclosed by the hole-like surfaces for $E > \xi_S$. E_{max} is determined in order to only take into account the holes added to the lower half-band. So we obtain for $T = 0$ K the density of free hole per CuO_2 plane, the Hall number $n_H = 1/(R_H\,e) = p_{h0}/V$. The scattering mechanism being probably the same for the electron and the hole orbits, which are very similar along the (1,1,0) direction, then we assume b = 1.

3   Results

The results of our calculations and their comparison with the experimental results are given in Figures 22, 23, 24, 25 . When the authors of the experiments give only the concentration x of doping atoms, and the critical temperature T_c we have to evaluate the actual hole doping p_{h0} using the universal phase diagram of T_c versus doping for high T_c superconductors [53].

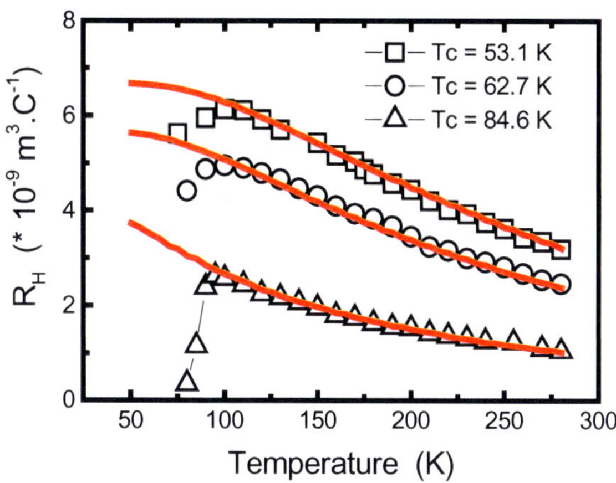

Figure 22

symbols: experimental $R_H(T)$ given by Matthey et al (Reference [50]) in $GdBa_2Cu_3O_{7-\delta}$
full lines: theoretical fits
theoretical hole doping level $p_{h0} = 0.10$ for the experimental Tc = 53.1 K
theoretical hole doping level $p_{h0} = 0.12$ for the experimental Tc = 62.7 K
theoretical hole doping level $p_{h0} = 0.16$ for the experimental Tc = 84.6 K
The calculations are made with : t = 0.18 eV , t' = 0.04328 eV , 2t'/t = 0.48, $\Gamma(\xi_S) = 0.2$

For the theoretical results of Figures 22, 23, 24 we use the following parameters : t = 0.18 eV, t' = 0.04328 eV, 2t'/t = 0.48, $\Gamma(\xi_S) = 0.2$. These values of t and t' means that the shape of the Fermi surfaces changes when we cross the critical doping $p_{h0} \approx 0.22$. This is also seen in the photoemission curves reported by Ino et al [7].

In Figure 23 , we can see the representation of the universal law $R_H(T)/R_H(T_0)$ versus T/T_0, where T_0 is defined experimentally by the fact that R_H becomes almost constant above this temperature [47-50]. In our model this temperature is given by $2k_BT_0 = \xi_F - \xi_S$, this shows that this universal behaviour is due to the 2D band structure, in which the shift $\xi_F - \xi_S$ is connected to the hole doping. This is very natural in our approach, because the factor $(\xi_F - \xi_S)/k_BT$ enters the Fermi-Dirac distribution.

We see that the agreement of our fits with the experiments are excellent. There is a small discrepancy between the values of our theoretical R_H and the experimental values. We think that this is due to the inhomogeneities in the material and to the way to carry out the R_H measurements. This can may be explained by the evaluation of the experimental volume V. We use in our calculation the unit cell volumes:

$V_{LSCO} \cong 189 \ 10^{-30}$ m^3 for LSCO and $V_{YBCO} \cong 174 \ 10^{-30}$ m^3 for YBCO.

The experimental value of R_H is determined by the geometrical aspect of the sample (the thickness in particular). This value is evaluated assuming that the current flow is homogeneous throughout the sample, this is not always true. We find a discrepancy between 1.5 and 2 in the case of YBCO [48] and GdBCO [50], a larger discrepancy is found in the case of LSCO [47]. In this later case, the authors find different R_H results for the same doping, with various compounds (single crystals and thin films).

Anyway, adjusting our values for R_H, at low temperature, we can fit many experimental results, for the three different compounds. We also use a rigid band model, where the bandwidth does not change with the doping. This is not exactly the case as shown in the photoemission experiments [51], but the effect is small and does not change our conclusions.

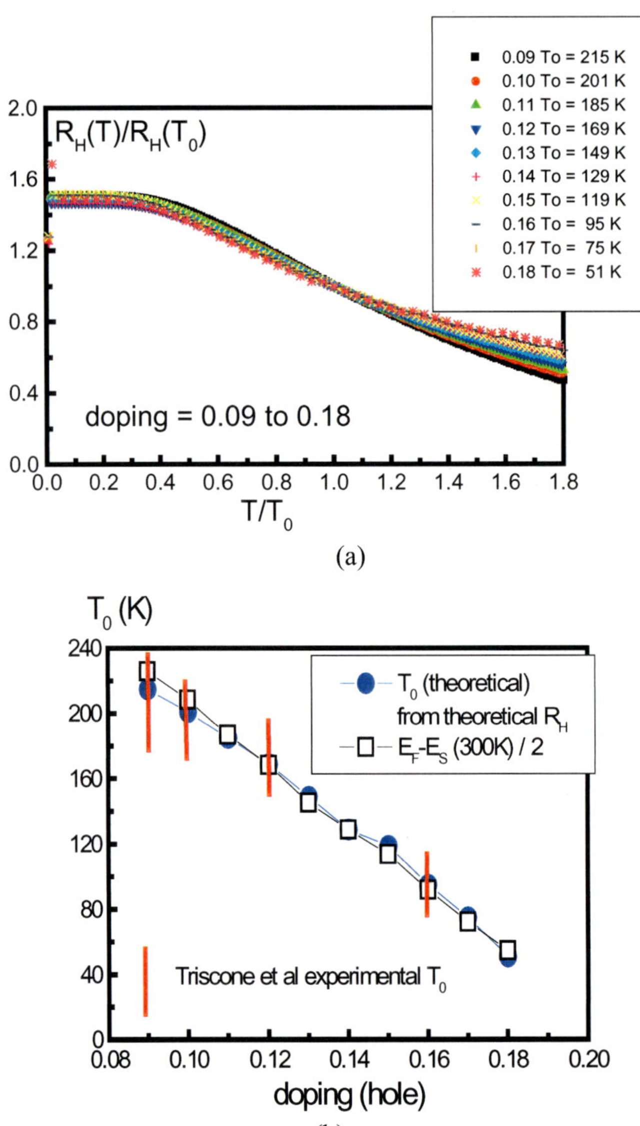

Figure 23: a) : universal law $R_H(T)/R_H(T_0)$ versus T/T_0 for various hole doping levels, from 0.09 to 0.18. b) : calculated T_0, $2k_BT_0 = \xi_F - \xi_S$, compared with the experimental T_0 given by Matthey et al (Reference [50]).

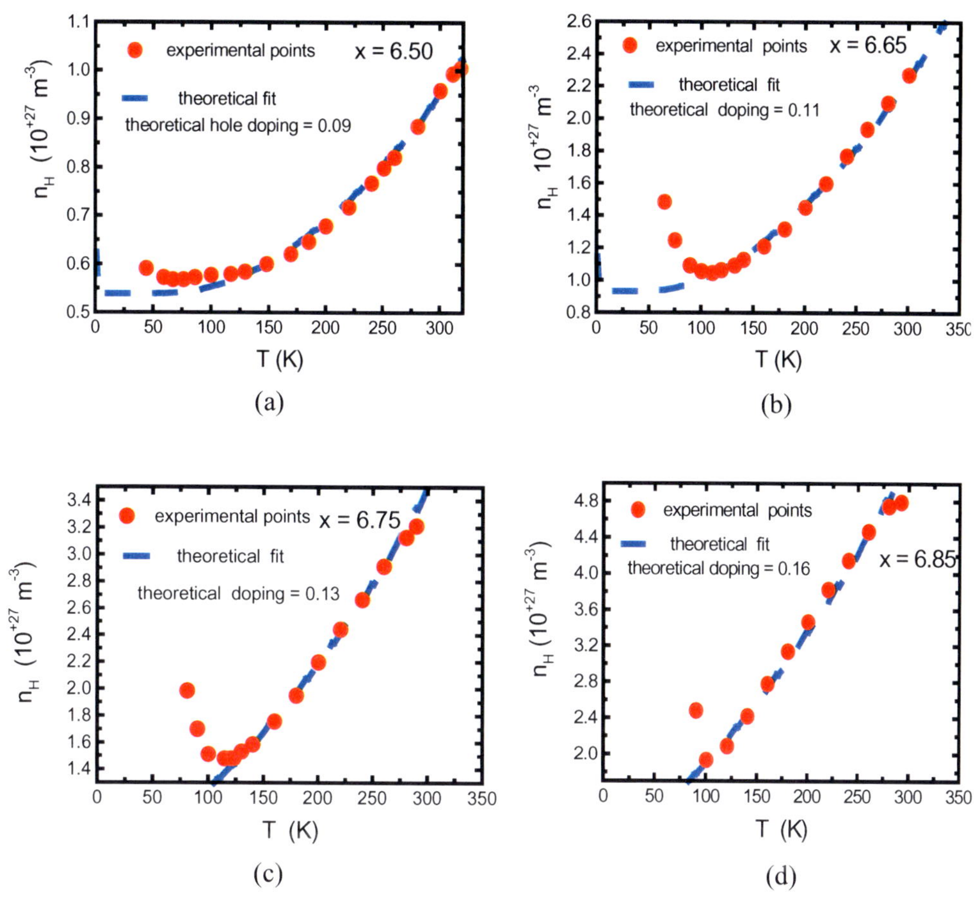

Figure 24:

Filled circles: experimental $n_H(T) = V/(R_H\,e)$ given by Wuyts et al (Reference [48]) in $YBa_2Cu_3O_x$. Dashed lines: theoretical fits

(a)   x=6.50, theoretical hole level = 0.09
(b)   x=6.65, theoretical hole level = 0.11
(c)   x=6.75, theoretical hole level = 0.13
(d)   x=6.85, theoretical hole level = 0.16

The calculations are made with: $t = 0.18$ eV, $t' = 0.04328$ eV, $2t'/t = 0.48$, $\Gamma(\xi_S) = 0.2$ then we obtain the same universal law as in Figure 23a, expressed in $n_H(T)/n_H(T_0)$.

From overdoped to lightly underdoped samples the upturns, at low temperature, in the experimental curves are due to the occurrence of the superconductivity transition.

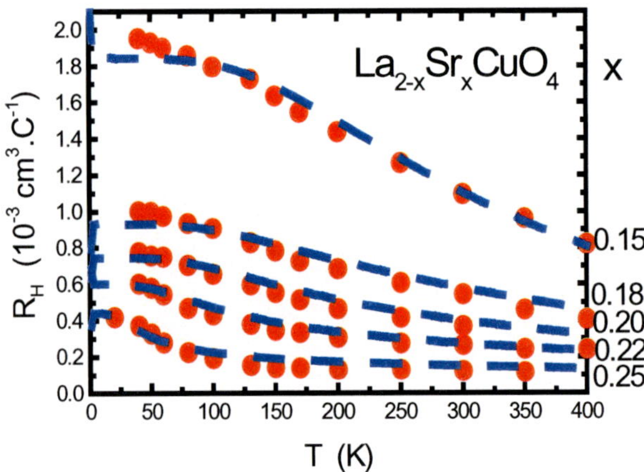

Figure 25: Filled circles: experimental R$_H$(T) given by Hwang et al (Reference [47]), in polycrystalline La$_{2-x}$Sr$_x$CuO$_4$, for x= 0.15, 0.18, 0.20, 0.22, 0.25. Dashed lines: theoretical fits, the theoretical hole levels as the same as the experimental.

The calculations are made with : t = 0.23 eV, t' = 0.06 eV, 2t'/t = 0.52, $\Gamma(\xi_S)$ = 0.1

4   Theoretical Results and Discussion

We use a theoretical band structure closed to the observed experimental one, but not in the fine details. We take a rigid band structure not varying with the doping, but we know that this variation occurs. Here we make our study with the ratio of transfer integrals of transfer closed to 2t'/t = 0.48 in order to obtain this special doping ph$_0$ $\approx$ 0.22 when $\xi_F = \xi_S$ as in our previous studies, leading to convincing results (see previous chapters and Reference [12]).

In Figures 22-25 , we give the best fits with the parameters that we need for this. The value of Γ maybe is too big because with our choice of t and t' the curvatures of the C.E.C. are not so pronounced as in reality.

But the aim of this chapter is to demonstrate that the temperature dependence of the Hall coefficient is due to the effect of the distribution of the hole carriers in the electron-like energy levels and in the hole-like energy levels with increasing temperature. The results of our model do not change appreciably if we change slightly our set of parameters.

Γ itself could change with the doping when the band structure varies. Near the optimum hole doped and overdoped systems Γ could decrease due to bigger curvatures of the C.E.C. In Figure 26 , we show the effect of the decreasing of Γ for a slightly overdoped system. This account for the behaviour of R$_H$(T) in the optimum and slightly overdoped samples, where R$_H$(T) is very flat and its value is very low closed to zero, and even can goes under zero at low or high temperature [47-49,51]. Theoretically this is due to the proximity of ξ_F and ξ_S.

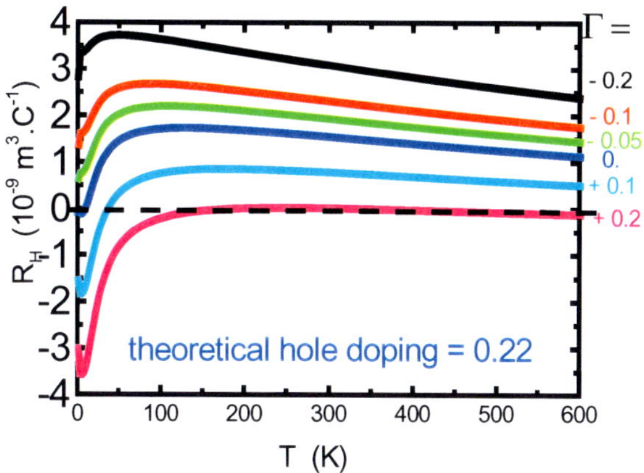

Figure 26: Theoretical curves of $R_H(T)$.showing the effect of Γ with , from the top to the bottom, the values: –0.2, -0.1, -0.05, 0, +0.1, +0.2. The calculations are made with: $t = 0.23$ eV, $t' = 0.0553$ eV, $2t'/t = 0.48$, $V_{YBCO} \cong 174 \; 10^{-30}$ m^3, for a theoretical hole doping = 0.22. As the curvatures of the orbits increase, Γ goes from positive to negative value. This leads to very low (even negative) values to higher positive values at low T, for $R_H(T)$ in the optimum and overdoped samples.

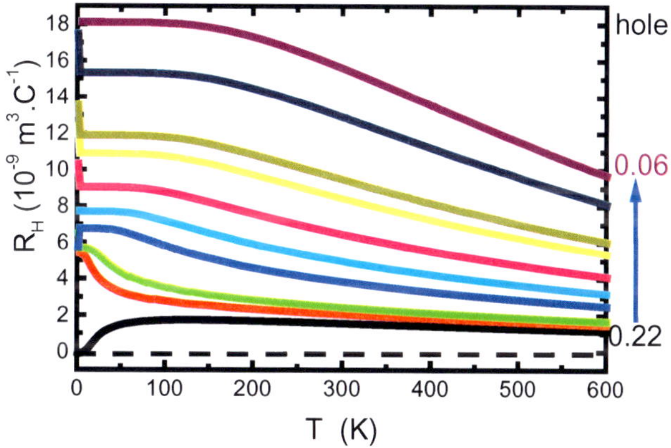

Figure 27: Theoretical curves of $R_H(T)$ with the same fit parameters of Figure 26 and with $\Gamma(\xi_S) = 0$. From the bottom to the top the hole dopings are the following : 0.22, 0.20, 0.19, 0.16, 0.14, 0.12, 0.10, 0.09, 0.07, 0.06.

In Figure 27 , we show the theoretical $R_H(T)$ curves for a set of doping, using the same fit parameters as in Figure 26 , letting $\Gamma(\xi_S) = 0$. We can see that the general behaviour of $R_H(T)$ is kept.

For very underdoped samples, near the metal-insulator transition our approach is no longer valid. We propose an explanation for the downturns observed in $R_H(T)$ [50,52] based on the localization of the carriers above an energy E_{loc} (see Figure 28).

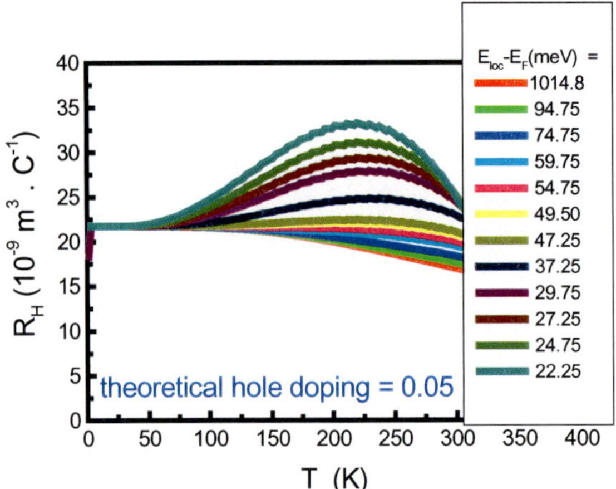

Figure 28: A localization level (E_{loc}) is introduced in the model to take into account the proximity of the metal-insulator transition in very underdoped sample.

The calculations are made for a hole doping of 0.05, with: t = 0.18 eV, t' = 0.04328 eV, 2t'/t = 0.48, $\Gamma(\xi_S) = 0$, $V_{YBCO} \cong 174\ 10^{-30}\ m^3$.

From the bottom to the top of the Figure 28 $E_{loc} - \xi_F$ varies from +1015 meV, that means no localization, to +22 meV, effective localization. We see that a strong maxima appears when the localization increases. This is due to the loss of localized particles, which do not contribute to transport.

In conclusion we find that the electronic structure of CuO_2 planes, with hole-like and electron-like orbits can explain the values of R_H for the high Tc cuprates in the normal state and its temperature behaviour [54], this conclusion is reinforced by the fact that we obtain a representation of the experimental universal law $R_H(T)/R_H(T_0)$ versus T/T_0.

Conclusion

In conclusion, we have proven that the Van Hove scenario explains many physical properties of the HTSC cuprates both in the normal and superconducting states. The existence of saddle points (VHs) close to the Fermi level is now well established by many experiments. This fact must be taken into account in any physical description of the properties of high T_c superconductors.

References

[1]   J. G. Bednorz and K. A. Müller, *Z. A. Phys.* **B 64**, 189 (1986).

[2]   L. Van Hove, *Phys. Rev.* **89**, 1189 (1953).

[3]   J. Labbé, J. Bok, *Europhys. Lett.* **3**, 1225 (1987).

[4]   H. Ding, J. C. Campuzano, K. Gofron, C. Gu, R. Liu, B. W. Veal and G. Jennings. *Phys. Rev. B* **50**, 1333 (1994).

[5] Z. X. Shen, W.E. Spicer, D.M. King, D.S. Dessau, B.O. Wells, *Science* **267**, 343 (1995).

[6] Jian Ma, C. Quitmann, R.J. Kelley, P. Alméras, H. Berger, G. Margaritondoni, M. Onellion. *Phys. Rev. B* **51**, 3832 (1995).

[7] Ino, C. Kim, M. Nakamura, Y. Yoshida, T. Mizokawa, A. Fujimori, Z. X. Shen, T. Takeshita, H. Eisaki, and S. Uchida, *Phys. Rev. B* **65**, 094504 (2002).

[8] T. Yoshida, X. J. Zhou, T. Sasagawa, W. L. Yang, P. V. Bogdanov, A. Lanzara, Z. Hussain, T. Mizokawa, A. Fujimori, H. Eisaki, Z.-X. Shen, T. Kakeshita, and S. Uchida, *Phys. Rev. Lett.* **91**, 027001 (2003)

[9] M. Abrecht, D. Ariosa, D. Cloetta, S. Mitrovic, M. Onellion, X. X. Xi, G. Margaritondo, and D. Pavuna, *Phys. Rev. Lett.* **91**, 057002 (2003)

[10] J. Bouvier and J. Bok, *Physica C* **249**, 117 (1995).

[11] J. Bouvier and J. Bok, *Physica C* **288**, 217 (1997).

[12] J. Bouvier, J. Bok. "Superconductivity in cuprates, the van Hove scenario: a review." In "The Gap Symmetry and Fluctuations in HTSC." Edited by Bok et al., Plenum Press, New York, 37 (1998).

[13] J. Labbé and J. Bok, *Europhys. Lett.* **3**, 1225 (1987).

[14] J. Bok and L. Force. *Physica C* **185**, 1449 (1991).

[15] L. Force and J. Bok. *Solid Stat. Comm.* **85**, 975 (1993).

[16] J. Bouvier and J. Bok, *J. of Superc.* **6**, 673 (1997).

[17] P. C. Pattnaik C. L. Kane D. M. Newns and C. C. Tsuei. *Phys. Rev. B* **45**, 5714 (1992).

[18] J. Bok, J. Bouvier, *Physica C* **403**, 263 (2004).

[19] B.L. ALTSHULER and A.G. ARONOV, (Electron-Electron Interaction In Disordered Conductors, in : Electron-Electron Interaction In Disordered Systems, eds. A.L. Efros and M. Pollak. Elsevier Science Publishers B.V. (1985).

[20] J. Bouvier, J. Bok, H. Kim, G. Trotter, M. Osofsky, *Physica C* **364**, 471 (2001).

[21] C.C. Tsuei, D.M. Newns, C.C. Chi et P.C. Pattnaik, *Phys. Rev. Lett.* **65**, 2724 (1990).

[22] C.G. Olson, R. Liu, D. W. Lynch R. S. List, A. J. Arko B. W. Veal, Y. C. Chang, P. Z. Jiang, and A. P. Paulikas, *Phys. Rev. B* **42**, 381 (1990).

[23] Z. Schlesinger et al, *Phys. Rev. Lett.* **67**, 2741 (1991).

[24] Y. Kubo, Y. Shimakawa, T. Manako, and H. Igarashi, *Phys. Rev. B* **43**, 7875 (1991).

[25] P. W. Anderson and P. Morel, *Phys. Rev.* **125**, 4 (1962).

[26] T. Yokoya, A. Chainini, T. Takashahi, H. Katayama-Yoshida, M. Kasai and Y. Tokura, *Physica C* **263**, 505 (1996).

[27] M.L. Cohen and P.W. Anderson in *Superconductivity in d and f band Metals*, edited by D.H. Douglass (A.I.P. New York) (1972) .

[28] V. Ginzburg, *Comtemporary Physics* **33**, 15 (1992).

[29] Y. Koike, Yoshihiro Iwabuchi, Syoichi Hosoya, Norio Kobayashi and Tetsuo Fukase, *Physica C* **159**, 105 (1989).

[30] M. Onellion, R.J. Kelley, D.M. Poirier, C.G. Olson and C. Kendziora, preprint "Superconducting energy gap versus transition temperature in $Bi_2Sr_2CaCu_2O_{8+x}$",

[31] R.J. Kelley, Jian Mia, C. Quitmann *Phys. Rev. B* **50**, 590 (1994).

[32] R.J. Kelley, C. Quitmann, M. Onellion, H. Berger, P. Almeras and G. Margaritondo, *Science* **271** (1996) 1255.

[33] J. Bok and J. Bouvier, Physica C **274**, 1 (1997).

[34] Ch. Renner and O. Fisher, *Phys. Rev. B* **51**, 9208 (1995).

[35] O. Riou, M. Charalambous, P. Gandit, J. Chaussy, P. Lejay and W.N. Hardy *"Disymmetry of critical exponents in YBCO"* published in *LT 21 proceedings*.

[36] Marcenat, R. Calemczuk and A. Carrington, *"Specific heat of cuprate superconductors near T_c"* published in *"Coherence in high temperature superconductors"* edited by G. Deutscher and A. Revcolevski World Scientific Publishing Company (1996).

[37] J. Bok and J. Labbé, *C.R. Acad. Sci Paris* **305**, 555 (1987).

[38] J. W. Loram, K.A. Mirza, J.M. Wade, J.R. Cooper and W.Y. Liang, *Physica C* **235**, 134 (1994).

[39] J.W. Loram, K.A. Mirza, J.R. Cooper, J.L. Tallon. published in the *proceedings of M^2S-HTSC-V*, Feb 28-Mar 4, 1997, Beijing, China. *Physica C: Superconductivity, Volumes 282-287, Part 3, August 1997, Pages 1405-1406*

[40] G.V.M. Williams,J.L Tallon,E.M. Haines,R. Michalak, and R. Dupree, *Phys. Rev Lett.* **78**, 721 (1997).

[41] J.R. Cooper and J. W. Loram, *J. Phys. I France* **6**, 2237 (1996).

[42] T.Timusk and B. Statt, *Rep. Prog. Phys.* **62**, 61 (1999).

[43] J. Bouvier end J. Bok, *J. of Supercond.* **10**, 673 (1997).

[44] Vobornik, H. Berger, D. Pavuna, M. Onellion, G. Margaritondo, F. Rullier-Albenque, L. Forró, and M. Grioni, *Phys. Rev. Lett.* **82**, 3128 (1999).

[45] O. Fisher *et al* - P. Müller *et al* - D. Rubio *et al*, *J. of Superconductivity* **13** (2000).

[46] T. Cren, D. Roditchev, W. Sacks, and J. Klein, *Phys. Rev. Lett.* **84** (2000) 147.

[47] M. S. OsofskyR. J. Soulen, Jr. , J. H. Claassen, B. Nadgorny, J. S. Horwitz, G. Trotter and H. Kim, *Physica C* **364-365** (2001) 427.

[48] H. Y. Hwang, B. Batlogg, H. Takagi, H. L. Kao, J. Kwo, R. J. Cava, J. J. Krajewski and W. F. Peck, *Phys. Rev. Lett* **16**, 2636 (1994).

[49] Wuyts, V. V. Moshchalkov and Y. Bruynseraede, *Phys. Rev. B* **53**, 9418 (1996).

[50] Z. Konstantinovic, Z. Z. Li and H. Raffy, *Physica C* **341-348**, 859 (2000).

[51] Matthey, S. Gariglio, B. Giovannini and J.-M. Triscone, *Phys. Rev. B* **64**, 024513 (2001).

[52] F. Balakirev, J. B. Betts, A. Migliori, S. Ono, Y. Ando, G. S. Boebinger, *Nature* **424**, 912 (2003).

[53] N. P. Ong, *Phys. Rev. B* **43** (1991) 193.

[54] J. Bobroff, H. Alloul, S. Ouazi, P. Mendels, A. Mahajan, N. Blanchard, G. Collin, V. Guillen, and J.-F. Marucco, *Phys. Rev. Lett* **89**, 157002 (2002).

[55] J. Bok, J. Bouvier. *Physica C* : Superconductivity **403**, 263 (2004).

In: New Topics in Superconductivity Research
Editor: Barry P. Martins pp. 107-156

ISBN 1-59454-985-0
© 2006 Nova Science Publishers, Inc.

Chapter 4

STUDIES OF YBCO ELECTROMAGNETIC PROPERTIES FOR HIGH-TEMPERATURE SUPERCONDUCTOR MAGLEV TECHNOLOGY

Honghai Song, *Jiasu Wang, Suyu Wang,*
Zhongyou Ren, Xiaorong Wang
Applied Superconductivity Lab, Southwest Jiaotong
University Mailstop152#, 610031 Chengdu, China†
Oliver de Haas, Gunter Fuchs, Ludwig Schultz
IFW Dresden, P.O. Box 270016, 01171 Dresden, Germany

Abstract

Melt textured $YBa_2Cu_3O_{7-\delta}$ superconductor has been widely used in the eld of high temperature superconductor (HTS) Maglev technology, such as the ywheel energy storage system and the transportation system. The induced (shielded) current may ow at large density without loss, circulating in large single-grained superconductors. So they can be used as permanent magnet, but with much higher magnetic elds. However, before good engineering designs for these applications can be derived, a deeper understanding of the magnetic behavior of YBCO superconductor must be obtained. Therefore, the studies on the electromagnetic properties of HTS YBCO bulks are reported for Maglev technology in this chapter. Both experimental and computational results have been discussed in terms of Electromagnetic Properties of Bulk High Temperature Superconductor for HTS Maglev Technology. It was found that not only growth sector boundaries (GSB) between the ve growth sectors (GS) but also superconduction property variations in these growth sectors contribute to inhomogeneities of bulk YBCO. Experiments were designed to investigate the macroscopic anisotropy of magnetization critical current density of bulk YBCO. While the eld is kept constant at 1.0 T, the ratio increases as the temperature decreasing from 85 K to 20 K. Although levitation force has linear relationship with the applied eld in the case of symmetrical, such a linear relationship disappears once the applied eld becomes unsymmetrical. However, levitation stiffness has linear relationship with the associated

---

*E-mail address: hhsong@home.swjtu.edu.cn, interhhsong@163.com
†Homepage: http://asclab.swjtu.edu.cn

levitation force, whether the applied field is symmetrical or unsymmetrical. The multiple seeded melt growth (MSMG) bulk has grain boundary (GB), but it still can be regarded as single larger grain bulk in the perpendicular mode due to the inter-grain critical currents flowing across GBs, and it has much larger levitation force than the stacked bulk array. During the lateral movement, the decay of levitation force is dependent on both the maximum lateral displacement and the movement cycle times, while the guidance force hysteresis curve does not change after the first cycle. Moreover, A variational approach was presented for the studies on the field dependence of the critical current density in YBCO Superconductor. When the anisotropy ratio into account in the HTS computation modelling, the calculated levitation forces between superconductor and magnet agree with the experimental ones. This work may be helpful to the system optimization and may provide scientific analysis for the HTS Maglev system design.

PACS 84.71.Ba; 85.25.Am; 74.81.-g

Keywords: Exponential model; Levitation force; Applied fields, field dependence, variational formulations, symmetrical and unsymmetrical field, current profile

1   Introduction

Since the discovery of high temperature superconductor (HTS), materials advances in HTS is promoting the HTS applications. In terms of both magnetic and mechanical properties of top seeded melt textured (TSMT) bulk samples, much progress has been achieved in the recent years [4, 2, 1, 3, 5]. Bulk HTS is being widely used in many fields, such as flywheels [6], levitation transportation [9, 7], electromagnetic launcher [10], linear motor [11], and so on [12]. One of the major applications of HTS $YBa_2Cu_3O_{7-\delta}$ is non-contacted superconducting Maglev, such as Maglev bearing [16, 17] and transportation system [18, 19, 8, 25].

The accomplishment of the world-first commercial Maglev line using TR-07 electromagnet system (EMS) in Shanghai demonstrates the great potential of Maglev technology in transportation industry [14]. However, the system employs conventional copper wire, with small levitation gap (8 mm) and complicated control requirement. Starting in 1997 [15], the Yamanashi test line recorded the highest relative speed up to 1003 km/h in 1999, but the low temperature superconductor magnet (SCM) system needs high-reliability cooling system resulting in great cost. In the field of transportation, the HTS maglev vehicle above the permanent magnet guideway (PMG) was proposed as one of the most promising applications of bulk HTS material [18].

Up to now, there are three people-carrying HTS test vehicles in the world [26, 27], pictured in Fig.1. As with the HTS Maglev test vehicle developed in Southwest Jiaotong University, China, there are 8 onboard superconducting equipments, each having 43 YBCO bulks [9, 28]. When five people stand on the vehicle, the total weight is 530 kg, and the net levitation gap is more than 20 mm. The levitation force of the entire Maglev vehicle with total 344 bulks at the gap of 15 mm is about 8940 N [9] and the guidance force is 2908 N when field cooled at the gap of 26 mm and the lateral displacement (LD) of the vehicle is 20 mm [24]. So far, in past 5 years, over 40 000 passengers have taken the vehicle, and it has travelled back and forth about 500 km. But the levitation performance is as before, which proves the HTS Maglev vehicle has the long-term stability [7].

Supposed that both the superconductor and the applied eld are uniform along the guideway length direction, the HTS/PMG model can be built up in the Cartesian coordinate, as shown in Fig.2. The shadow part with arrow is for the PMG and the arrow direction notes the magnetization direction. The shadow part without arrow is for the HTS. LG is the levitation gap in short, indicating the position of the levitated superconductor. Three postulate conditions are listed in the following:

- The surface current (Meissner current) will be neglected since the applied eld is much more than the low critical eld H_{c1}.

- The displacement currents will neither be considered since the eld is one kinds of the quasi-static magnetic elds.

- All the constitutive laws describe superconductor at a macroscopic level rather than mesoscopic or microscopic level, i.e. without the necessity of solving Ginzburg-Landau equations.

Generally the superconductor is eld cooled above the PMG and lowered to a certain working height [20]. During the movement the change of the applied eld will induce currents in the superconductor [21]. Simultaneously there will be Lorentz force resulting from the magnetic interaction between the induced current and the applied eld. If a balance is reached between the total weight and the levitation force, the stable levitation can be achieved [22].

Both experimental and theoretical studies on the HTS levitation performance above the PMG are under extensive investigation [38, 39, 21, 22]. It is important to note that in the highly anisotropic YBCO superconductor the critical current density resulting from the ux-line pinning depends on the angle between the local direction of the magnetic induction and the c axis which is normal to the ab plane of the sample in typical experiments [84]. In this chapter, we also present a calculation of levitation force considering the both magnitude and angle dependence of the applied eld. This work may provide scienti c basis for the design of the present HTS maglev train system from both sides of science and technology.

2 Experiment and Discussion

2.1 Evaluation of Inhomogeneities of Bulk $YBa_2Cu_3O_{7-\delta}$ Superconductors Using Pulsed Field Magnetization

With the scaling up fabrication and application to larger quantities for industrial applications [8], it is necessary to evaluate the quality of such bulks using nondestructive methods. As regards characterizing bulk high temperature superconductor (HTS), the most common methods are levitation force measurement and mapping of the trapped eld. The latter yields much more information on the spatial variation of the superconducting properties [13].

When it comes to magnetization method, pulsed eld magnetization (PFM) has been regarded as an effective method for quality test of HTS bulks, since it is more sensitive to sample inhomogeneities than ordinary static eld magnetization process, such as eld

(a) In China, 2000, photoed by Prof. Wang, J.S.

(b) In Russia, 2004, from Prof. Kovalev, L.K.

(c) In Germany, 2004, photoed by Dr. De Haas, O.

Figure 1: The research and development of the people-carrying HTS test vehicles in the world

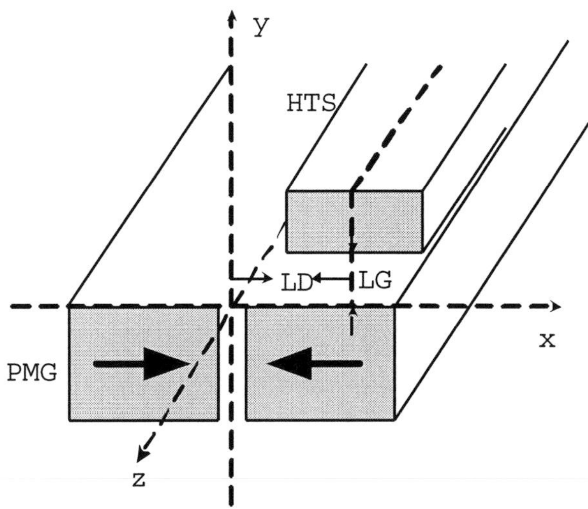

Figure 2: The schematic for HTS bulk and PMG system.

cooling magnetization (FCM) or zero field cooling magnetization (ZFCM) [29]. Not only from crystal center to the edge but also from upper to lower inhomogeneities in large bulk superconductors were observed using this magnetization method [30]. Various parameters of magnetic pulses (i.e. the shapes, durations and amplitudes) [31, 32], and pulse sequences [33] were discussed and experimented.

Although much work has been done by Ioth in Refs. [31] and [34], the problem whether high PFM field will cause damage or cracking inside bulk samples has not been brought forward. However, the mechanical tensile stress of YBCO bulk has the limit, so cracking does happen during activation process resulting from micro-cracks inside [35]. Consequently the magnitude of PFM can not be applied as great as possible. We are wondering if there is such an amplitude of magnetic pulses exceeding the fully penetration field, which is most sensitive for inhomogeneities in a general YBCO bulk. If such a magnitude is found, the PFM method will become both effective and economic. In order to solve these questions four bulk pieces of a YBCO cylinders will be employed in our experiments using the PFM evaluation method. This work is undertaken in view of large scale industrial applications in Magnetic Levitation transportation system [8].

Four pieces of single domain melt textured bulk samples (ϕ30 mm$\times$15 mm) were fabricated by the General Research Institute for Nonferrous Metals (Beijing), the details of which were described elsewhere [36]. They were denoted as BJ1, BJ2, BJ3 and BJ4 corresponding to their original order (005, 018, 022 and 250). Their average force densities are 13.3, 11.0, 13.1 and 12.4 N/cm^2 while interacting with a permanent magnet (ϕ35 mm$\times$15 mm) in the coaxial configuration system.

The experiment setup (JCK Electromagnet, China) is similar to that in Ref.[31]. Firstly the bulk samples were magnetized in zero field cooling (ZFC), immersed in liquid nitrogen, surrounded by copper solenoid coils. Through feeding a pulsed current to a solenoid winding (ϕ51mm$\times$80mm inside), we changed the amplitudes of pulsed magnetic fields at 77 K. The duration time was about 5 ms from the beginning to the end of the pulse field. The pulsed fields ranging from 50 V to 870 V have a linear relationship for different pulsed fields, corresponding to peaks of the magnetic field from 0.4 T to 5.1 T. On the other hand, bulk BJ3 was field-cooled (FC) in liquid nitrogen, 6 mm above the PMG with the seeding surface downwards. The magnetic field above the PMG is not homogeneous, and the fields at the seed and at the back surface centers are about 0.7 T and 0.4 T, respectively.

After the superconductor sample was field activated at 77 K, the axial component of its trapped magnetic flux density B_T was measured by a horizontally scanning Hall element sensor (Lake Shore **450** Gauss-meter). Unless otherwise stated, it moved stepwise at a height of 1.0 mm above the top seed surface of the bulk samples, the scanning area being about 0.25 mm^2.

The applied field dependencies of both bulk sample BJ1 and BJ3 at 77 K in PFM are shown in Fig. 3. The maximum trapped flux density begins to decrease when the applied field exceeds a certain value and levels off as the applied field increases. These results agree well with the conclusions in Ref [31]. The maxima for the two bulks are 1.79 T and 2.08 T respectively, which is slightly different from the reported applied field of 1.9 T [31]. The reason is that the sample size and the average superconducting properties of the bulk samples used in our experiments are significantly similar to those of the cylindrical bulk samples (ϕ34 mm$\times$14 mm) employed in Ref. [31].

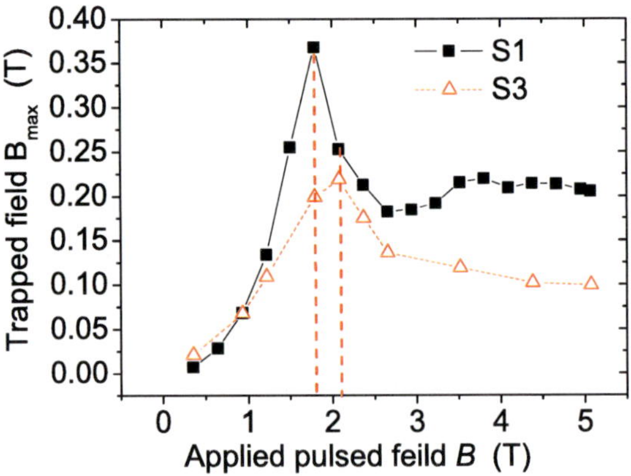

Figure 3: The applied field dependence of the maximum magnetic field. Notice: the scanning gap is 0.5 mm above the seed surface of bulk BJ1, and 1mm for the bulk sample BJ3.

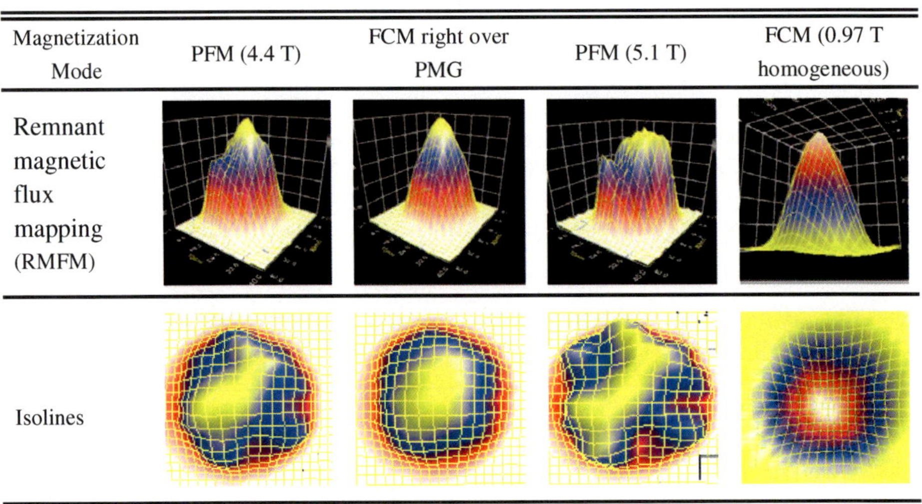

Figure 4: The remnant magnetic flux mapping (RMFM) of the bulk BJ3 using different magnetization modes.

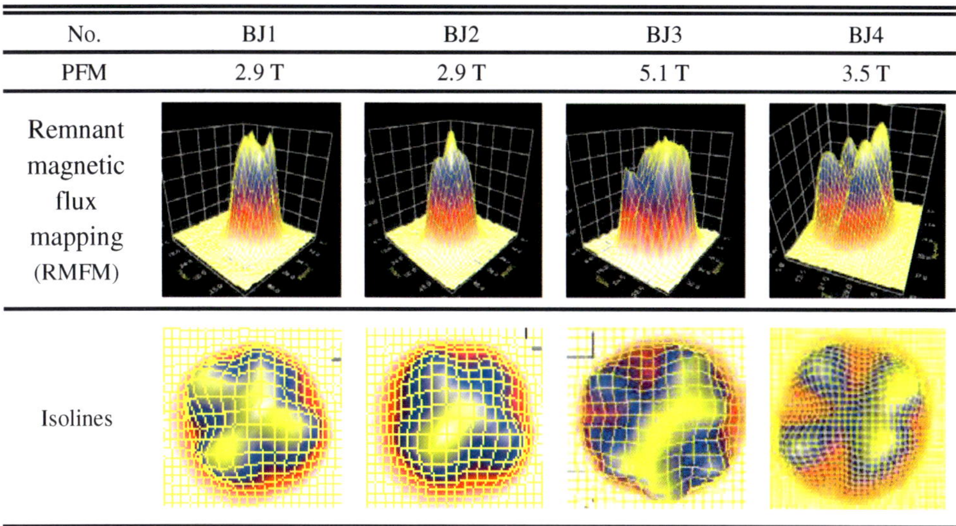

No.	BJ1	BJ2	BJ3	BJ4
PFM	2.9 T	2.9 T	5.1 T	3.5 T
Remnant magnetic flux mapping (RMFM)				
Isolines				

Figure 5: The optimum sensitive pulsed field magnetization (PFM) voltages of the bulk BJ1, BJ2, BJ3 and BJ4.

As reported in Ref. [35], the cracking or damage is more likely to happen during magnetization, when the radial stress resulting from the interaction of the trapped field and the critical current is beyond the limit of the mechanical tensile strength of bulk samples. A series of magnetization processes were applied to activate sample BJ3 as shown in Fig. 4. Since the fully penetrated fields are about 0.65 and 1.9 T in the cases of FCM and PFM respectively, we used the following magnetization processes: PFM with the amplitude of 4.4 T, FCM right over the PMG with seed surface downwards [8], PFM with the amplitude of 5.1 T, and FCM in the homogenous field of 0.972 T.

Trapped fields corresponding to other applied fields were scanned and the associated mapping results were defined as remnant magnetic flux mapping (RMFM). For bulk BJ3 the RMFM data in both inhomogeneous and homogeneous fields show single peaks, which proves that no cracking or damage was observed within the sensitivity of the mapping method even at the maximum PFM field of 5.1 T at 77 K. However, the RMFM data in both cases of PFM are very different compared to those in the cases of FCM, which confirms that PFM is more sensitive to bulk inhomogeneities.

Since the PFM field is applied for very short time, the magnetic flux penetrates and is expulsed from YBCO bulk samples at very high speed. Part of the fluxes pinned at weaker pinning centers will expulse during the retreat process. The rapid change of applied field will induce AC losses inside the samples, thus weakening the pinning capabilities, helping more flux being expulsed. Therefore the comparison between the strong and weak pinning centers is strengthened. Moreover, the density of the 211 phrases inclusions, i.e. pinning centers responsible for the J_c values, does not change across the a-a growth sector boundaries (GSB), but differs essentially at the intersections of a-c GSB [13]. This explains why the PFM method can detect bulk inhomogeneities.

Furthermore, the remnant magnetic flux mapping (RMFM) of sample BJ3 are somewhat different in the two cases of PFM in Fig. 4. This implies that different applied fields have different sensitivities to the inhomogeneities for an identical sample. The question arises, whether there is most sensitive PF amplitude for inhomogeneity assessment of general YBCO bulks, i.e. is there an optimum PFM field from the industrial application point?

Since no cracking will happen at the maximum pulsed field at 77 K, the PFM field may be freely applied to magnetize the four pieces of YBCO bulks. In the following, each of the four bulks was respectively magnetized by a series of pulsed voltage ranging from 300 V to 870 V with an increment of 50 V. In Fig. 5, the most sensitive PFM amplitude of the four bulks was 2.9 T (500 V), 2.9 T (500 V), 5.1 T (870 V) and 3.5T (600 V) respectively, corresponding to BJ1, BJ2, BJ3, and BJ4. The corresponding RMFMs exhibit different profiles, in particular, four obvious peaks as to bulk BJ4. The sensitive PFM voltages (or PFM fields) for different bulks are not uniform, since the superconducting qualities of these YBCO bulks are not uniform.

However, all the isolines in Fig. 5 present X-like distribution, which confirms that the bulk inhomogeneities are mainly caused by related GSB. The trapped field of bulk BJ1 is symmetrically distributed about its axis. But the standard "cross-wheel", where rungs have the form of a dovetail [13], does not appear. This indicates that critical currents do not decrease as much as the increasing distance from the seed crystal. Moreover, two rungs upon the isolines of bulk BJ3 have the form of a dovetail, whilst the other two rungs are significantly different, which results from superconducting property variations in the five growth sectors (GS) [37]. Bulk BJ1 and BJ4 also have such property variations. Thus, not only GSB between GS but also superconducting property variations within growth sectors contributes to inhomogeneities of YBCO bulks.

Shortly, in the case of PFM mode the amplitude of pulsed applied fields should exceed 1.9 T to guarantee sensibility to the inhomogeneity of YBCO bulks. Optimum sensitive pulsed fields are always higher than 1.9 T but have different amplitudes for different bulk samples, due to their different microstructures. It was found that not only GSB between the five growth sectors but also superconducting property variations within GSs contribute to inhomogeneities of YBCO bulk samples.

2.2   Macroscopic Studies on the Anisotropic Properties of High-Temperature Superconductor $YBa_2Cu_3O_{7-\delta}$ Bulk

Owing to the layer structure of HTS YBCO, the critical current density J_c in ab plane is much more than that along c axis. For instance, Murakami et. al. found that the J_c anisotropy ratio of melt textured YBCO is about three at 77 K [4]. However most of the measured sample cut from the big bulk was fabricated as small as possible to avoid exceeding the maximum magnetic moment in the conventional vibration sample magnetometer (VSM) system [40, 41, 42]. Recent studies indicate that the magnetic interaction between HTS and applied field is very complicated including lots of microscopic phenomena, such as vortex locking, vortex kink, geometrical locking and so on [43, 44]. Consequently, the geometry or size of the superconductor sample may have effect on the magnetization critical current density [45, 46]. Therefore the anisotropy ratio from the small sample may not equal to that from the big bulk which is employed in the present Maglev system. In addition, it is impossible to directly measure such a big bulk in the VSM system.

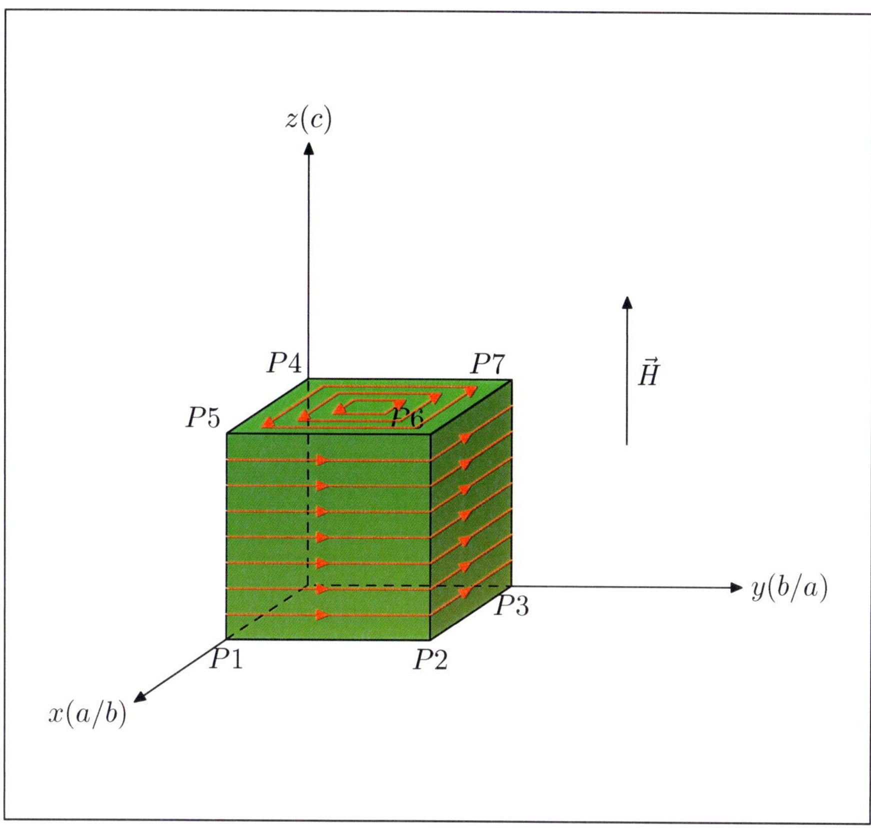

Figure 6: The flow paths of magnetization currents in the case of the applied field parallel to the c axis ($\vec{H} \parallel c$). The cubic bulk (#1) with the seeding surface upward is illustrated in the Cartesian coordinate and its 8 points are denoted from P_0 to P_7 respectively.

On the other hand, the trapped field mapping has been widely used, as an non-destructive method to evaluate sample quality. Xing et. al. have discussed the method that the shielded (or trapped) current distribution can be calculated inversely from the field mapping using Kim-Anderson model [47]. Moreover this method has been extended from the film to bulk superconductor by Iliescu et. al [48]. So it is possible to calculate the currents along c axis and in ab plane through changing the direction of the applied field. The anisotropy of critical current density may be obtained for the big bulk.

A preliminary part of this work was made with three YBCO melt textured samples separated from the large bulk which was synthesized by the top-seeding melt texture growth technique. The details of this procedure were described elsewhere [50]. One of the samples is a cubic bulk (#1) with $10.0 \times 10.0 \times 10.0$ mm^3 in size and the other two small samples are $4.9 \times 1.1 \times 0.4$ mm^3 (#2) and $1.0 \times 3.9 \times 0.8$ mm^3 (#3) along the a, b and c axis respectively.

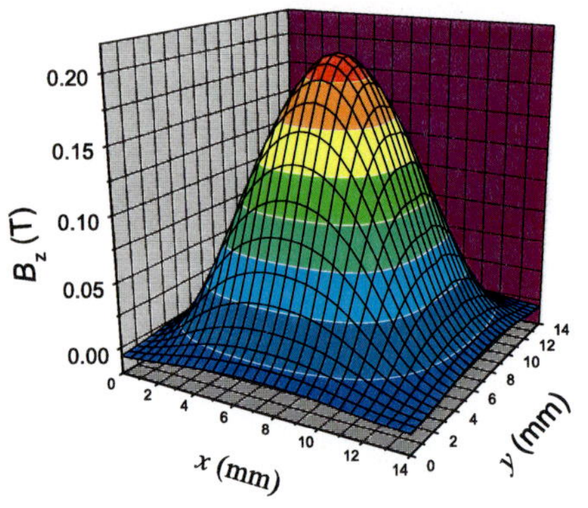

(a) The profile of trapped magnetic field 1 mm above the seeding surface $P_4 P_5 P_6 P_7$.

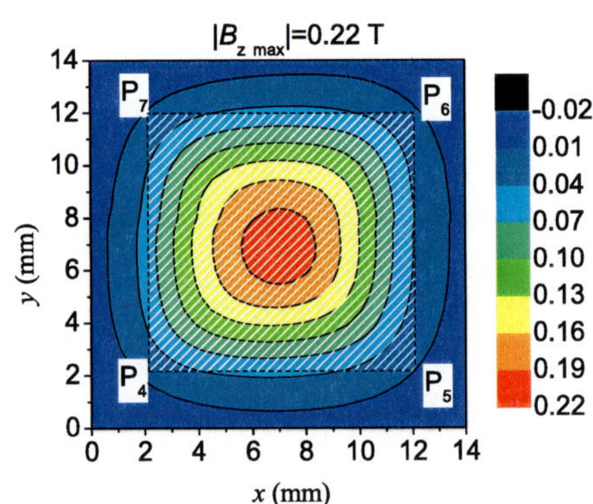

(b) The trapped magnetic flux distributions.

Figure 7: The trapped magnetic field of the cubic sample (#1) at 77 K in the case of $\vec{H} \parallel c$. The part with tilted line indicates the position of the scanned sample.

2.2.1   Anisotropy Ratio of Critical Current Densities from the Cubic Bulk

First, the cubic bulk with seeding surface upward was magnetized in a superconducting 8 T magnet system at 77 K (in liquid nitrogen). The maximum of the applied field was 3.0 T, which is more than the fully penetrated field in the case of ZFC [51]. The magnetic field was applied and removed with its direction parallel to c axis and the sweeping rate is 1 T/min. The induced currents circulate in ab plane perpendicular to c axis as shown in Fig. 6. Subsequently, the field distribution was scanned 1 mm above the surface of the sample using an axial Hall sensor in liquid nitrogen. Fig. 7 presents the field distribution above the seeding surface of the cubic bulk, i.e. the $P_4 P_5 P_6 P_7$ plane. The scanning area was 14×14 mm^2 and the scanning precision was 0.25 mm^2. Both the single peak and concentric contour of the trapped magnetic field are typical for the field mapping, with the maximum of the trapped field 0.22 T which indicates the #1 sample is a quasi-single domain [52].

For comparison, the cubic bulk was then magnetized with the applied field along ab plane rather than parallel to c axis. Fig. 8(a) shows the paths of the circulating currents induced in this case. Part of them is along the c axis and the other is in ab plane. So the field above the $P_0 P_1 P_5 P_4$ plane instead of the $P_4 P_5 P_6 P_7$ plane has been scanned. Although the other setting parameters have not been changed, the maximum of the absolute trapped field is only 0.07 T. This confirms that the critical current along c axis is much less than that in ab plane.

Subsequently, the current maps associated to the trapped field maps were computed using Kim-Anderson model [47]. The program basically solved the inverse Biot-Savart problem in a "quasi" 3D mode, i.e. the current density circulation J_c is considered to be planar and uniform along the sample thickness (c axis). As discussed in Ref. [41], if we suppose it is isotropic in ab planes, in principle there are three components of the critical current density J_c. They are $J_c^{ab,c}$, $J_c^{ab,ab}$ and $J_c^{c,ab}$, where $J_c^{ab,c}$ represents the current

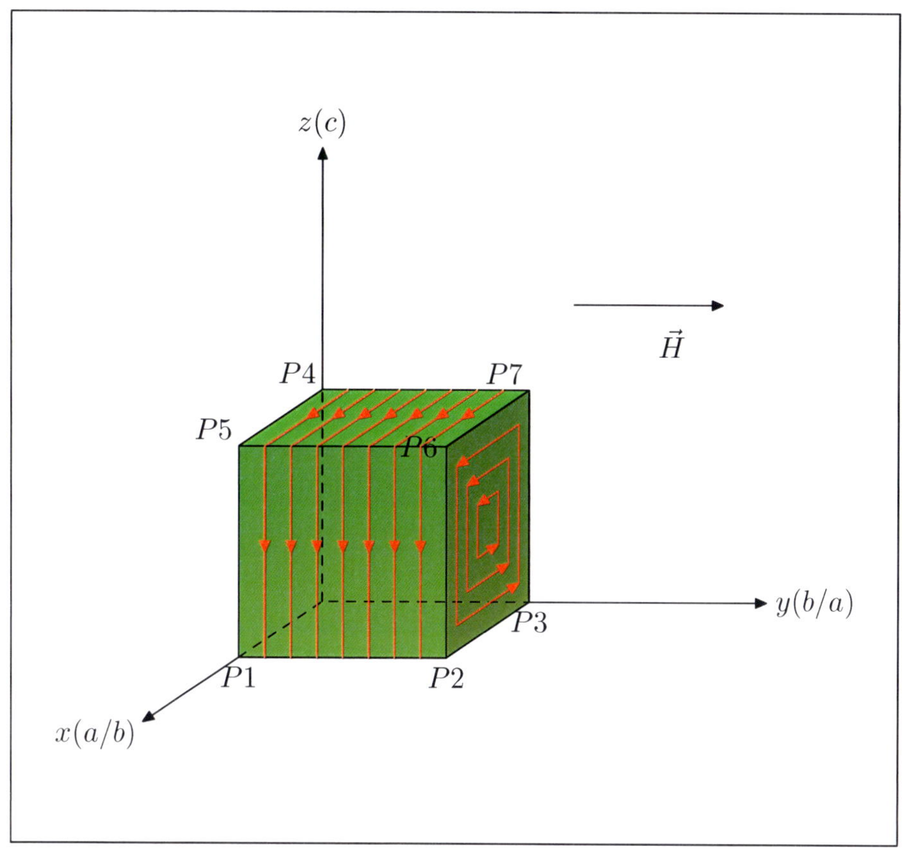

(a) The flow paths of magnetization currents in the case of $\vec{H} \parallel ab$.

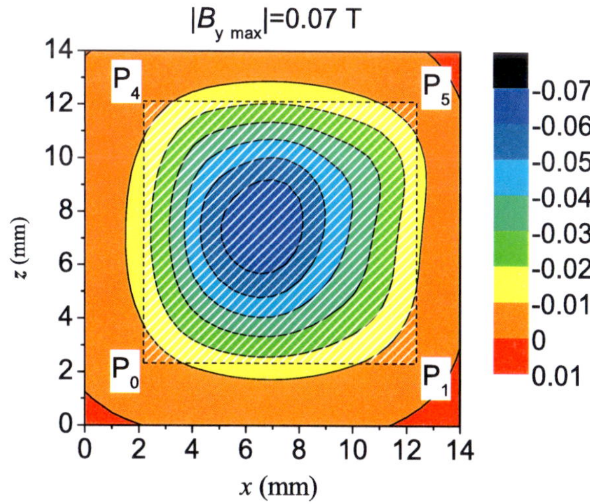

(b) The trapped field 1 mm above the $P_0 P_1 P_5 P_4$ plane.

Figure 8: The trapped magnetic field of the cubic sample (#1) at 77 K in the case of the applied field parallel to the ab plane ($\vec{H} \parallel ab$).

1

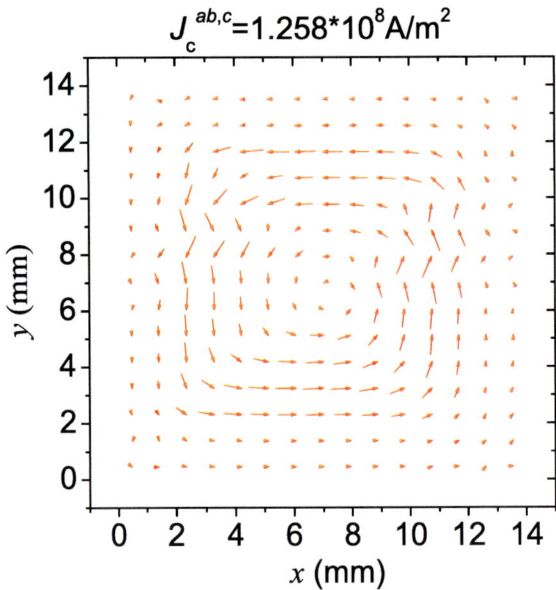

(a) The calculated current distribution in the case of $\vec{H} \parallel c$

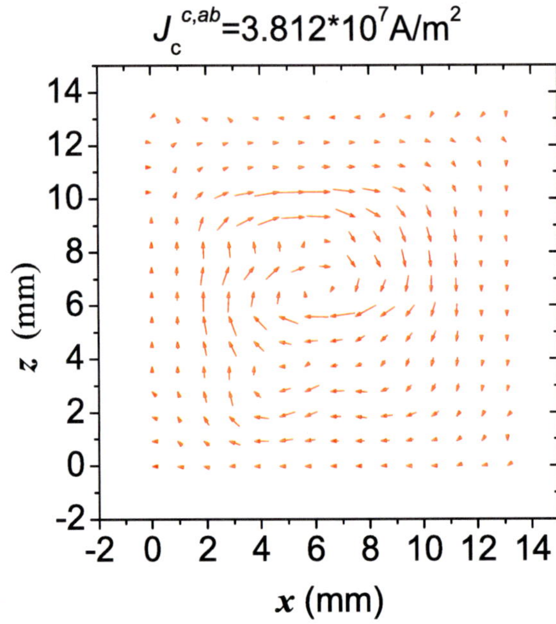

(b) The calculated current distribution in the case of $\vec{H} \parallel ab$

Figure 9: The comparison of the calculated current distributions between the two different cases. The length and direction of the arrows indicate the magnitude and direction of the local currents overall the sample.

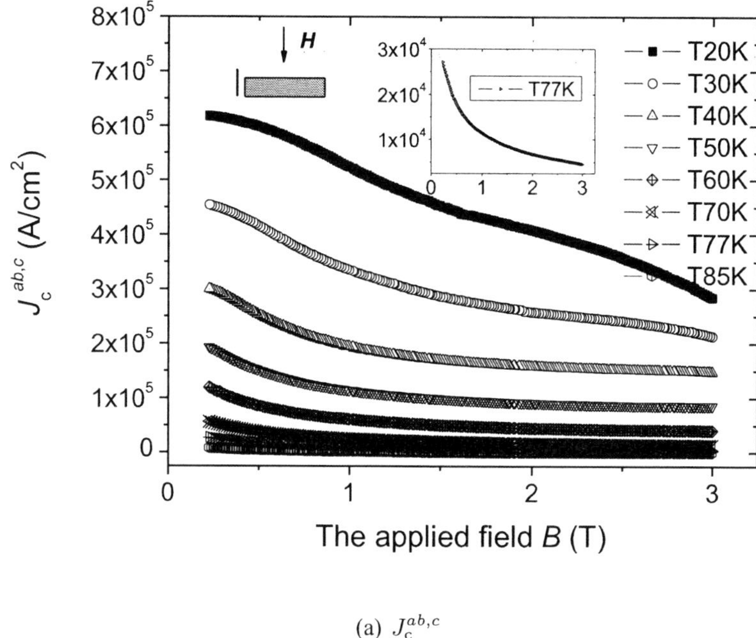

(a) $J_c^{ab,c}$

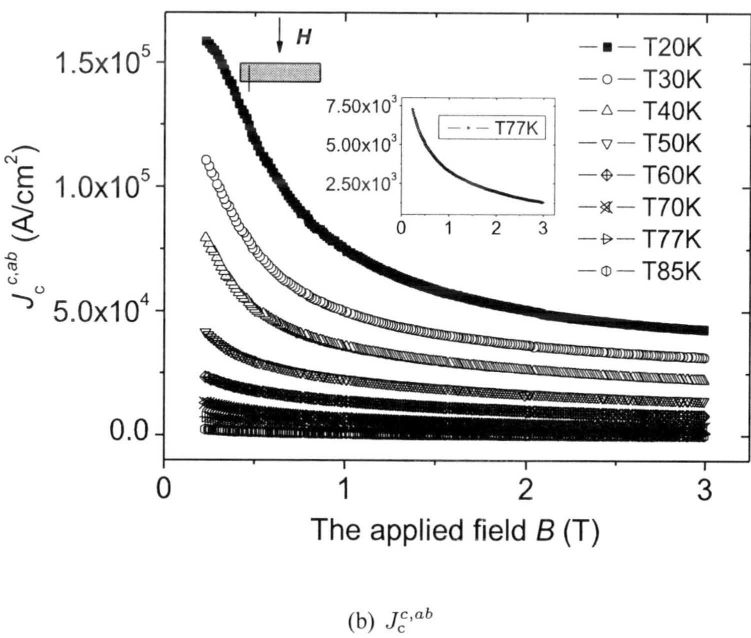

(b) $J_c^{c,ab}$

Figure 10: The eld and temperature dependence of the critical current densities with the applied eld perpendicular to their lengths in terms of the two small samples (#2 and #3). The insets are for the temperature of 77 K.

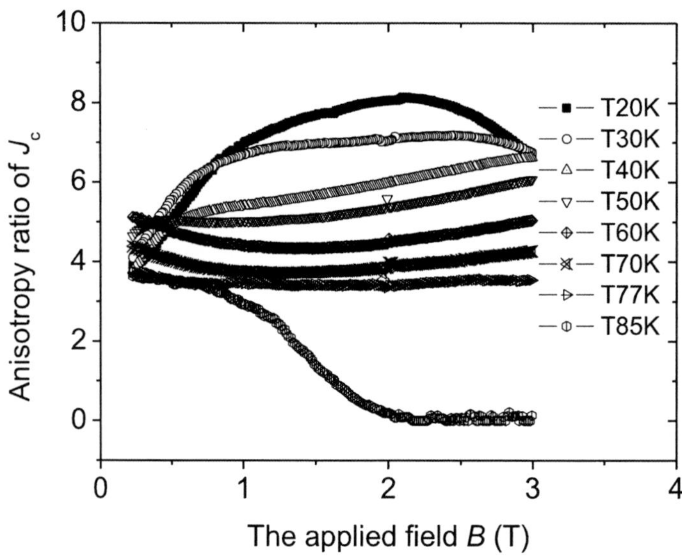

Figure 11: The temperature dependence of anisotropy ratio γ, $\gamma = J_c^{ab,c}/J_c^{c,ab}$.

owing in ab plane with the applied eld in c direction, $J_c^{ab,ab}$ and $J_c^{c,ab}$ corresponds to the currents owing in and across ab plane with the applied eld parallel to ab plane.

Experimental and theoretical work proved that there are inhomogeneities resulting from growth sector boundaries in single domain melt textured YBCO bulk [37]. Therefore the currents calculated from the eld mapping should be regarded as approximate values. Fig. 9(a) shows the distributions of the calculated currents in the case of $\vec{H} \parallel c$. The length and direction of the arrows indicate the magnitude and direction of the local currents overall the sample. Approximately the whole current loop is rectangle shaped, which is consistent with the shape of the cubic bulk. The little disorder indicates that the bulk is slightly inhomogeneous. The maximum current density in this case was regarded as $J_c^{ab,c}$, roughly 1.258×10^8 A/m^2.

Moreover, the current distribution corresponding to the case of the applied eld in ab plane was calculated and mapped in Fig. 9(b). It is obvious that there was higher current density in the upper part of the bulk than in the lower part. This indicates that the sample (#1) is inhomogeneous from the seeding surface to the bottom surface due to the growth sector. The critical current densities in ab plane and c direction are 5.47×10^7 A/m^2 and 3.81×10^7 A/m^2 in this case of magnetization, i.e. $J_c^{ab,ab}$ and $J_c^{c,ab}$. Therefore the ratio $J_c^{ab,c}/J_c^{c,ab}$ is about 3.3, which is de ned as the anisotropy ratio of J_c for the cubic bulk (#1) in the experiment.

2.2.2   Anisotropy Ratio of Critical Current Densities from the Small Samples

On the other hand, we also fabricated the other two small samples (#2 and #3). Their magnetization measurements $M(B)$ were carried out using a vibrating sample magnetometer in the applied field up to 4 T at a constant sweep rate of 160 mT/s. Generally, the current density was calculated from the magnetization differences between the upper and lower branches of hysteresis loops ($\triangle M$). The anisotropy between c axis and ab plane has been taken into account using the extended Bean model developed by Gyorgy et al [53]. The direction of the field is always perpendicular to the length of both the two small samples. As for the sample #2, the applied field is in c direction, so the $J_c^{ab,c}$ in the ab plane can be calculated from $\triangle M$ in the following formula

$$\triangle M \;=\; \frac{J_c^{ab,c} \cdot t_{ab}}{20}[1 - \frac{t_{ab}}{3l_{ab}}] \tag{1}$$

where l_{ab} and t_{ab} are 0.49 cm and 0.04 cm, i.e. the size of the sample #2 along the ab plane and c axis respectively. $\triangle M$ is in emu and the current density in A/cm^2.

Although the applied field was still perpendicular to the length of the sample #3, it was in the ab plane rather than parallel to the c axis. The $J_c^{c,ab}$ along the c axis can be derived using the subsequent formula

$$\triangle M \;=\; \frac{J_c^{c,ab} \cdot t_{ab}}{20}[1 - \frac{t_{ab}}{3l_c} \frac{J_c^{c,ab}}{J_c^{ab,ab}}] \tag{2}$$

where l_c and t_{ab} are 0.39 cm and 0.1 cm, i.e. the size of the sample #3 along c axis and ab plane respectively. Note that in this geometrical configuration $l_c > t_{ab}$ and that $J_c^{ab,ab} > J_c^{c,ab}$, yielding

$$\triangle M \;=\; \frac{J_c^{c,ab} \cdot t_{ab}}{20}, \tag{3}$$

so in this case the critical current density is limited by the c axis.

Moreover, the temperature was extended from 77 K to different temperatures, such as 85 K, 70 K, 60 K, 50 K, 40 K, 30 K and 20 K respectively. Both $J_c^{ab,c}$ and $J_c^{c,ab}$ were calculated from their $M(B)$ as shown in Fig. 10(a) and Fig. 10(b). It is obvious that the critical current density is increasing as the temperature decreases. However, at the constant temperature, the $J_c^{ab,c}$ is much higher than $J_c^{c,ab}$ and the field dependence of the former is much less than the latter.

The critical current density in ab plane and along c axis is compared directly and the corresponding anisotropy ratio is obtained in Fig. 11. It is interesting to find that the anisotropy ratio at 77 K is about 3.2, which is very close to that from the big cubic bulk using the field mapping. The difference between them may result from the inhomogeneities of the big samples. The independence of the ratio at 77 K of the applied field is surprising.

However, at the other temperatures, the field dependence of the ratio is significantly nonlinear. Particularly at 85 K, the ratio decreases from 3.3 with the zero field to about zero when the field is more than 2 T. For the one at 20 K, the ratio has single peak at the field around 2 T. Since the magnetic field above the PMG in the maglev system is around 1.0

T, we concentrate the attentions on the anisotropy ratio for the field less than 1.0 T. It was found that the ratio decreases from 7.1 to 2.5 with the temperature decreasing from 20 K to 85 K. So it is reasonable to regard as 2D the superconductivity at lower temperature [54].

But the superconductivity become 3D as the temperature rises to 77 K. This phenomenon can be explained by the anisotropic GL(Gainsburg Landau) function for high-temperature superconductivity derived by Blatter et al [49]. As the temperature increasing, the coherence length along c axis become longer. When the length is more than the distance between the superconducting CuO layer in the atomic structure, the vortex state will change from 2D to 3D. Therefore at 77 K, the superconductivity is neither isotropic, nor is it absolutely 2D. It is necessary to take the anisotropic ratio into account in the computer modelling for magnetic interactions between HTS bulks and the field above the PMG in the maglev system.

In short, two kinds of experiments were designed to investigate the anisotropy of magnetization critical current density of (HTS) YBCO. It was found that the anisotropy ratios of critical current density in both cases are about 3.5 independent of the applied field at 77 K, and the anisotropy ratio increases as the temperature decreases from 85 K to 20 K. Furthermore the comparison between calculated and experimental results proves that it is necessary to take the anisotropy ratio into account in the HTS computation modelling.

2.3 The Relationship between Levitation Force and Stiffness in Symmetrical and Unsymmetrical Applied Fields

In order to make the best use of the PM guideway, optimize the arrangement of superconductor bulks, and improve the levitation performance of the HTS vehicle, it is necessary to study levitation force distribution of a YBCO bulk over the PM guideway. Much work [55, 38] has been done on levitation force, while little work on levitation stiffness which is also an important parameter as to levitation performance. It was found that the slope of the minor loop rather than the slope of the major hysteresis loop gives levitation stiffness [56]. Levitation stiffness were measured and calculated [57] in terms of the decoupled grains model [58], and a good quantitative agreement between theoretical curve and experimental data was obtained. Alvaro Sanchez and et al [59] also introduced an analytical expression of levitation stiffness, and found that it depends on penetration field of bulk superconductors. In most cases, the relative position of a single superconductor and a single magnet is symmetrically fixed on the same line [58, 59, 60].

However, the width of the PM guideway in the experiment is several times as wide as that of a single rectangle bulk. When the bulk is just above the center of the PM guideway, the applied field is symmetrical about the bulk axis. Once the bulk is away from this center position, the applied field is no longer symmetrical. Since the applied field has two cases for a given bulk, symmetrical or unsymmetrical about its axis, the analytical expression of the stiffness in symmetrical applied field [58, 59] may not be applied directly. In this paper minor loops upon major hysteresis loop will be measured, according to which we calculate levitation stiffness in both symmetrical and unsymmetrical applied field. Both the levitation force and stiffness are dependent on the applied field, so it is necessary and possible to correlate the two parameters. This work will have double aims: to deepen the understanding of the general interaction between bulk HTS and permanent magnet in both

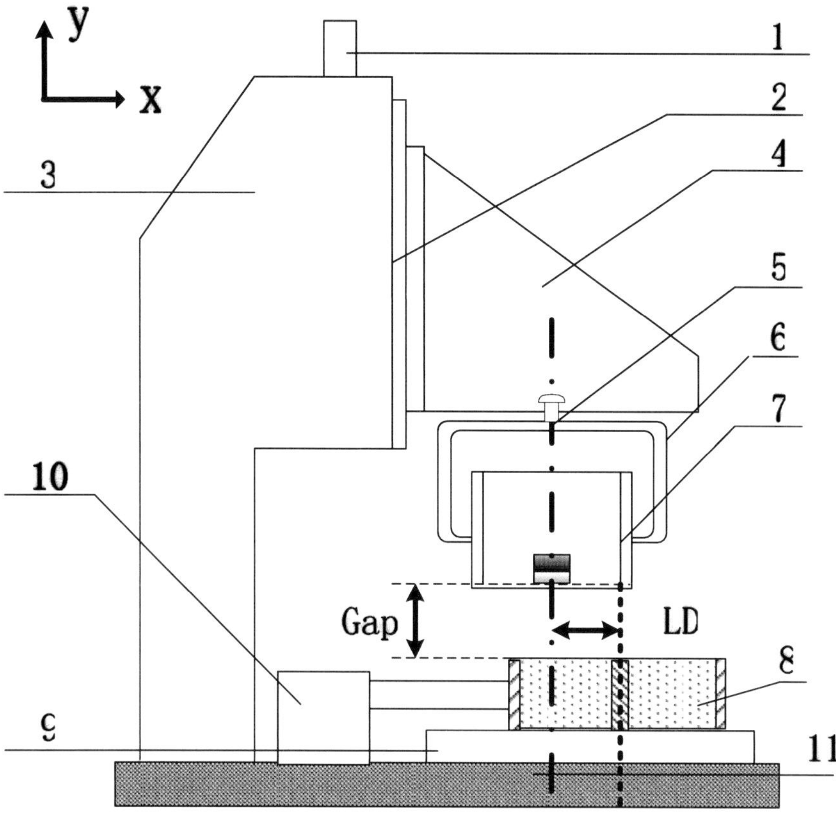

Figure 12: Scheme of HTS Maglev measurement system: 1 servo motor 2 vertical guided way 3 vertical column 4 cant lever 5 vertical sensor 6 fix frame of vessel 7 liquid nitrogen vessel 8 PM guideway 9 horizontal drive platform 10 step motor 11 base.

symmetrical and unsymmetrical applied fields and to provide a systematic framework for optimization of the full scale Maglev vehicle system.

Fig.12 illustrates the HTS measurement system [60], and PM guideway can be moved by the step-motor with the velocity of 1 mm/s along the x axis. The bulk is fixed in the cylinder vessel, which can be held up and down with the mechanical arms operated by the servomotor at the speed of 2 mm/s along y axis. Perpendicular to xy plane, inward is the direction of z axis, which is parallel to the longitudinal direction of the PM guideway. It is supposed that the center axis of the cylinder vessel intersects with the PM guideway surface plane at the grid origin (x=0 mm, y=0 mm, z=0 mm) in Cartesian co-ordinates. In the following, LD is the relative distance from the bulk center axis to the center symmetry plane of the PM guideway parallel to the xy plane, and levitation gap is defined as the net distance from the bulk seeded surface to the PM guideway surface, as shown in Fig.12. The rectangle-shaped single-domain melt-textured bulk (30 mm $\times$35.6 mm $\times$15.5 mm) is fabricated by IFW, Dresden, Germany. There are 20 levitation force-gap (LFG) curves in ZFC at different LDs, which are increased by 5 mm each time from 0 to 95 mm.

On the other hand, levitation stiffness over the PM guideway is studied. According

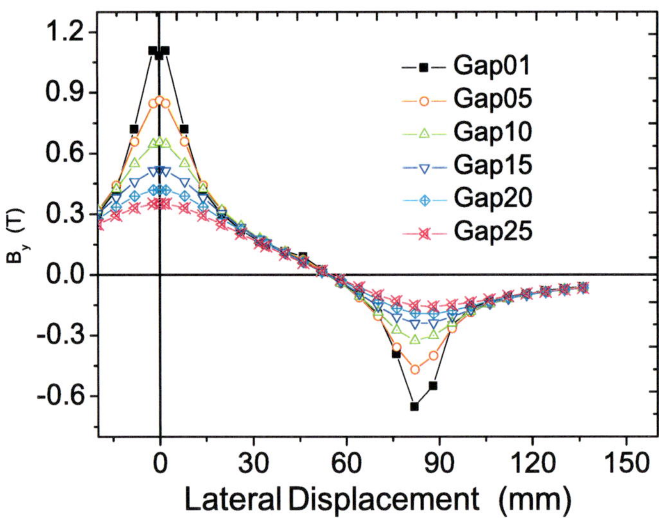

Figure 13: Gap dependence of magnetic field distribution along Z-axis over the PM guideway. Gap01 denotes the magnetic field distribution is at the gap of 1 mm, the same to Gap05, Gap10, Gap15, Gap20 and Gap25.

to the practical operation of the full-scale HTS Maglev vehicle, we measured the minor loop in the case of FC. The procedure in details is as following. Firstly, we initialize HTS measurement system, and then start its function of the levitation force. The PM guideway will be moved to a set displacement, and a single bulk goes down and stops at the height of 40 mm over the PM guideway, where the YBCO bulk is field cooled. Ten minutes later, we reset the measurement range of the gap (from 40 to 20 mm), and obtain the relationship between levitation force and gap, which is defined as Step1. Consequently, Step2 is from 22 to 18 mm and backward to 22 mm, Step3 from 22 to 5 mm and back forth to 22 mm, Step4 from 22 to 18 mm and backward to 22 mm again. Levitation stiffness is calculated from minor loops at Step 2 and Step 4 within the range from 22 to 18 mm.

Such minor loops at the displacement of 0, 10, 15, 20, 25, 45, 70, and 95 mm were measured. According to [57], the stiffness is given by the formula

$$k_{\text{Lev}} = \frac{\partial F_{\text{Lev}}(y)}{\partial y} = \frac{\triangle F_{\text{Lev}}}{\triangle y}, \tag{4}$$

where $F_{\text{Lev}}(y)$ is for the vertical levitation force along the y axis, $\triangle F_{\text{Lev}}$ and $\triangle y$ are for the absolute value of levitation force change upon the minor loop and the loop interval (2mm).

Therefore, the stiffness can be calculated by the transformed formula

$$k_{\text{Lev}} = \frac{|F_{\text{Lev}}(y + \triangle y) - F_{\text{Lev}}(y - \triangle y).|}{2 \times \triangle y} \quad (\text{UNIT} : \text{N/mm}) \tag{5}$$

Fig.13 displays the magnetic field distribution over half of the PM guideway in the xy plane, which is uniform along the z axis, i.e. the longitude direction. Two peaks are

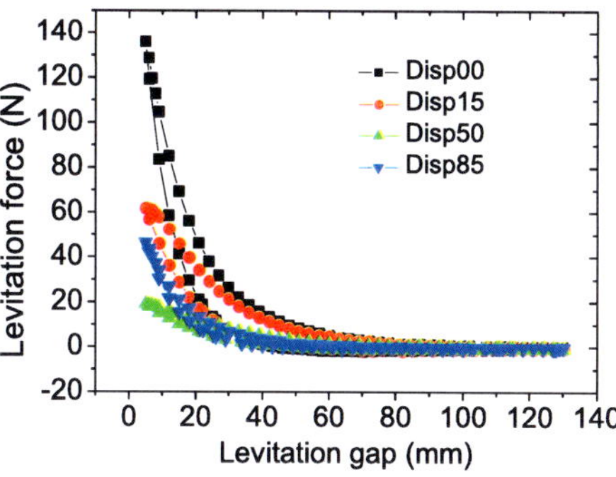

(a) LFG curves at the displacements of 0, 15, 50 and 85 mm.

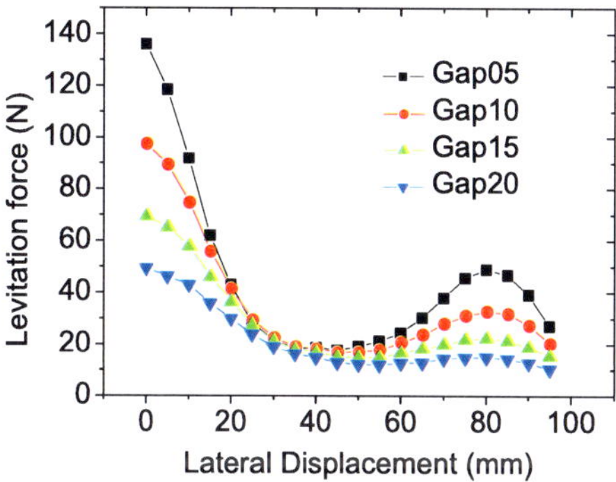

(b) Gap dependence of levitation force distribution over the half of the PM guideway. Gap05 denotes the levitation force distribution at the gap of 5 mm.

Figure 14: Levitation force hysteresis loops and its lateral distribution at different heights.

$+1.1$ and -0.73 T at the displacement of 0 and 85 mm, where the magnetic fluxes are concentrated. When the bulk is just above the center of the PM guideway, its LD is 0 mm and the applied field is symmetrical about the bulk axis. However, once the bulk is away from the right position, the applied fields at the other displacements become unsymmetrical.

In Fig.14(a) we compare LFG curves in ZFC and their LDs are 0, 15, 50, and 85 mm. The levitation force of the bulk at the displacement of 0 mm is noticeably larger than those at the other displacements for a given levitation gap. When the bulk arrives at 5 mm, the force at the displacement of 0 mm is 136 N, roughly twice larger than 46.5 N at the displacement of 15 mm, and six times larger than the one at the displacement of 50 mm (19.2 N). Since the processing of superconductor and the history of motion are identical, the levitation force is significantly dependent on the applied field distributions [61].

Fig.14(b) presents the levitation force distributions at the same gaps of 5, 10, 15 and 20 mm. The distributions also have two peak values, and they are around the displacement of 0 and 80 mm. For a given gap the levitation force is the smallest at the displacement of

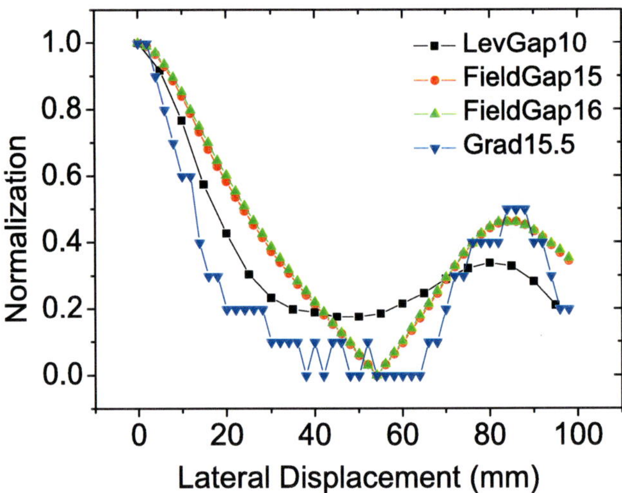

Figure 15: Relationship between the levitation force and the associated applied field. Square symbol is for the levitation force distribution at the gap of 10 mm; circle and triangle (upwards) symbols are for the absolute applied field at the gap of 15 and 16 mm; and triangle (downwards) is for the absolute applied field gradient at the gap of 15.5 mm. (Notice: the thickness of the rectangle bulk is about 15 mm.)

50 mm in Fig.14(b), where the magnetic field along z-axis is roughly zero in Fig.4. It was found that levitation forces of a single bulk exponentially increasing with the gap decreasing and has linear relationship with the applied field at LD of 0 mm [62]. This conclusion was obtained in symmetrical applied field. However, the applied field becomes unsymmetrical once the bulk moves away from the center plane of the PM guideway. So it is necessary to find new force-field relationship in unsymmetrical applied field.

When the geometric center point of the YBCO bulk is at the gap of 15.5 mm the associated applied field distributions is at the gap of 15 and 16 mm, and the field gradient is at the gap of 15.5 mm as displayed in Fig.15 The whole trends of their distributions are similar, but they do not agree well with each other. The variation of the levitation force is smaller than that of the applied field or its associated gradient. As for small HTS bulk levitation force is proportional to the gradient of applied magnetic field. However, the size of the rectangle bulk is comparable to the magnetic field distribution and the levitation force is an integral parameter considering total volume effect. Consequently the approach is difficult to apply to the large bulk in the present case. In short, levitation force does not have linear relationship with neither the applied field nor its gradient when the field becomes unsymmetrical.

Now that it is difficult to find linear relationship between the levitation forces and the unsymmetrical field, we may analyze problems on the basis of the levitation force distribution in Fig.14(b), which may be directly helpful for the further design of the vehicle system.

As to the practical operation of the vehicle, the field-cooling gap and subsequent operation gap are around 40 and 20 mm. Levitation stiffness at the gap of 20 mm is essential

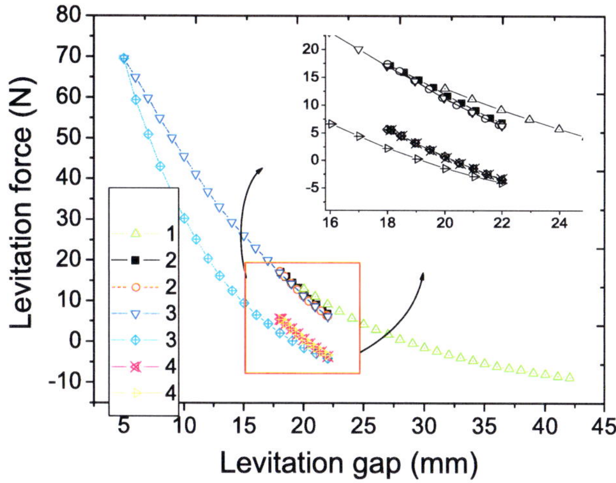

Figure 16: Levitation force vs gap for the rectangle bulk at the displacement of 5 mm. Inset is for the minor hysteresis loops from gap of 22 to 18 mm.

for the practical operation. For example, the measured process at LD of 5 mm is shown in Fig.16. Firstly the bulk is in FC at the gap of 40 mm, where the applied field is not symmetrical about the axis of the rectangle bulk. At Step2 and Step4, the minor loops are obtained on upon the major hysteresis loop. Associated levitation stiffness upon the descending and ascending processes at Step2 is calculated, 2.63 and 2.75 N/mm respectively. Those upon the descending and ascending processes in Step4 are 2.28 and 2.18 N/mm. As to the same minor loop, the stiffness upon the descending process is slightly different with the stiffness upon the ascending process. The stiffness upon the descending process of the major hysteresis loop is larger than that upon the ascending process of the major hysteresis loop. This phenomenon agrees well with the conclusion in [57] that levitation stiffness shows an even stronger variation as that shown by the levitation force itself.

The stiffness at the other LDs was measured with the same measurement procedure. The solid square in Fig.17 is for the normalization of the calculated stiffness at the displacements of 0, 5, 10, 15, 20, 25, 45, 70, and 95 mm respectively. Once LD is more than zero, the applied fields become unsymmetrical about the bulk axis.

In reference [63], it was found that the stiffness appeared to be directly proportional to the levitation force, though not for the whole range of forces and not for all cases. However their applied field is symmetrical, so it is necessary to find the relationship between levitation force and stiffness in unsymmetrical applied field. Although the levitation force is measured in ZFC, the levitation stiffness in FC, both of them are dependent on the associated applied field. Therefore it is possible to correlate the two parameters. The distributions of levitation force normalization at different gaps are also displayed in Fig.17, with three joints, at the displacements of 0 mm, 70 mm, and 95 mm respectively. This indicates that as the gap increasing, the variation of the levitation force distribution decreases.

Although the applied fields are not symmetrical any more, the lateral distribution of

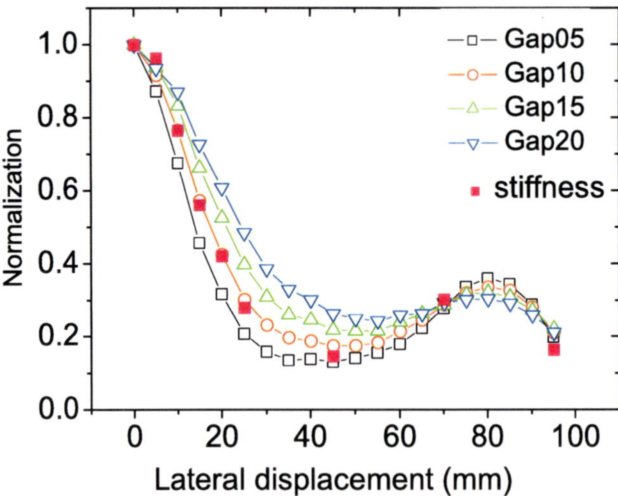

Figure 17: Relationship between levitation force in ZFC case and levitation stiffness in FC case. Open symbols are for gap dependence of the normalized levitation force distribution. Solid rectangle dot is for the normalized stiffness at different displacements.

the normalized stiffness agrees with that of the levitation force at the gap of 10 mm. This phenomenon is consistent with the theory in [59]. Why the levitation stiffness agrees with the normalization distribution of the levitation forces at the gap of 10 mm rather than 15 or 20 mm? Both the levitation force and stiffness are determined by the induced current in the penetration depth. Their current distributions are determined by the local applied field and the motion of the bulk. During the motion up and down upon the minor loop the current will decrease and then increase, or vice versa. The drastic change of current will increase the change of the levitation force. That is the reason why levitation stiffness changes more drastically than levitation force. Moreover, on the basis of this quantitative relationship, we calculate levitation stiffness at the other LDs. They are 30, 35, 40, 50, 55, 60, 65, 75, 80, 85 and 90 mm respectively. Both the calculated and measured data are presented in Table.1.

In conclusions, both the levitation force and stiffness are dependent on the displacement of a single YBCO bulk over the PM guideway. Except for the applied field is symmetrical about the bulk axis, the levitation forces in unsymmetrical magnetic fields do not have linear relationship with the associated applied field or its gradient. Moreover levitation stiffness in FC is directly proportional to the levitation force in ZFC in both symmetrical and unsymmetrical applied field. On the basis of the quantitative relationship, we have calculated levitation stiffness at the other LDs according to the measured stiffness. This data is important for the design of HTS Maglev system.

Table 1: Levitation stiffness of the rectangular bulk at the gap of 20mm with the FC height of 40 mm.

No.[1]	Disp(mm)	Stiffness(N/mm)	No.	Disp(mm)	Stiffness(N/mm)
1	0	2.29	2	50	0.32
3	5	2.21	4	55	0.37
5	10	1.76	6	60	0.45
7	15	1.29	8	65	0.55
9	20	0.98	10	70	0.69
11	25	0.64	12	75	0.74
13	30	0.48	14	80	0.76
15	35	0.38	16	85	0.71
17	40	0.35	18	90	0.58
19	45	0.34	20	95	0.38

2.4 Magnetic Interactions between Multiple Seeded YBCO Bulks and the Permanent Magnet Guideway

Since size of shielding current loop depends on bulk grain size [65], it is necessary to enlarge the grain size of the superconductor to increase levitation force. However, it is signi cantly dif cult to directly grow high quality YBCO bulks with a size of larger 30 cm. Another disadvantage of such TSMG method is the long processing time that is attributed to the slow growth rate of 123 grains [68]. Schatzler [66] and Jee [67] proposed multiple seeded melt growth (MSMG) as the way to fabricate larger and well textured YBCO bulks. Multiple-seeding technique is preferable and promising to overcome the limitation of size available within a moderate processing time [68].

However, as to MSMG YBCO bulks, poor connection between two adjacent Y123 grains results from the residual melt-forming phrases near grain boundary (GB) [69, 70]. Schatzle [66] also reported the decrease of the critical currents at grain boundaries. Liquid phrase in GBs and accompanying misorientation angles are main causes for the poor connection between adjacent grains, which leads to a decrease of critical current density. So it is not clear that how macroscopic current loops will become larger with the increasing size of MSMG bulks.

Moreover, there are few reported results on magnetic interaction between MSMG bulks and the PMG. The concentrating eld created by the PMG is uniform along its length direction while inhomogeneous along the width direction with great magnetic gradient. In this paper, trapped eld, levitation and guidance force of MSMG bulks have been studied in two modes: parallel and perpendicular between the length directions of the PMG and that of the large bulks. Furthermore they are correlated since the trapping and shielding capabilities of bulk superconductor are consistent and both of them result from the ux pinning.

All the TSMG bulk samples fabricated by IFW, Dresden, Germany [66], are divided into two series and numbered from bulk IFW01 to IFW04 in Talbe2. Generally a three seeded bulk includes three grains (domains), so it is possible to separate it into three equal parts. Each trisected part corresponds to one growth domain with one seeding crystal. There are three pieces of trisected bulks in Series A cut from three seeded bulks. They are numbered

Table 2: Specifications of YBCO Bulks in the Experiments.

Series	No.	Number of seeds	Dimensions($a \times b \times c$(mm))
A	IFW01, IFW02, IFW03	1	$35 \times 30 \times 15$
B	S04	3	$90 \times 36 \times 15$

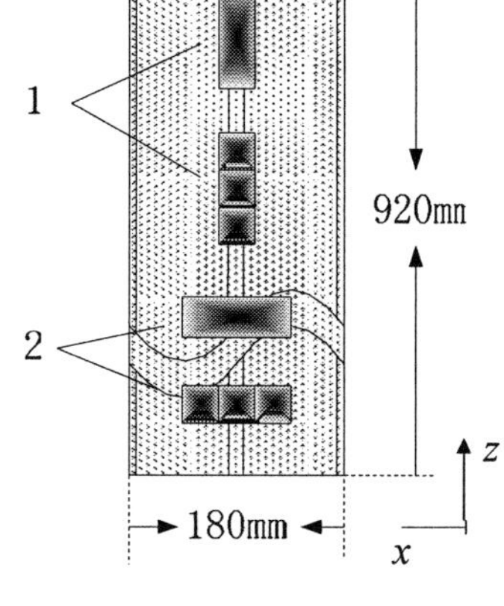

(a) The magnetic flux distribution of the PMG in the vertical xy plane.

(b) The displacement of bulks in the horizontal xz plane.

Figure 18: Case 1 and 2 are the two displacement methods: parallel and perpendicular between length directions of large bulk (or bulk array) and that of the PMG.

from bulk IFW01 to IFW03 and stacked up as one bulk array ($90 \times 35 \times 15$ mm^3). Moreover, there is a piece of MSMG bulk in Series B with the dimension ($90 \times 36 \times 15$ mm^3). On the basis of similar geometrical dimensions, their levitation and guidance forces over the PMG will be studied.

The magnetic field over the PMG is assumed to be continuous and uniform along the length direction. However, according to the magnetization direction, the concentrating field has great magnetic gradient along the width direction. Fig.18(a) presents the magnetic flux distribution in the vertical plane. It is obvious that the field is nonuniform along the width direction Therefore, as to the placement of the MSMG bulk or the stacked array, there are two modes: parallel and perpendicular between the length directions of the PMG and the large bulks. 1 and 2 are for the parallel and perpendicular modes as illustrated in the lower part of Fig.18(b).

Fig.12 shows the scheme of the HTS Maglev system, in which YBCO bulk is fixed in the cylinder vessel with liquid nitrogen, which can be moved both vertically and horizontally with the mechanical arms under the control of the measuring system [60]. We will use Cartesian co-ordinates with the origin located at the center of the surface of the PMG. Axis z and x correspond to its length and width direction in the horizontal plane of the PMG, and y axis is for the vertical direction. In the following LD is the relative distance from the bulk geometrical center axis to y axis, and levitation gap is defined as the relative distance from the seeded surface to the PMG surface.

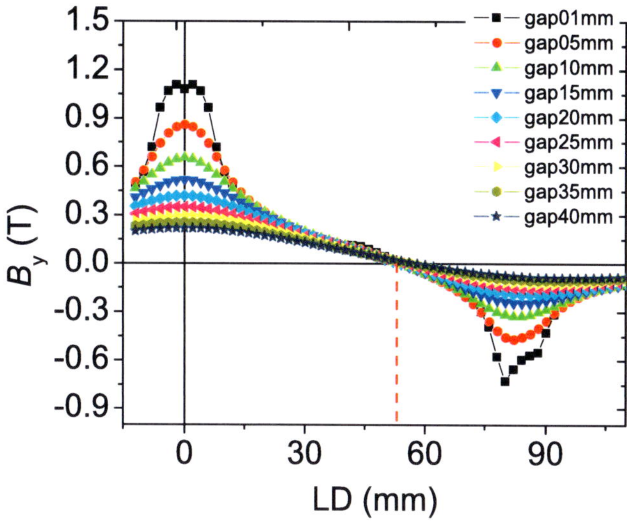

Figure 19: Gap dependence of the distribution of the y component magnetic flux density in the xy vertical plane.

Both levitation and guidance force are important parameters for the HTS maglev vehicle. Vertical levitation forces are measured in the case of vertical movement. After the specimen bulk is cooled down to liquid nitrogen temperature in ZFC, it descends from 130 mm to 5 mm and ascends backward (along y axis) to the start point of the vertical movement [60]. On the other hand, guidance forces are recorded in the case of horizontal movement. Firstly, the specimen bulk is cooled rightly above the center of the PMG and the FC height is 20 mm as shown in Fig.12. Then it moves along the x axis from a maximum LD (+20 mm) and finally returns to the start point of this lateral movement [28]. The speed is 2 mm/s in the cases of both vertical and horizontal movement, so the interaction between PMG and MSMG maybe be regarded as quasi-static.

Moreover, since the mapping of the trapped field may yield much information on the spatial variations of the superconducting properties related to HTS bulks [29], an axial component of the trapped magnetic flux density B_y is measured by a scanning horizontally Hall element sensor (Lake Shore 450 Gaussmeter) after the specimen bulk is FC with the displacement in Fig.18. Unless otherwise stated, the FC height is always 6 mm above the PMG surface and measurement height is 1.0 mm above the bulk seeded surface. For bulk array with dimension $90 \times 36 \times 15$ mm^3, the area is set as 90 mm$\times$60 mm.

First, we measured the applied field distribution (along the x direction) at different gaps over the half of the PMG in Fig.19. Most of the magnetic fluxes are concentrated at the center where the gap dependence of the magnetic field is obvious. The maximum value is up to 1.2 T close to the surface of the PMG. However, when the LD is 52 mm the magnetic field drops to zero, which is the turning point of the magnetic field. Owing to the symmetrical structure the whole magnetic field is assumed to symmetrical within

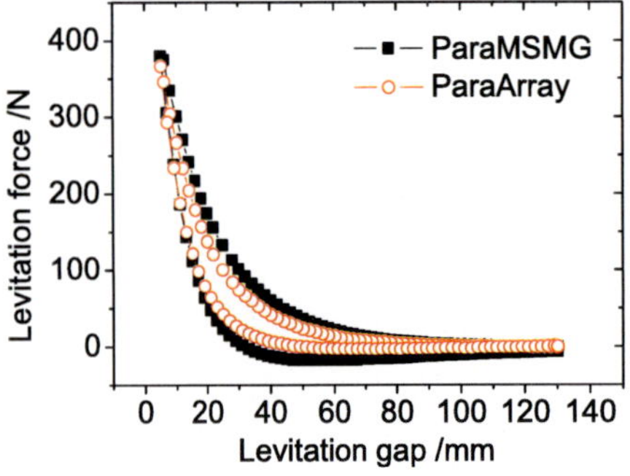

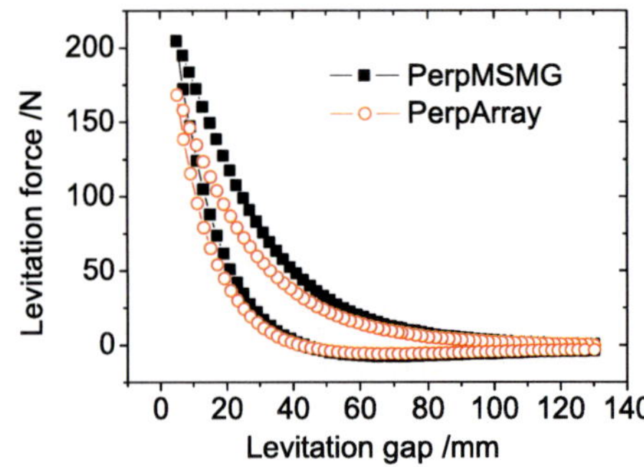

(a) Levitation force curves of the MSMG bulk and the stacked array consisting of bulk IFW01, IFW02 and IFW03 in the parallel mode. Closed dot is for the MSMG bulk and open dot is for the stacked bulk array.

(b) Levitation force curves of the MSMG bulk and the stacked array consisting of bulk IFW01, IFW02 and IFW03 in the perpendicular mode. Closed dot is for the MSMG bulk and open dot is for the stacked bulk array.

Figure 20: Levitation force

engineering error.

Furthermore, levitation force curves of the MSMG bulk and the stacked array in parallel and perpendicular mode are compared. In the parallel mode hysteresis loss of the MSMG bulk is more obvious than the stacked array in Fig.20(a), and the weak-link GB give rise to hysteresis behavior. However, their maximum levitation forces (MLvF) are slightly different. The difference between them is 7.0 N with the percentage of 4.68%. But the difference rises up to 36.1 N with the percentage of 21.40% in the perpendicular mode though their maximum forces drop to 204.8 N and 168.7 N respectively as shown in Fig.20(b).

There is no critical current flowing through the boundaries of the stacked array. However, the currents flowing through GBs may exist as to the MSMG bulk though they are much depressed. Therefore macroscopic currents J_c in MSMG bulks may be divided into two kinds, i.e. intra-grain J_c circulating in each seeding grain corresponding to each crystal and inter-grain J_c flowing across GBs. Due to the distribution of the applied field the induced currents will drastically fall off in the perpendicular mode, which makes the role of inter-grain currents stand out. This may be the reason why the force difference between their maximum forces increases. This also implies the average size of macro current loops of the MSMG bulk is larger that of the stacked array in the perpendicular mode owning to macro inter-grain currents.

According to Ref.[31], trapped fields are consistent with the distribution of the critical currents. In order to understand such magnetic interactions, we present trapped fields of the MSMG bulk and the stacked array (consisting of bulk IFW01, IFW02 and IFW03) in both parallel and perpendicular modes. Their corresponding peak profiles (along the trace perpendicular to boundaries) are presented in Fig.21. In the parallel mode the peak profile upon the trapped field of the MSMG bulk have three peaks, which confirms that critical currents are depressed at GBs. But the distances between peak and valley of the stacked array are extremely larger than those of the MSMG bulk in the parallel mode. Moreover the trapped magnetic flux densities B_y are normalized to the maximum B_y, we have isolines of

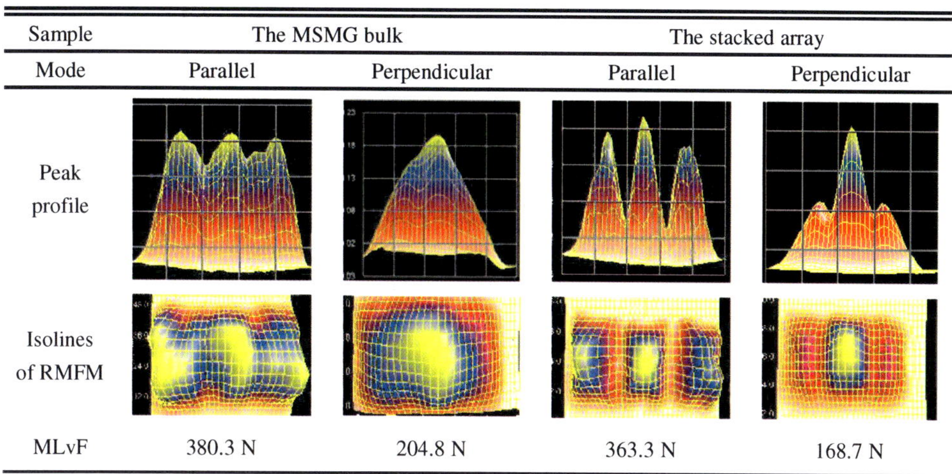

Sample	The MSMG bulk		The stacked array	
Mode	Parallel	Perpendicular	Parallel	Perpendicular
Peak profile				
Isolines of RMFM				
MLvF	380.3 N	204.8 N	363.3 N	168.7 N

Figure 21: Trapped field of the MSMG bulk and the stacked bulk array in both parallel and perpendicular modes.

their trapped fields. It is obvious that connectivity of the trapped field of the MSMG bulk is better than the stacked array. These prove that there is an amount of inter-grain J_c flowing through GBs which connects the intra-grain J_c in the two adjacent grains of the MSMG bulk. This is consistent with the conclusions that the trapped field is significantly different from zero despite the drop [66].

As for the perpendicular mode, intra-grain J_c flows in two side grains will become less due to the decrease of the applied field. There is single peak upon the trapped field of the MSMG bulk, while three peaks of the stacked array. It is possible to predict that the intra-grain J_c in the side grains of the MSMG bulk is not beyond the maximum current transportation capabilities of GBs, i.e. all those intra-grain J_c circulating in the side grains may flow through GBs of the MSMT bulk. So it can be regarded as single larger grain bulk in the perpendicular mode. During the measurement process of levitation force, inter-grain J_c in perpendicular mode is less than that in the parallel mode, but the ratio between macro inter-grain and intra-grain J_c in the perpendicular mode becomes larger. In other words the role of inter-grain currents stands out in this mode and average size of shielding current loop is larger than that of the stacked array. This is the reason why the MSMG bulk has larger levitation force than the stacked array in perpendicular mode.

On the other hand, guidance force is responsible for lateral stability of maglev vehicle. Guidance forces of the MSMG bulk and the stacked array are compared in both parallel and perpendicular modes (Fig.22(a) and Fig.22(b)). Their maximum guidance forces are 70.1 N and 56.3 N in the parallel mode, while 69.3 N and 56.2 N in the perpendicular mode. The difference between them is 13.8 N and 13.1 N in the two different modes. Thanks to inter-grain critical J_c guidance forces of the MSMG bulk are larger than those of the stacked array in the two modes. On the other hand, although both the placement and move-ment direction of the MSMG bulk are extremely different in the parallel and perpendicular modes, their maximum guidance forces nearly do not change. This phenomenon seems to be controversial with the conclusion that guidance force is dependent on the trapped field of

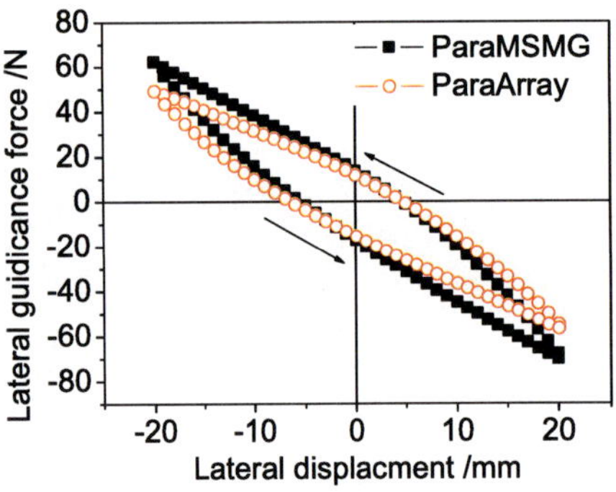
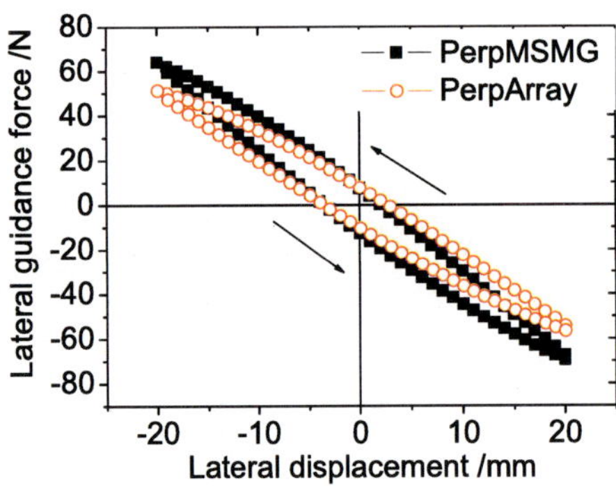

(a) Guidance forces of the MSMG bulk and the stacked bulk array in the parallel mode.

(b) Guidance forces of the MSMG bulk and the stacked bulk array in the perpendicular mode.

Figure 22: Guidance force:Closed dot is for the MSMG bulk and open dot is for the stacked bulk array.

the HTS bulk [71], because the trapped fields in the different modes are significantly different according to the displacement in Figure.1. In fact guidance force is total pinning force determined by both the trapped field and the applied field during the magnetic interaction process.

Summarily, we employ one piece of MSMG bulk and a stacked array consisting of three separated bulks in our experiments. They have similar dimensions and the maximum levitation forces of them are slightly different in parallel mode. Due to inter-grain critical currents flowing through GBs, the MSMS bulk can be regarded as single larger grain bulk in the perpendicular mode, so it has larger levitation force than the stacked bulk array. Guidance force is the total pinning force determined by both the trapped field and the applied field during the magnetic interaction process. The maximum guidance forces of the MSMG bulk are almost the same in parallel and perpendicular modes. However, guidance forces of the MSMG bulk are always larger than the stacked array in the two modes. In addition, hysteresis loss is reduced in perpendicular mode. MSMG YBCO bulk superconductor is preferable to optimize the present HTS-PMG Magnetic levitation system.

2.5   Influence of the Lateral Movement on the Levitation and Guidance Force in the High-Temperature Superconductor Maglev System

As two key parameters for the Maglev system, vertical levitation force and lateral guidance force used to be studied separately [16]. In fact, they are two components of the integral Lorentz force resulting from the interaction between the induced currents and the applied magnetic field [39]. This paper will optimize the system field-cooling height (FH) and working-levitation height (WH) combining the two parameters from the point of the practical application [72].

Initially the bulk superconductors onboard the HTS Maglev system were fixed over the PMG, and the current distance between the bottom of superconductor and the surface of the PMG was defined as FH. Dozens of minutes after being cooled in liquid nitrogen,

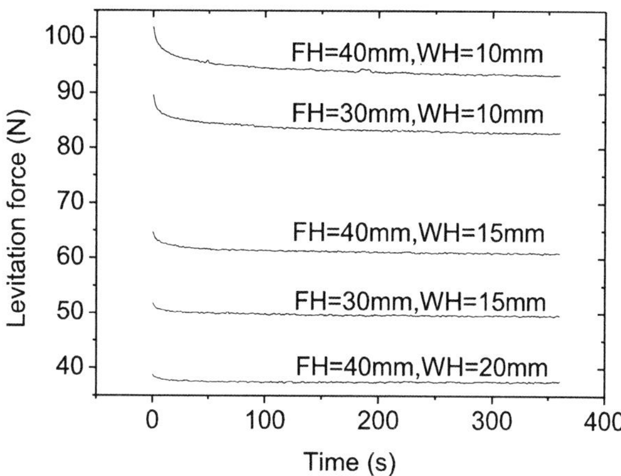

Figure 23: The relaxation of vertical levitation force for different eld-cooling height (FH) and working-levitation height (WH).

the bulk superconductors were dropped from the eld-cooling position to the nal working height where the vehicle body was levitated. At this moment the distance between the bottom of superconductor and the surface of the PMG was regarded as WH. It has been accepted that the bulk superconductor is not in a really stable state after the drop; that is, the levitation force has a logarithmal decrease with time [73, 74, 75]. This phenomenon de ned as levitation force relaxation resulting from magnetic relaxation [76], has been investigated experimentally and theoretically. But most of their systems composed of the superconductor and magnet were coaxial [77] and there was no movement during the measurement.

However in the practical operation of the HTS Maglev vehicle, even if the vehicle body is right above the PMG at the beginning , it will be off the right position when it goes along the curved part of the PMG [78]. Moreover the mechanical-electrical coupling will also cause the vibration along the lateral direction. Therefore, it is possible that the LD will take place, resulting in changes of levitation and guidance forces, while there are few reports about the in uence of lateral movement on the vertical levitation force.

What's more, both the vertical levitation force and lateral guidance force can be measured at the same time in these experiments. The HTS sample was a YBCO monolith, 90 mm long, 36 mm wide, and 15 mm thick. It was cooled in liquid nitrogen, with the seeded surface downward and its length direction perpendicular to the length of the PMG. After the sample was completely cooled at a certain FH, it dropped downwards because of gravitation. Once the sample reached the WH position, the relaxation measurement for levitation force was started, 360s later, the relaxation measurement was ended and the sample was moved along the lateral direction, parallel to the surface of the PMG. There were four maximum lateral displacements (MLD) such as 3, 6, 9 and 12 mm, respectively.

Fig.23 presents ve curves for levitation force versus time, i.e., the levitation force relaxation for about 360s. Whatever the WH and FH were, there is a logarithmic relationship between the levitation force and the time, and the force had little change in the end of re-

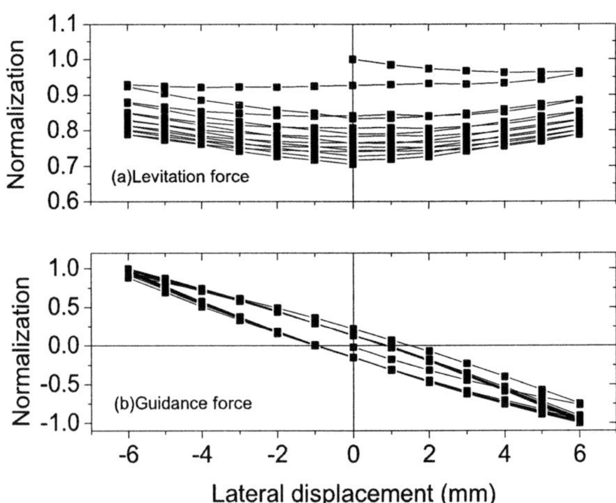

Figure 24: Normalization curves for both levitation and guidance force during the lateral cycle movement with the maximum absolute displacement 6 mm after the force relaxation measurement.

laxation measurement. For WH=10 mm, the levitation force with FH=40 mm was more than that with WH=30 mm. The greater the drop between the FH and WH, the greater the levitation force. However, Hull et al. found that the rotational loss is dependent on the drop between the WH and FH [79]. It is necessary to minimize their drop to lessen the moving loss at high speed for the HTS Maglev system. Moreover, for the same FH (40 or 30 mm), the force in the case of WH=10 mm was about twice than in the case of WH=15 mm. Therefore, the smaller the WH, the greater the levitation force, when FH is constant.

Once the force relaxation measurement was ended, the sample was horizontally moved at the certain WH perpendicular to the length direction of the PMG. For each cycle the sequence was: rstly, from the center to the end of the right, and then back across the center, continuing to the end of the left, and back to the center. As shown in Fig.24, there were three times in the rst cycle for the sample across the center, i.e., the right position above the PMG. The levitation force signi cantly decreased to 85% for the third time compared to the initial one, which results from hysteretic property of a hard Type-II superconductor. Since the practical vehicle will go across and across the center plane of the PMG, the go-and-back cycle was repeated in the experiments. The levitation force continuously decreased but the rate of the decrease was gradually reduced.

Fig.25(a) collected the levitation force when the sample went across the center position during the continuous cycles. Whatever the FH was, the levitation force was the most with WH=10 mm, while its decrease was also the most drastic. The levitation force with FH=40 mm started from 93.5 N, and decreased to 81.5 and 78.0 N when the sample went second and third across the center, and lowered to 70.0 N seven cycles later. The same happened to the case of FH=30 mm and WH=10 mm with the force decreasing from 84.0 to 60 N. However, the force slowly decreased with the WH=15 mm, and the force decay for FH=40 mm and FH=30 mm were about 13.0 and 11.0 N, respectively. Although the force

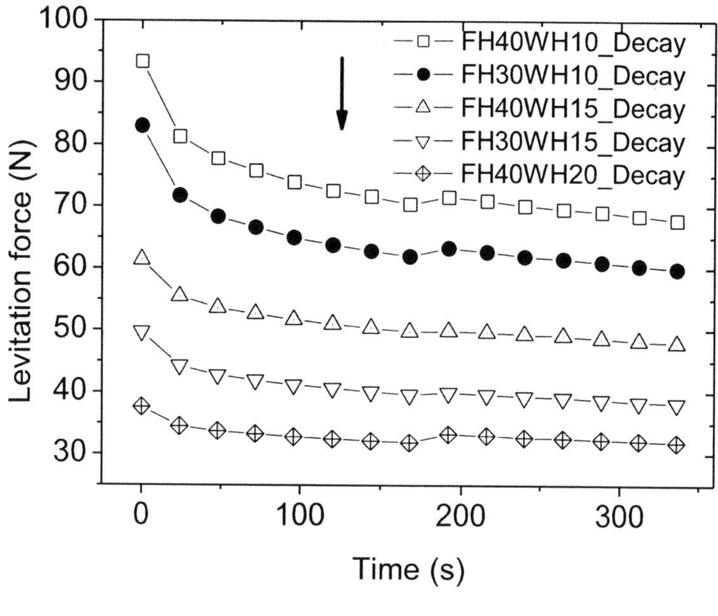

(a)

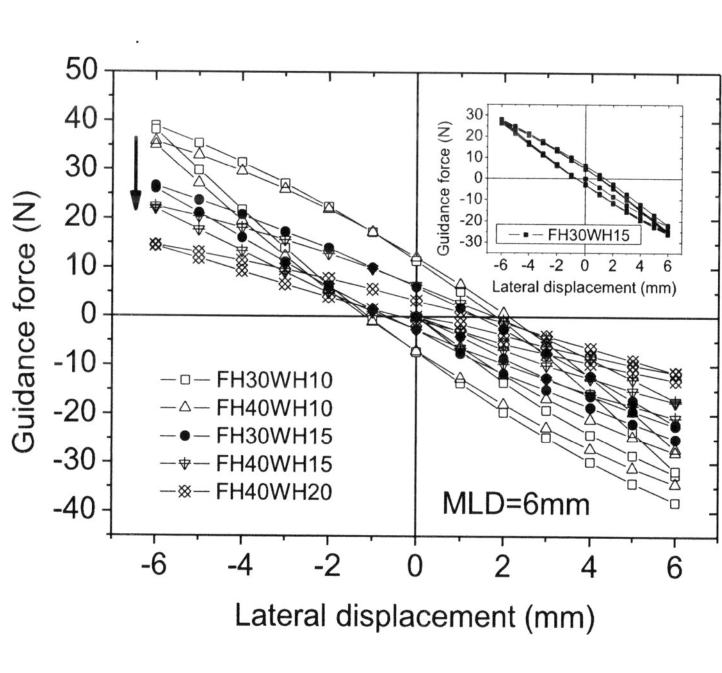

(b)

Figure 25: Levitation and guidance force for different FH and WH with the MLD at 6 mm.

decay was not so much for the case of FH=40 mm with WH=20 mm, the corresponding levitation force is also smaller.

As another key parameter, the guidance force was also obtained during the cycle movement as shown in the lower part of Fig.24. Fig.25(b) shows the rst cycle of the guidance force curves for different FH and WH. At the beginning the force was zero, and there was a lateral force when it went back across the center. The force curve corresponding to the rst cycle stands out compared to the others, so there is little in uence of lateral movement on the guidance force after the rst cycle, which indicates that the WH has the main effect on the guidance force rather than FH. Although the guidance force is much higher for WH=10 mm, such small WH will bring the higher requirement to the practical engineering resulting in higher cost. Owing to the less levitation and guidance force, the WH=20 mm will be out of consideration. Consequently, the WH=15 mm and FH=30 mm is preferred in the present HTS/PMG system. Additionally, the absolute irreversible force for WH=15 mm was about 6.5 N, when the sample went across the center, less than that for WH=10 mm, as shown in the inset of Fig.25.

Furthermore, the MLD was extended from 6 to 3 , 9, and 12 mm. Fig.26(a) presents the rst cycle of the guidance force curves for the different MLD. The curve from the center to the right end and back to the center is enlarged in the inset of Fig.26(a). The starting branches for different MLD (i.e., 3, 6, 9 and 12 mm) overlap each other from the center to the left end, corresponding to the maximum absolute force 12.0, 22.5, 29.8 and 35.8 N respectively. And the hysteresis loss increased as the MLD increased.

Fig.26(b) presents the levitation forces for different MLD across the center during the cycling lateral movement. Since this series of experiments was carried out directly after the force relaxation measurement and the FH and WH were always 30 and 15 mm, the starting forces were roughly identical, 50 N. However, the time consumption in each cycle was different for the different MLD. The greater the MLD is, the less time for the sample to go across the center, So the number of the recorded data points was different as shown in Fig.26(b) and their total numbers were 46, 23, 15 and 12 respectively, within the same measurement period. As the time of the movement cycle increased, the levitation force decreased, which had the same phenomenon with Fig.25(b). The decrease of the levitation force became more drastic as the MLD increased up to 9 mm. However, there was a little difference between the MLD=9 mm and MLD=12 mm, which indicates that the decrease will not be much increased once the MLD is more than 9 mm.

In summary, the in uences of FH and WH on both the levitation and guidance forces were discussed and the optimum FH and WH are proposed as 30 and 15 mm, respectively, for the present HTS/PMG system. Although the levitation force became stable at end of the relaxation measurement, the force drastically decreased once the sample was horizontally moved. Additionally, it was found that levitation force decay right above the PMG is dependent on both the MLD and the cycle times, but, the guidance force after the rst cycle movement is independent of the cycle times.

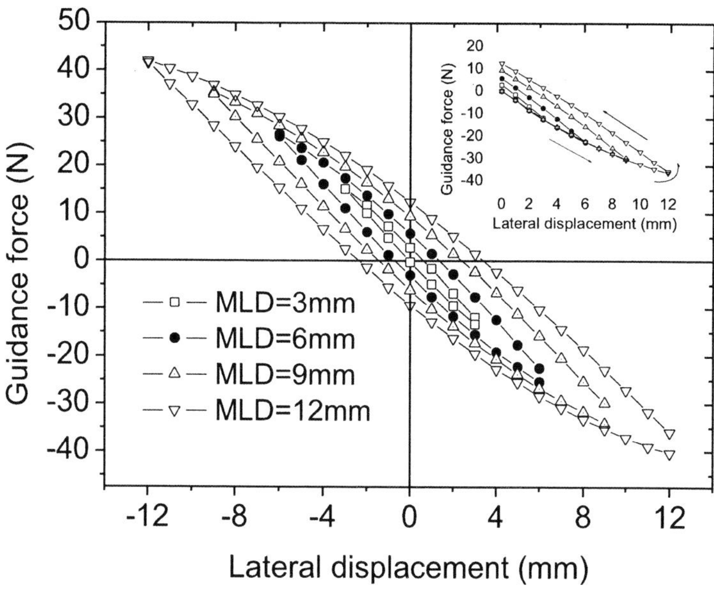

(a)

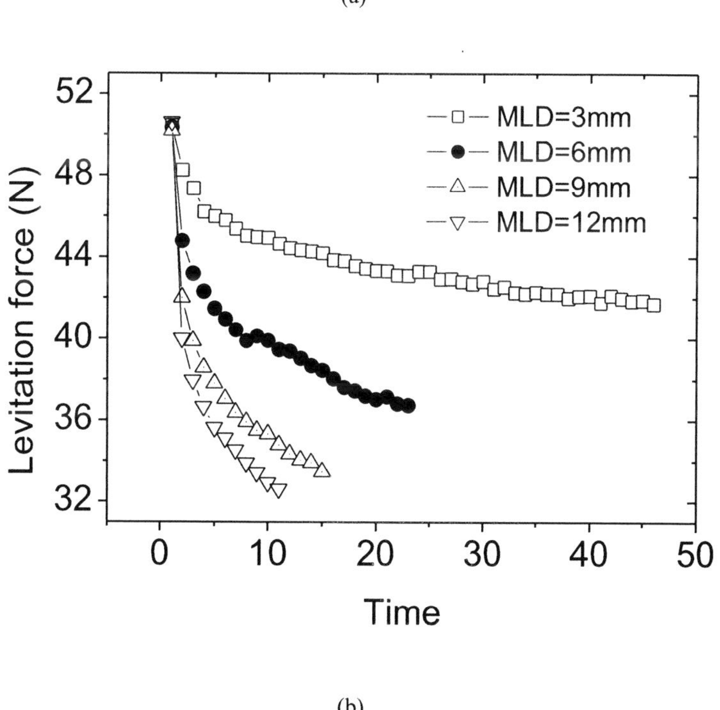

(b)

Figure 26: The MLD dependence of levitation and guidance with FH=30 mm and WH=15 mm.

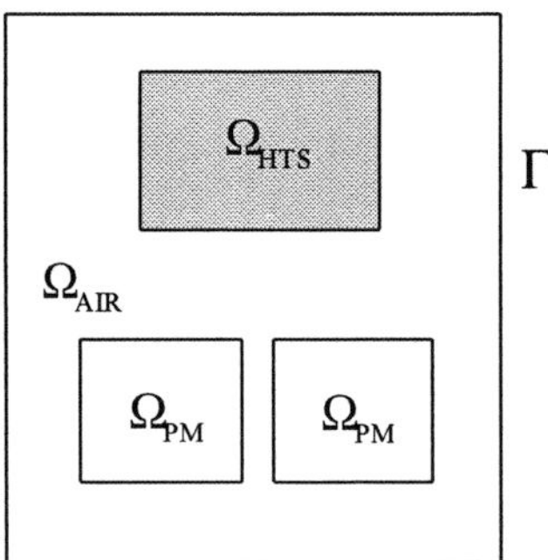

Figure 27: The total computational domain is Ω consisting of Ω_{HTS}, Ω_{PM} and Ω_{AIR}, which are for HTS, PMG and the air respectively.

3 Numerical Calculation

3.1 Introduction

Most of the levitation systems are made up of a single high temperature superconductors (HTS) bulk and a permanent magnet sharing the same axis, i.e. co-axial system [80].The magnetic eld above the PMG is continuous and uniform along the length direction, however, such a eld has great magnetic gradient along the width direction. Therefore the applied elds are divided into two cases: symmetrical and unsymmetrical about the HTS symmetry axis [39]. The magnitude dependence of the critical current density $J_c(B)$ should be considered. Chen found that the calculated magnetization curves of Kim or Exponential model are more accurate than those of Bean model [81, 82].

To optimize the design of the Maglev transportation system, it is necessary to predict the magnetic interaction between the HTS and the PMG. To study this electromagnetic behavior, a variation Formulation and numerical solution has been developed by L. Prigozhin [23]. By using this approach, the in uence of the geometry of HTS and PMG on the guidance force has been investigated [39].

The magnetic eld is uniform along the length of the permanent magnet guideway, while it is inhomogeneous, with great eld gradient along the width direction [9, 7, 83]. If the superconductor is levitated rightly above the guideway, the applied eld is symmetrical about the center plane of the superconductor. Once the superconductor is off the right-above position, the applied eld will become unsymmetrical. In other words, the applied eld is neither parallel to c axis nor to ab plane, but titled.

3.1.1 Variational Formulations

There are different methods to calculate the levitation force such as Maxwell stress, virtual work or Lorentz force. Also there are different model for describing the E-J relationship, such as critical state model, ux ow and ux creep, or two uid models. As to the HTS/PMG system, the changing of the applied eld will be regarded as quasi-static, and the integral magnetic force will be obtained from Lorentz force.

First of all, we will give a brief presentation of transition to the variation formulation in terms of the modelling of HTS critical state proposed by Prigozhin, and please see [85] for the mathematical problem in details. The macroscopic magnetic phenomenon in the computation domain Ω can be described with the Maxwell equations:

$$\nabla \times \vec{E} = -\frac{\partial \vec{B}}{\partial t} \tag{6}$$

$$-\vec{J} = \nabla \times \vec{H} \tag{7}$$

and one of the continuous laws is $\vec{B} = \mu_0 \vec{H}$, where μ_0 is the permeability of the vacuum. The external current density in Ω_{AIR} is assumed to be:

$$-\vec{J}_e = \nabla \times \vec{H}, \quad \nabla \cdot \vec{J}_e = 0 \quad \text{in} \quad \Omega_{\mathrm{AIR}}, \tag{8}$$

In terms of the superconducting currents owing through the superconductor, there is

$$\vec{E} = \rho \vec{J} \quad \text{in} \quad \Omega_{\mathrm{HTS}}, \tag{9}$$

where $\rho(x, t) \geq 0$ is an unknown nonnegative function. Recently Bossavit proposed a modi cation of Ohm's law with effective resistivity, i.e. ρ has two state, one is superconducting state (the current density J is less than or equal to J_c) and another is normal state (the current density J is more than J_c) [86]. After specifying the initial and boundary conditions and taking into account that the tangential components of $\vec{E}$ and $\vec{\varphi}$ are continuous on Γ, we obtain the variational relation

$$\mu_0 \int_0^T \int_\Omega \frac{\partial \vec{H}}{\partial t} \cdot \vec{\psi} + \int_{\Omega_{\mathrm{HTS}}}^T \int_\Omega \nabla \times \vec{H} \cdot \nabla \times \vec{\psi} = 0, \tag{10}$$

which is valid for all test functions $\vec{\varphi}$, from $\vec{V} = \{\vec{\varphi}(x, t) | \nabla \times \vec{\varphi} \text{ in } \Omega_{\mathrm{HTS}}, [\vec{\psi}_\Gamma] = 0, \text{ on } \Gamma\}$. So there are only two unknowns: the magnetic eld and the effective resistivity, without the electric eld.

Now we introduce a new variable, $\vec{h} = \vec{H} - \vec{H}_e$, satisfying

$$\nabla \times \vec{h} = 0 \quad \text{in} \quad \Omega_{\mathrm{AIR}}, \tag{11}$$

$$|\nabla \times \vec{h}| \leq J_c(|\vec{h} + \vec{H}_e|) \quad \text{in} \quad \Omega_{\mathrm{HTS}}, \tag{12}$$

$$[\vec{h}_\Gamma] = 0 \quad \text{on} \quad \Gamma. \tag{13}$$

Let us de ne the set of functions

$$K(\vec{h}) = \{\vec{\varphi} \in \vec{V} | \nabla \times \vec{h}|^2 \leq J_c^2(|\vec{h} + \vec{H}_e|)\}, \tag{14}$$

which is dependent on $\vec{h}$. Moreover, since

$$\nabla \times \vec{h} = \nabla \times \vec{H}, \quad |\nabla \times \vec{H}| \leq J_c(\vec{H}) \quad \text{in} \quad \Omega_{\mathrm{HTS}}. \tag{15}$$

The formulation may be yielded as

$$\int_0^T \int_{\Omega_{\mathrm{HTS}}} \rho |\nabla \times \vec{h}|^2 = \int_0^T \int_{\Omega_{\mathrm{HTS}}} J_c^2(|\vec{h} + \vec{H}_e|. \tag{16}$$

Correlating the above four functions (10), (14) - (16), and considering $\vec{\varphi} - \vec{h} \in \vec{V}$ for the function $\vec{\varphi}$ from $K(\vec{h})$, we obtain

$$\mu_0 \int_0^T \int_\Omega \frac{\{\partial \vec{h} + \vec{H}_e\}}{\partial t} \cdot (\vec{\varphi} - \vec{h}$$

$$= -\int_0^T \int_{\Omega_{\text{HTS}}} \rho \nabla \times \vec{h} \cdot \nabla \times (\vec{\varphi} - \vec{h})$$

$$\geq \int_0^T \int_{\Omega_{\text{HTS}}} \rho(J_{\text{c}}^2(|\vec{h} + \vec{H}_e| - |\nabla \times \vec{\varphi}||\nabla \times \vec{h}|). \tag{17}$$

This proves that $\vec{h}$ is a solution to the problem:

$$\begin{cases} \text{Find function } \vec{h} \in K(\vec{h}) \text{ such that} \\ \left(\frac{\{\partial \vec{h} + \vec{H}_e\}}{\partial t}, (\vec{\varphi} - \vec{h}) \right) \geq 0 \quad \text{for any} \quad \vec{\varphi} \in K(\vec{h}) \\ \vec{h}|_{t=0} = \vec{h}_0, \end{cases} \tag{18}$$

where $(\vec{u}, \vec{w}) = \int_0^T \int_\Omega (\vec{u} \cdot \vec{w})$ is the scalar product of two vector functions.

Let us assume that both HTS bulk and the applied eld above the PMG are homogeneous along the length direction, so the plane eld at certain height z has only the x and y components,

$$\vec{H} = H_x \hat{x} + H_y \hat{y} \tag{19}$$

So far, the current owing in superconductor has only the z component, which can be expressed as

$$\vec{J} = \begin{bmatrix} \hat{x} & \hat{y} & \hat{z} \\ \frac{\partial}{\partial x} & \frac{\partial}{\partial y} & \frac{\partial}{\partial z} \\ H_x & H_y & H_z \end{bmatrix} = \left(\frac{\partial H_y}{\partial x} - \frac{\partial H_x}{\partial y} \right) \hat{z} \tag{20}$$

It is obvious that the currents. For two-dimensional problems, the inequality can be presented in scalar formulations, and the above inequality variational formulations may be rewritten in terms of the critical current using the convolution algorithm, as described in [23].

3.1.2   Field Dependence of Critical Current

The variational formulation that has been presented, changes the front-tracking problem to free boundary problem, which is applicable to the whole computation domain and valid for any material of the problem. In terms of the electromagnetic characteristics, critical state model was used to describe the E-J nonlinear relationship [23].

$$J_{\text{c}}(B) = y_1 + A_1 e^{-B/t_1} \tag{21}$$

where $A_1 = 2.158 \times 10^8$, $t_1 = 0.245$, $y_0 = 2.158 \times 10^8$. Such dependence of J_{c} on the applied eld magnitude will be considered in the computer modelling.

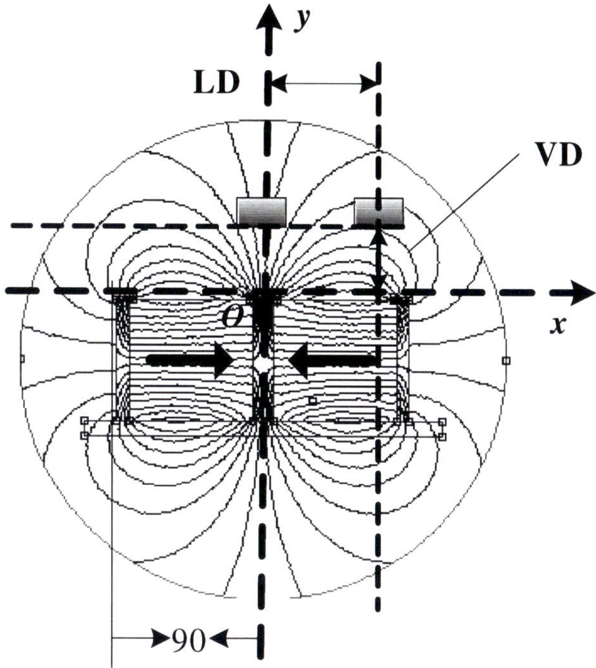

(a) Two dimensional (2D) magnetic flux distribution of the PMG. The rectangle is for HTS bulk.

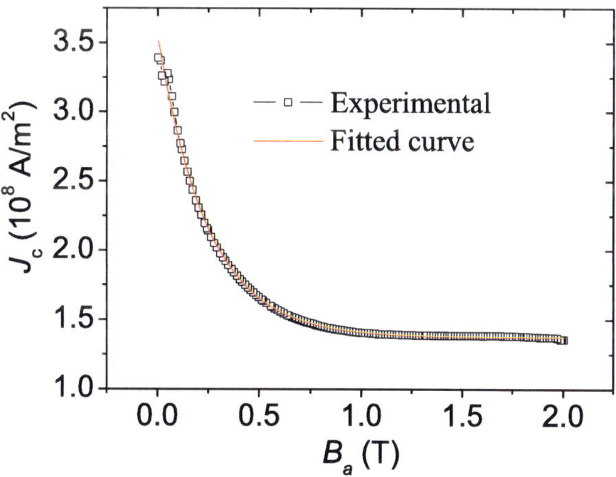

(b) The fitted function of experimental $J_c(B)$ curve.

Figure 28: The field dependence of J_c of YBCO bulk above the PMG. LD and VD are lateral and vertical displacement in short.

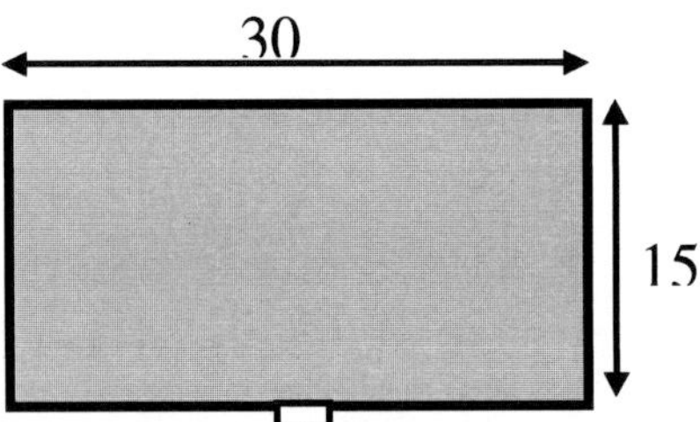

Figure 29: The schematic of the cross section in vertical xy plane.(Unit:cm)

Fig. 28(a) presents the magnetic flux distribution in the vertical xy plane. It is obvious that the field is non-uniform along the width direction. As reported in the maximum field is up to 1.17 T close to the center of the PMG surface [9]. On the other hand, $J_c(B)$ curve at 77 K is measured by **IFW** Dresden as illustrated in Fig. 28(b). The experimental results are fitted with the field ranging from 0 T to 2.0 T. The fitted function goes as the following,

However, neither Bean model or Kim model [87], consider the angle dependence of critical current on the applied field. Recently Sawamura has numerically calculated the levitation force in the co-axial magnet and superconductor system [88], supposing the critical current density is dependent on both magnitude $|\vec{B}|$ and angle θ of the local applied field which follows the equation

$$J_c(B,\theta) \;=\; J_c(B) \times F(\theta) \tag{22}$$

where θ is the angle between the applied field and the normal component of the ab layer of YBCO. $J_c(B)$ and $F(\theta)$ are for the influence of magnitude and angle on the critical current density respectively. Moreover Savamura extended the Morris research work and found that there is such angle dependence $F(\theta)$ for the applied field between 1 T and 2 T [89],

$$F(\theta) \;=\; (\alpha \cos^2(\theta) + \sin^2(\theta))^{-0.75}, \tag{23}$$

where $\alpha = 2.2$. The magnitude and angle dependence of the critical current density will be taken into account in the numerical calculations.

Prigozhin proposed that the constraint should be imposed by the forced transport currents [88]. In the infinitely long superconductor, all the currents are induced by the change of the external field, without any extracted transport current, so the net current in the across section has to be zero, i.e.

$$\int_{\Omega_{\text{HTS}}} J \;=\; I_{\text{transport}}(t) = 0. \tag{24}$$

The vertical cross section is 30 mm wide and 15 mm high, with 450 elements as illustrated in Fig.29. In the numerical analysis, the current distribution is calculated by the Quadratic Programming [39].

The levitation force is the vertical component of the macroscopic integral Lorentz force, calculated by

$$\vec{F}_{\text{Lev}} \;=\; \int_V \vec{J} \times \vec{B}_x dv, \tag{25}$$

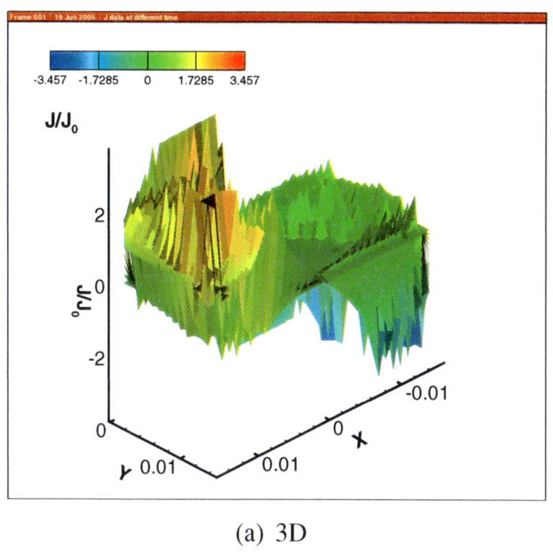

(a) 3D

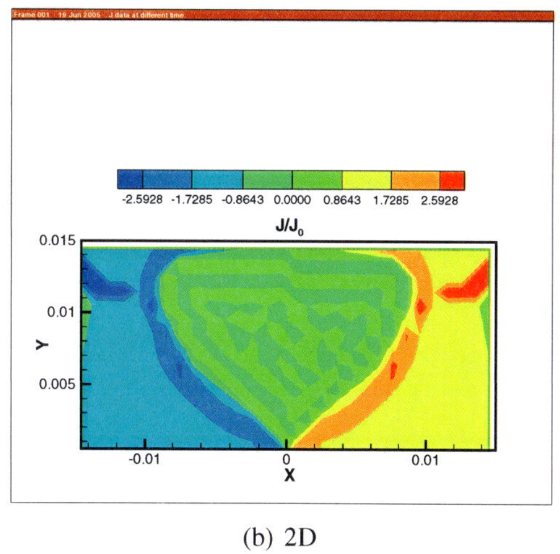

(b) 2D

Figure 30: The current distribution upon the vertical cross section in symmetrical applied field with LG=5 mm.

where $\vec{J}$ is supercurrent density distributed in the superconductor, $\vec{B}_x$ is the horizontal component of magnetic field by the PMG, V is the volume of the superconductor, and "Lev" is for levitation in short. The guidance force is the vertical component of the macroscopic integral Lorentz force, calculated by

$$\vec{F}_{\mathrm{Gui}} \;=\; \int_{V} \vec{J} \times \vec{B}_y dv, \tag{26}$$

where $\vec{J}$ is also supercurrent density distributed in the superconductor, $\vec{B}_y$ is the horizontal component of magnetic field by the PMG, V is the same with that in 25, and "Gui" is for guidance in short.

3.2 Calculation Results

3.2.1 Current Profiles

When the bulk is right above the PMG, the applied field is symmetrical about the axis of the HTS bulk. Once the bulk is away from the right position, the applied field becomes unsymmetrical. Supposed that the superconductor is roughly zero field cooled at the height of 130 mm above the PMG, Fig.30 and Fig.31 present the current distribution upon the vertical cross section in xy plane when the superconductor moves to the height of 5 mm, i.e. the smallest levitation gap. As shown in Fig.30 the current distribution is anti-symmetrical about the center plane of superconductor in the symmetrical applied field. However the current distribution becomes unsymmetrical when LD is 30 mm, resulting from the titled field in unsymmetrical applied field as depicted in Fig.31.

3.2.2 Levitation and Guidance Force

Fig.32 and Fig.33 compare the calculated and experimental levitation force in both symmetrical and unsymmetrical applied fields. The details of the measurements have been

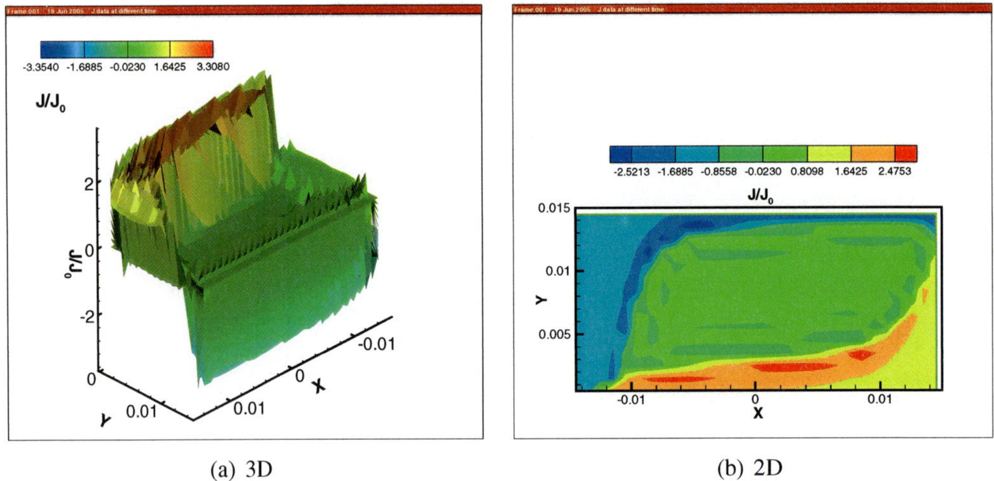

(a) 3D   (b) 2D

Figure 31: The current distribution upon the vertical cross section in unsymmetrical applied field with LG=5 mm.

described extensively in Ref.[38, 39]. In the experiments of this part, the sample #4 used is 36 mm long, 30 mm wide and 15 mm thick. After the sample was cooled in liquid nitrogen with zero field, the levitation force was real-time collected while the sample moves from 130 mm to 5 mm above the PMG and went back to the original position. When the sample vertically moved right above the PMG, the applied field is symmetrical and the relationship between the levitation force and gap is approximately linear. Once the bulk was off the center plane, the applied field became unsymmetrical. As shown in the inset of Fig. 35, there is much difference between two levitation force curves in the symmetrical and unsymmetrical fields. In the unsymmetrical applied field, the descending part upon the hysteresis curve is convex and the maximum force comes out at the vertical gap of about 10 mm.

We have two cases for the use of the critical exponential description: without and with the angle dependence. There are differences between the calculated results in these two cases, which indicates that the angle of the applied field does have the influence on the critical current distribution. Moreover, both the calculated results are about twice more than the experimental results. Qualitatively, the calculated results agree with the experimental results in the symmetrical applied field as the force normalization curves shown in Fig.32(b), however, there is both qualitative and quantitative difference between them in the unsymmetrical applied field, as shown in Fig.33.

The guidance force was carried out in FC 20 mm above the PMG. Keeping the measuring height (MH) at the gap of 20 mm, the sample was horizontally moved from the center to the displacement of -20 mm and across the center back to the displacement of 20 mm, and back to the starting position. The same happens to the guidance force in Fig.34(a) and Fig.34(b), i.e. the difference between results without and with angle dependence still exists and the calculated hysteresis curve does not agree with the experimental curve, because the experienced field for the HTS sample is changing during the lateral movement, from symmetrical to unsymmetrical or vice versa.

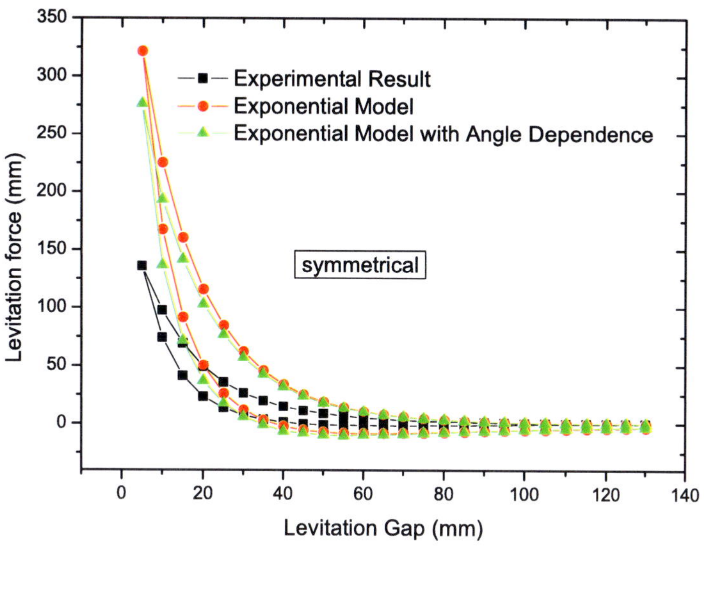

(a) Leviation force

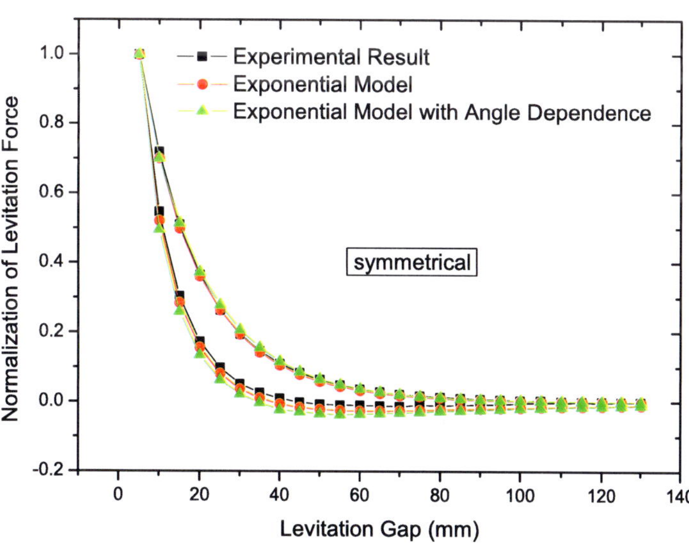

(b) Normalization of levitation force

Figure 32: Comparison between the calculated and experimental levitation force in the symmetrical field.

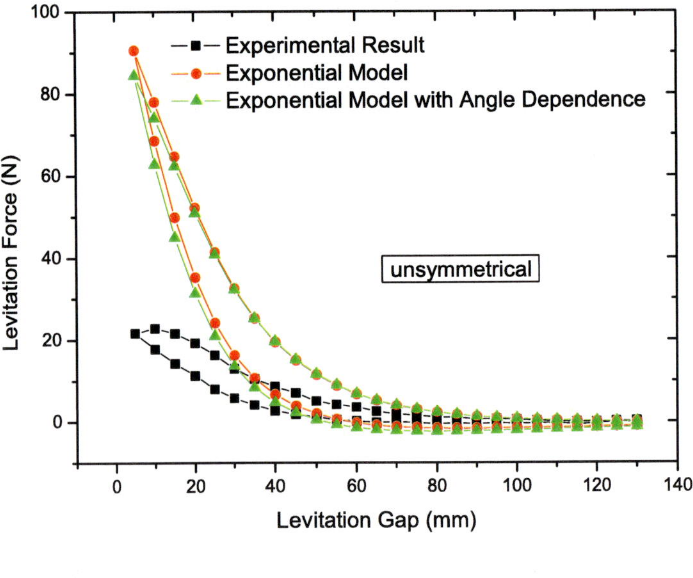

(a) Leviation force

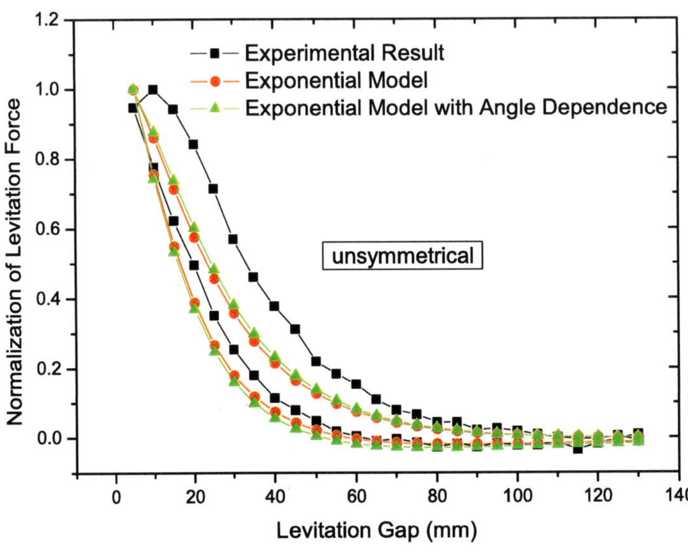

(b) Normalization of levitation force

Figure 33: Comparison between the calculated and experimental levitation force in the unsymmetrical field.

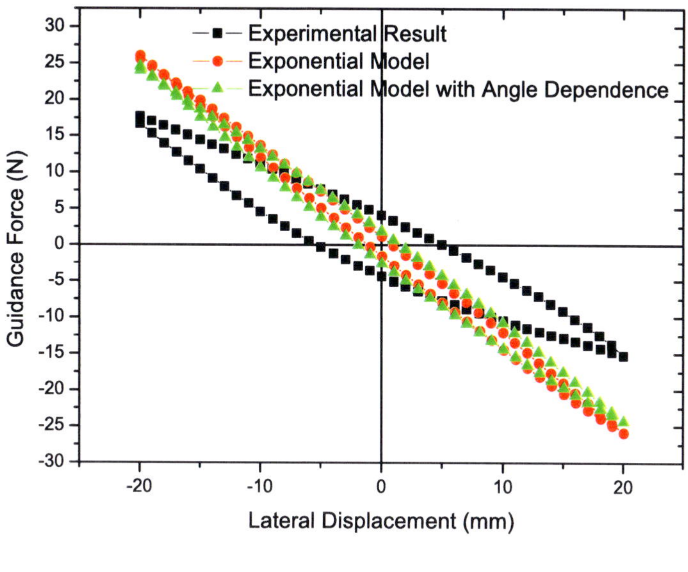

(a) Guidance force

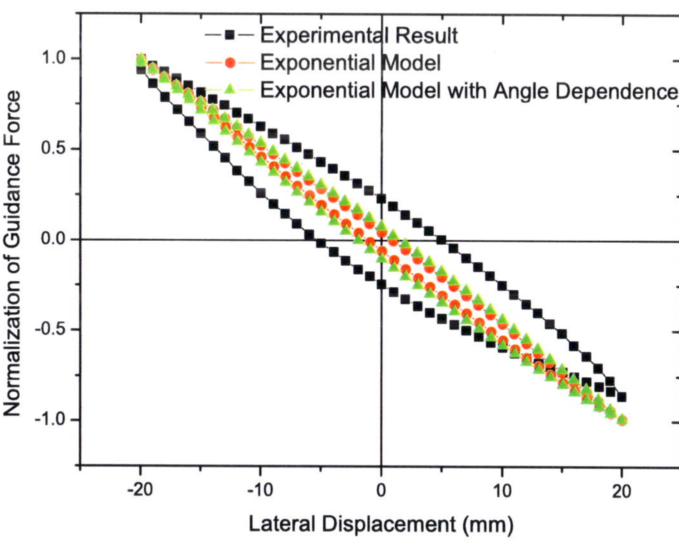

(b) Normalization of guidance force

Figure 34: Comparison between the calculated and experimental guidance force during the lateral movement.

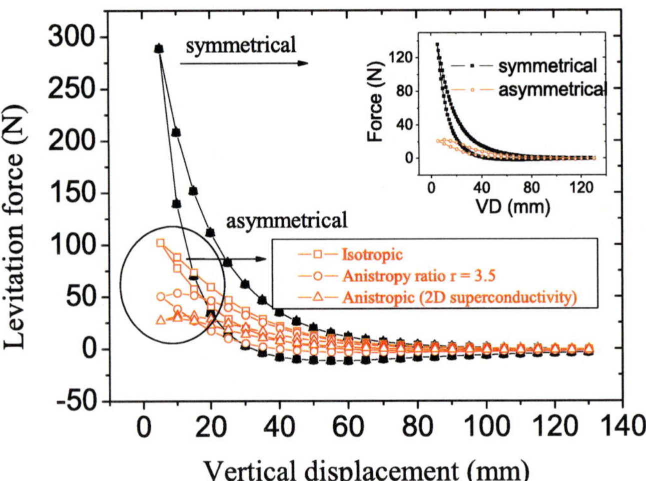

Figure 35: The comparison between calculated and experimental vertical levitation force in the case of zero field cooling.

3.2.3   Anisotropy

The superconductor is assumed to be isotropic by Prigozhin in Ref.[23, 87], which is different with the anisotropic properties of HTS. The qualitative disagreement between the calculated and experimental results in the unsymmetrical applied field may result from that the anisotropy of the critical current density. So we divided the across section into several layers to take account of the anisotropy, defined as $\Omega_{\text{HTS}-i}$ and the imposed constraints are changed to

$$\int_{\Omega_{\text{HTS}-i}} J \;=\; I_{\text{transport}-i}(t), \qquad \forall \quad i = 1, 2, \dots, k \qquad (27)$$

where k is the total number of sub-cross-sectional components with independently forced transport current.

Fig. 35 and 36 presents the comparison between the experimental and calculated results considering the anisotropic properties. There are three cases for the superconductivity of YBCO bulk in the calculations. First is isotropic, which was supposed with the Bean model by Prigozhin [87]. Next, the superconductivity is absolutely 2D, which has been discussed by J.R. Clem [54]. That is, there is no component along the c axis of critical current density, whatever the field direction and magnitude are. Third, the anisotropy ratio around 3.5 measured in the above experiments were applied, to describe the superconductivity of the YBCO superconductor. In the symmetrical applied field, the calculated forces are the same for the above three cases, i.e., there is qualitative agreement between calculated and experimental results. The reason is that the applied field right above the PMG is roughly perpendicular to its surface with few horizontal components and the currents mainly flow in the ab plane rather than along c axis.

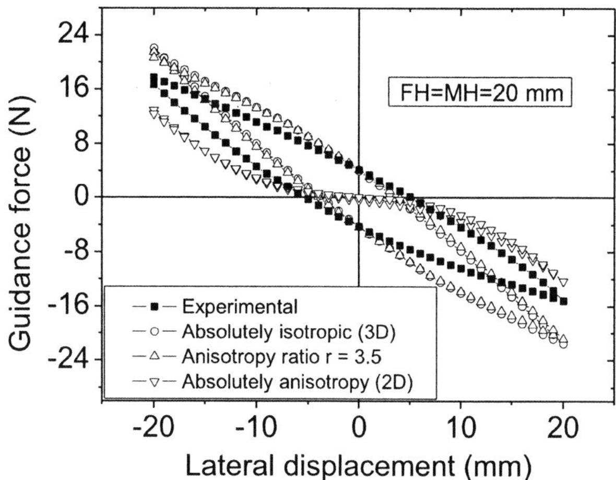

Figure 36: The comparison between calculated and experimental horizontal guidance force in the case of eld cooling.

However, in the unsymmetrical applied elds, the calculated results in the rst case (isotropic) does not agree with the experimental one. In the second case (absolutely 2D anisotropic), there is nearly no hysteresis upon the major loop of levitation force, which is far away from the intrinsic properties of hard superconductor. When the anisotropy ratio 3.5 is taken into account, the calculated results qualitatively agree with the experimental ones, but the quantitative disagreement still exists. The reason may be that the induced currents in the actual sample identically ow in three dimensions rather than one dimension in the two dimension model [94].

As shown in Fig. 36, there is no hysteresis upon the guidance force loop in the case of 2D superconductivity, which has distinguished difference with the experimental result. Such difference con rms the anisotropic superconductivity is 3D rather than 2D at 77 K. Although there is no difference between the other two cases (isotropic and 3D anisotropic), the case of isotropic is not t for levitation force calculation in the unsymmetrical applied eld. So, neither isotropic nor absolutely 2D anisotropic the superconductivity of YBCO at 77 K is. It is necessary to take the exact anisotropy ratio into the computer modelling for the present HTS Maglev vehicle system.

Conclusively, both the magnitude and angle dependence of critical current density on the applied eld have been considered in the numerical analysis. The comparisons prove that the eld angle does in uence the current distribution and magnetic interaction between HTS and PMG. If the anisotropy of critical current density is taken into account, the calculated results of the two dimension model qualitatively agree with the experimental results in the both symmetrical and unsymmetrical applied elds, while quantitative disagreement still exists. Therefore it is necessary to build up a three dimension model to describe the electromagnetic properties of YBCO superconductor for the further research.

4 Outlook

History has shown that every new advance in communication, from telegraph to television, has been met with an increased demand for travel [78], particularly, the safe, comfortable, environmental transportation with high speed up to several hundred kilometers per hour. Melt textured $YBa_2Cu_3O_{7-\delta}$ superconductor has been widely used in the eld of the HTS Maglev transportation system. The long-term stability since Jan. 2001 has be con rmed by the experiments on the rst people-carrying HTS test vehicle. Moreover, HTS wire technology is progressing from the present-generation BSSCO to the second-generation of coated conductor. The major challenge that remains before HTS sees signi cant use in Maglev technology is reduction of the cost of fabrication, but efforts are ongoing [91, 92, 93]! Actual industrial implementation of people-carrying HTS Maglev vehicle is in the foreseeable future.

Acknowledgment

Discussions with Zheng, J.; Wang, X.Z.; Beyer, C.; Ma, G.T.; Zhang, L.C.; Lu, Y.Y.; Liu, M.X.; Qin, Y.J.; Huang, Y.G.; Deng, Z.G. and Zhang, J.H. are much appreciated. The authors are grateful to Prigozhin, L. and Bossavit, A. for providing with the helpful references. The authors are grateful to Zhao, Y., Muller, K.H. and Krabbes, G. for providing with a great deal of bene cial discussions. We wish to acknowledge the support from the material staff in IFW Dresden Germany, providing and fabricating samples, offering suggestions and encouragement, testing measurement apparatus. Anyway, the authors would like to thank all the persons who contributed to the chapter. This work is supported by the National High Technology Research and Development Program of China (2005AA306150), the National Natural Science Foundation in China (50377036) and the Specialized Research Fund for the Doctoral Program of Higher Education (0613009). This work is also one part of the PPP project, which is funded by China Scholarship Council and Deutscher Akademischer Austauschdienst.

References

[1] Fuchs, G.; Gruss, S. and et.al. *Physica C*, 2002, 372-376, pp 1131-1133.

[2] Tomita, M.; Murakami, M. *Nature*, 2003, 421, pp 517-520.

[3] Leendes, A.; Walter, H. and et. al. *IEEE Trans. Appl. Supercond.*, 2001, 11, pp 3728-3731.

[4] Murakami M. *Melt Processed High-Temperature Superconductor*; World Scienti c Publisher: Singapore, 1992, pp 66.

[5] H Johansen, T. *Supercond. Sci. Technol.*, 2000, 13, pp R121CR137.

[6] Bakke, P. (2005). The Superconductivity Partnership Initiative (SPI), including the Boeing - Phase II HTS Flywheel Program [pdf].
URL: http://www.energetics.com/meetings/wire05/pdfs/session8/session8d.pdf.

[7]   Wang, J.S.; Wang, S.Y.; Zeng, Y.W.; Deng, C.Y.; Wang, X.R.; Song, H.H.; Wang, X.Z.; Zheng, J. *Supercond. Sci. Technol.*, 2005, 18, pp S215-S218

[8]   Wang, S.Y.; Wang, J.S. and et. al. *IEEE Trans. Appl. Supercond.*, 2003, 13, pp 2134.

[9]   Wang, J.S.; Wang S.Y.; et.al. *Physica C*, 2002, 378-381, pp 809.

[10] Putman, P.T.; Zhou, Y.X.; Fang, H.; Klawitter, A; Salama, K. it Supercond. Sci. Technol., 2005, 18, pp S6-S9.

[11] Stumberger, G.; Timur Aydemir, M.; ?arko, D.; A. Lipo, T. *IEEE Trans. Appl. Supercond.*, 2004, 14, pp 54-62.

[12] Hull, J.R. *Rep. Prog. Phys.*, 2003, 66, pp 1865-1886.

[13] Zhou, Y.X.; Fang, H.; Balachandran, U.; Salama, K. *IEEE Trans. Appl. Supercond.*, 2003, 13, pp 3072 -3075.

[14] see: http://www.transrapid.com.cn/

[15] Sakamoto, S.; Watanabe,H.; Takizawa,T.; Suzuki, E.; Terai, M. *IEEE Trans. Appl. Superconduct.*, 1997, 7, pp 3791-3796.

[16] Hull, J.R.; *Supercond. Sci. Technol.*, 2000, 13, pp R1.

[17] Ma, K.; Postrekhin, Y.; Chu, W.K. *Rev. Sci. Instrum.*, 2003, 74,pp 4989.

[18] Weh, H. in *Proc. 15th Int. Conf. on Magnet Technology Part 2 (Beijing)*; Lin, L.Z.; Shen, G. L.; Yan, L. G.; Ed.; Part 2; Science Press:Beijing; 1998, vol 2, pp 833-838.

[19] Fujimoto, H.; Kamijo, H.; et. al, *IEEE Trans. Appl. Superconduct.*, 1999, 9, pp 301.

[20] Song, H.H.; De Haas, O.; Beyer, C.; Krabbes, G.; Verges, P.; Schultz, L. *Appl. Phys. Lett.*, 2005, 86, pp 192506.

[21] Song, H.H.; De Haas, O.; Ren, Z.Y.; Wang, X.R.; Zheng, J.; Wang, X.Z.; Wang, S.Y.; Wang, J.S.; Zhao, Y. *Physica C*, 2004, 407, pp 82-87.

[22] Song, H.H.; Wang, J.S.; Wang, S.Y.; De Haas, O.; Ren, Z.Y.; Wang, X.R.; Wang, X.Z.; Zheng, J.; Zhao, Y. *Supercond. Sci. Technol.*, 2005, 18, pp S95-S98

[23] Prigozhin,L. *Europe Journal of Applied Mathematics*, 1996, 7, pp 237-247

[24] Wang, J.S. Wang, S.Y.; Ren, Z.Y.; Wang, X.R.; Zhu, M.; Jiang, H.; Song, H.H.; Wang, X.Z.; Zheng, J. *IEEE Trans. Appl. Superconduct.*, 2003, 13, pp 2154-2156.

[25] Wang, J.R.; Zhang, P.X. Zhang; Zhou, L.; Wu, M.Z.; Gawalek, W.; Gornert, P.; Weh, H. in *Proc. 15th Int. Conf. on Magnet Technology Part 2 (Beijing)*; Lin, L.Z.; Shen, G. L.; Yan, L. G.; Ed.; Part 2; Science Press:Beijing; 1998, vol 2,

[26] Wang, S.Y.; Wang, J.S.; Ren, Z.Y.; Zhu, M.; Jiang, H.; Wang, X.R.; Shen, X.M.; Song, H.H.*Physica C*, 2003, 386, pp 531-5.

[27] Schultz, L; De Haas, O.; Verges, P.; Beyer, C.; Rohlig, S.; Olsen, H.; Kuhn, L; Berger, D.; Noteboom, U. and Funk, U. *IEEE Trans. Appl. Superconduct.*, 2005, 15, pp 2301-2305.

[28] Wang, X.R.; Wang, J.S.; Wang, S.Y.; et.al. *Physica C*, 2003, 390, pp 113-115.

[29] Takebayashi, S.; Murakami, M. *Materials Science and Engineering*, 1999, B65, pp 17-25.

[30] Surzhenko, A.B. and et. al. *Physica C*, 2002, 327-376, pp 1212-1215.

[31] Itoh, Y.; Mizutan, U.; *Jpn. J. Appl. Phys.*, 1996, 35, pp 2114-2125.

[32] Sander, M.; Sutter, U.; Klaser, M.; it Super. Sci. Technol., 2002, 15, pp 748-753.

[33] Sander, M.; Sutter, U.; Koch, R.; Klaser, M.; it Super. Sci. Technol., 2001, 13, pp 841-845.

[34] Itoh, Y.; Yanagi, Y.; *J. Appl. Phys.*, 1997, 82,pp 5600-5611.

[35] Ren, Y.; Weinstein, R.; Liu, J.; Sawh, R.P.; Foster, C. 1995, *Physica C*, 251, pp 15-26.

[36] Xiao, L.; Ren, H.T.; Jiao, Y.L.; Zheng, M.H.; Chen, Y.X. *Physica C*, 2003, 386, pp 262-265.

[37] Diko, P.; Zmorayova, K. and et. al. *Physica C*, 2003, 384, pp 125-129.

[38] Ren, Z.Y.; Wang, J.S; et. al, *Physica C*, 2002, 378-381, pp 873-6.

[39] Wang, X.R.; Ren, Z.Y.; Song, H.H.; Wang, X.Z.; Zheng, Jun; Wang, S.Y.; Wang, J.S. and Zhao, Y. Supercond. Sci. Technol.(2005), 18, pp S99-S104.

[40] Gawalek, W.; Schuppel W.; et. al, *Supercond. Sci. Technol.*, 1992, 5, pp S407.

[41] Ding, S.Y.; Lin J.; et al,*Supercond. Sci. Technol.*, 1992, 5, pp S240.

[42] Salama, K.; Selvamanickam, V.; et. al,*Supercond. Sci. Technol.*, 1992, 5, pp S85.

[43] Zhukov, A.A.; Perkins, G.K.; Bugoslavskyand, Y.V.; Caplin, A.D.; *Phys. Rev. B* , 1997, 56, pp 2809.

[44] Zhukov A.A.; Perkins G. K.; et. al, *Phys. Rev. B* , 1997, 56, pp 3481.

[45] Daumling, M.; Larbalestier, D.C. *Phys. Rev. B* , 1991, 40, pp R9350.

[46] Fisher, L.M.; Kalinov, A.V. *Physica C*, 2001, 350, pp 152.

[47] Xing W.; Heinrich B.; et. al,*J. Appl. Phys.*, 1994, 76, pp 4244.

[48] Iliescu, S.; Sena, S.; et. al, *IEEE Trans. Appl. Supercond.*, 2003, 13, pp 3136.

[49] Blatter, G.; Geshkenbein, V.B.; Larkin, A.I. *Phys. Rev. Lett.*, 1992, 68, pp 875.

[50] Shlyk, L.; Krabbes, G.; et. al, *Appl. Phys. Lett.*, 2002, 81, pp 5000.

[51] Itoh, Y.; Yanagi, Y.; *J. Appl. Phys.*, 1997, 82, pp 5600.

[52] Fuchs, G.; Krabbes, G.; Mulle, r K.H.; et. al, *J. Low Temp. Phys.*, 2003, 133, pp 159.

[53] Gyorgy, E. *Appl. Phys. Lett.*, 1989, 55, pp 283.

[54] Clem, J.R. *Supercond. Sci. Technol.*, 1998, 11, pp 909.

[55] Wang, S.Y.; Wang, J.S.; Ren, Z.Y.; Jiang, H.; Zhu, M.; Tang, Q.X. *IEEE Trans. Appl. Supercond.*, 2001, 11, pp 1808-1811.

[56] Moon, F.C.; Weng, K.C.; et.al. *J. Appl. Phys.*, 1989, 66, pp 5643.

[57] Riise, A.B.; H Johansen, T.; Bratsberg, H. it Physica C, 1994, bf 234, pp 108-114.

[58] H Johansen, T.; Bratsberg H. *J. Appl. Phys.*, 1993, **74**, pp 15.

[59] Sanchez, A.; Navau, C. *Physica C*, 1996, 268, pp 46-52.

[60] Wang, J.S.; Wang, S.Y.; Ren, Z.Y.; Zhu, M.; Jiang, H.; Tang, Q.X. *IEEE Trans. Appl. Supercond.*, 2001, **11**, pp 1801-4.

[61] Nagashima, K.; Otati, T.; Murakami, M. *Physica C*, 1999, **328**, pp 137-44.

[62] Ren, Z.Y.; Wang, J.S.; Wang, S.Y.; Zhu, M.; Jiang, H.; Wang, X.R.; Shen, X.M. Song, H.H. *Chinese Journal of Low Temperature Physics*, 2002, **24**, pp 293-297.

[63] Chang, P.Z.; Moon, F.C. *J. Appl. Phys.*, 1990, 67, pp 1.

[64] Murakami,M. *Appl. Supercond.*, 1, 1993, pp 1157

[65] Yang, W.M.; Zhou, L.; Feng, Y.; Zhang, P.X.; Zhang, C.P.; Yu, Z.M.; Tang, X.D.; Nicolsky, R.; Andrade Jr. R. *Cryogenics*, 2002, 42, pp 589-592.

[66] Schatzle, P.; Krabbes, G.; Stover, G.; Fuchs, G.; Schlafer,D. *Supercond. Sci.Technol.*, 1999, 12, pp 69.

[67] Jee, Y.A.; Kim, C.-J.; Sung, T.H.; Hong, G.W. *Supercond. Sci. Technol.*, 2000, 13, pp 195

[68] Gombos, M.; Vechione, A.; et. al. *IEEE Trans. Appl. Supercond.*, 2001, **11**, pp 3517-3520.

[69] Hilgenkamp, H.; Mannhart, J. *Reviews of Modern Physics*,2002, 74, pp 485-529.

[70] Joong, C.; Kim, H.J.; Joo, J.H.; Hong, G.W.; et.al. *Physica C*, 2000, **336**, pp 233-238.

[71] Hikihara, T.; Isorumi, G. *Physcica C*, 1996, 270, pp 68-74.

[72] Navau, C.; Sanchez,A.; Pardo, E.; Chen, D.X. *Supercond. Sci. Technol.*, 2004, pp 828.

[73] Riise, A.; Johansen, T.; Bratsberg, H. *Appl. Phys. Lett.*, 1992, pp 2294.

[74] Smolyak, B.; Perel'sshtein, G.; Ermakov, G. *Tehnical Physics Letter*, 2001, pp 21.

[75] Postrekhin, E.; Ma, K.; Ye, H.; Chu, W. *IEEE Trans. Appl. Superconduct.*, 2001, 11, pp 1984.

[76] Yeshurun, Y.; Malozemoff, A.P.; Shaulov, A. *Reviews of Modern Physics*, 1996, 68, pp 911-941.

[77] Qin, M.; Li, G.; Dou, S. *Phys. Rev. B.*, 2002, 66, pp 024516.

[78] Moon, F.C. *Superconducting levitation: Applications to Bearings and Magnetic Transportation*; JOHN WILEY: New York, 1994; pp 213.

[79] Hull, J.; Mulcahy, T.; Labataille, J. *Appl. Phys. Lett.*, 1997, 70, pp 655.

[80] Li, G.; Qin, M. J.; et. al. *IEEE Trans. on Appl. Supercon.*, 2003, 13, pp 2142.

[81] Chen, D.X.; et. al. *J. Appl. Phys.*, 1990, 67, pp 3430.

[82] Chen, D.X.; et. al. *Phys. Rev. B*, 1990, 41, pp 9510.

[83] Song, H.H.; Ren, Z.Y.; Zheng, J.; Wang, X.Z.; Wang, S.Y.; Wang, J.S. *International Journal of Modern Physics B*, 2004, 19, pp 395-397

[84] Mikitik, G. P.; Brandt, E.H. *Phys. Rev. B*, 2000, 62, pp 6800.

[85] Prigozhin, L.; *Europ. J. Appl. Math*, 1996, 7, pp 237.

[86] Bossavit, A. *IEEE Trans. Magn.*, 1994, 30, pp 3363.

[87] Prigozhin, L. *IEEE Trans. Appl. Supercond.*, 1997, 7, pp 3866-3873.

[88] Sawamura, M.; Tsuchimoto, M. *Japan Journal of Industrial and applied mathmetics*, 2000, 17(2), pp 199-208.

[89] Morris, R.C.; Coleman R.V.; Bhandari R. *Phys. Rev. B*, 1972, 5, pp 895.

[90] Visual Numerics Inc. IMSL(R) Math/Library, 1999, 1,2, pp 982-984 (With Visual Compaq FORTRAN Compilor)

[91] Wang, J.S.; Wang, S.Y.; Zeng, Y.W., Deng, C.Y., Song, H.H., Zheng, J., Wang, X.Z.; Huang, H.Y.; Li, F; *IEEE Trans. Appl. Supercond.*, 2005, 15, pp 2273-2276.

[92] Wang, S.Y.; Zheng, J.; Song, H.H.; Wang, X.Z.; Wang, J.S.*IEEE Trans. Appl. Supercond.*, 2005, 15, pp 2277-2280.

[93] Haugan, T.; Barnes, P.N.; Wheeler, R.; Meisenkothen, F.; Sumption, M. *Nature*, 2004, 430, pp 867-870.

[94] Grilli, F.; Stavrev, S.; Le Floch, Y.; Costa-Bouzo, M.; Vinot, E.; Klutsch, I.; Meunier, G.; Tixador, P.; Dutoit, B. *IEEE Trans. Appl. Supercond.*, 2005, 15, pp 17-25.

In: New Topics in Superconductivity Research
Editor: Barry P. Martins, pp. 157-193

ISBN: 1-59454-985-0
© 2006 Nova Science Publishers, Inc.

Chapter 5

STUDY OF HIGH TEMPERATURE SUPERCONDUCTOR

Liang FangYing[1]

School of Mechanics, Architecture and Civil Engineering, China University of Mining
and Technology, 100085 Beijing. P.R. China.
Institute of Science and Technology for OPTO-Electronic Information of Yan Tai
University, 264005 Yan Tai, P.R. China, March 8, 2005

Abstract

This chapter discusses five important issues:

1. **Anomalous transport characteristics of high temperature superconductors and Josephson currents**

The electric currents of superconductor and electrical field are relation of direct proportion; the currents and magnetic field are relation of inverse ratio. In a special condition, the Josephon currents has anomalous characteristic.

2. **Thermodynamic properties of high temperature superconductor**

A new systematic calculation of the specific heat contributions of vortex liquids and solids is presented. Three derivatives of the free energy with respect to the temperature of superconductor, the entropy, the specific heat, the temperature of superconductor derivative of the specific heat are continuous across the phase transition.

3. **The study of characteristics of superconductive rings**

The current of superconductive rings is change with jump in theory. The magnetic field of superconductive rings is quantization. If increasing magnetic field, the order parameter is gradually decreasing, leads to a decrease of the size of the jump of the flux in the vorticity. In a special condition, if the outer magnetic field is gathering, the sign of supercurrent can reversal.

4. **Study of thermodynamic properties of type I superconductive films**

The specific heat of the type I superconductive films

$$C_V = 2a_0^3 \left\{ 5\left(ba_0 - \frac{2k_B \rho_0^{3/2}}{3L_0} \right) T^3 + 6\left(1 - 2ba_0 T_{c0} + \frac{k_B T_{C0} \rho_0^{3/2}}{L_0} \right) T^2 \right.$$

---

[1] E-mail address yangming88@eastday.com

$$-9\left(T_{C0} - ba_0 T_{C0}^2 + \frac{k_B T_{C0}^2 \rho_0^{3/2}}{3L_0}\right)T + 3T_{C0}^2 - 2ba_0 T_{C0}^3 + \frac{k_B T_{C0}^3 \rho_0^{3/2}}{3L_0}\right\}.$$

5. Study of high temperature superconductor under pressure

When outer pressure is a constant on superconductor, the pressure intensity with the temperature is the relation of quadratic curve. The temperature is increasing with the pressure intensity. When outer pressure on superconductor is not a constant, the external pressure intensity has a relation of partial differential equation with the temperature of superconductivity. As increasing the external pressure intensity, the temperature is rising. The critical temperature is decreasing quasi-linearly with applied hydrostatic pressure for superconductor, and observed negative pressure coefficient of the critical temperature of superconductor. In another special case, we obtain the critical temperature increases quasi-linearly with applied pressure on superconductor.

1 Anomalous Transport Characteristics of High Temperature Superconductors and Josephson Currents[1]

Abstract

The electric currents of superconductor and electrical field are relation of direct proportion; the currents and magnetic field are relation of inverse ratio. In a special condition, the Josephon currents has anomalous characteristic.

Keywords: Superconductivity, Junction, Josephson Current.
PACS numbers: 74.60.Jg 74.80.Dm 74.72.-h 74.50.+r

1.1 Introduce

The study of vortex motion in type II superconductors continues to attract the attention of theorists and experimentalist alike, due in part to the rather unusual mixed state transport properties of the high temperature superconductors (HTSC)[2-11]. One of many vexing of these properties is the anomalous behavior of the transport properties, which is observed to change sign in the superconductivity mixed state. However, it is important for the direction of Josephon junctions of anisotropic (layer) superconductors. The transport properties of Josephson currents and artificial craft have been a very challenged for theorists and experimentalist, applied technology. Previous works have had a lot of work for study superconductors and study Josephon junctions[3-11]. Here, we consider the time dependant Ginzburg-Landau modified model and the model of Lawrence-Doniach to take into account the transport characteristics of high temperature superconductor and Josephson junctions[13,14,15]. We make use of the models and evaluate the vortex motion equations. We focus our discussion on the anomalous characteristics for high temperature superconductor currents and Josephson currents.

1.2 Model and Modified TDGL Equations

Because of high temperature superconductor can be regarded as stacker of the region of strong-superconductor alternates with region of weak-superconductor. The study of high temperature superconductors can be fitted by the time-dependant Ginzburg-Landau (TDGL) theory. Layered model of HTSC are rationale of deal with layered system. Our starting point is modified model of TDGL; we suppose order parameter $\Psi_n(\mathbf{r},t)$ is the same in every layer, this deviation of order parameters is constant between each layer. Substitute wave function $\Psi_n(\mathbf{r},t)$ in each layer for $\Psi(\mathbf{r},t)$ of continuous complex order parameter in G-L theory (here,

$\Psi(\mathbf{r},t) = \sum_n \Psi_n(\mathbf{r},t), \mathbf{z}_n = ns, n$ is layer's number and positive round number, s is

distance between two layers.). The Ginzburg-Landau free energy of a layered super-conductor[13,14,15] is

$$F = \int d^2\mathbf{r}\; s\sum_n \left\{ a|\Psi_n(\mathbf{r},t)|^2 + \frac{1}{2}b|\Psi_n(\mathbf{r},t)|^4 + \frac{\hbar^2}{2m}\left|(-i\frac{\partial}{\partial r} - \frac{2e}{c}\mathbf{A})\Psi_n(\mathbf{r},t)\right|^2 \right.$$

$$\left. + \frac{1}{2Ms^2}\left| \Psi_{n+1}(\mathbf{r},t)\exp[-\frac{2ie}{c}\int_{ns}^{(n+1)s}d\mathbf{z}\mathbf{A}_z] - \Psi_n \right|^2 + \frac{\mathbf{H}^2}{8\pi} \right\}, \tag{1}$$

TDGL Equation is
$$\left[\partial_t + \frac{i\,\tilde{\mu}}{\hbar}\right]\Psi = -\Gamma\frac{\delta F}{\delta\Psi^*}. \tag{2}$$

Here, $a = a_0(T/T_C - 1)$, a_0 being a constant; T expresses temperature, while T_C is critical temperature; b is a constant[15]. The parameters $a.b.M$ are expressed through the microscopic characteristics. $e = 1.6\times10^{-19}$ coulomb; the z axis is along c axis of the crystal; the two-dimensional vectors $\mathbf{r} = (\mathrm{x,y})$ are components in the ab plane; m is the effective mass of quasi-particles moving in the ab plane; M is the effective Ginzburg-Landau mass; $\mathbf{A}$ is the vector potential, $\mathbf{H}$ is magnetic field, $\mathbf{B} = \langle\mathbf{h}\rangle$ is induction field with $\mathbf{h} = \nabla\times\mathbf{A}$ the microscopic magnetic induction field. $\Gamma = \Gamma_1 + i\Gamma_2$ is a complex dimensionless relaxation rate; $\tilde{\mu}$ is the total chemical potential, $\tilde{\mu} = \mu + 2e\Phi + \delta F/\delta n_s$, Φ is electric potential, μ is chemical potential energy, $\delta F/\delta n_s$ is supercurrent kinetic-energy, $n_s = |\Psi|^2$ is supercurrent density; the order parameter $\Psi(\mathbf{r},t) = \sum_n \Psi_n(\mathbf{r},t)$; n is the layer number. The order parameter Ψ_n depends on the layer number n and the point $\mathbf{r}$ on it. We neglect the anisotropy of Ψ_n in the ab plane[13,14,15]. The Ψ_n is usually expressed for $\Psi_n = f_n\exp[i\chi_n(\mathbf{r},t)]$. ($f_n$ is an amplitude. Note that a moving vortex does not possess cylindrical symmetry, so that the phase variable χ_n is equal to the angular variable only near the center of the vortex.).

Now, we defining a dimensionless order-parameter relaxation time $\gamma = \gamma_1 + i\gamma_2$ $= [\Gamma_1 - i(1+\Gamma_2)]/[\Gamma_1^2 + (1+\Gamma_2)^2]$, we make use of modified model of TDGL equations and take L-D free energy into the equations. Our order-parameter equations of motion becomes[17]

$$\hbar\gamma\left[\partial_t + i\frac{2e}{\hbar}\tilde{\Phi}\right]\Psi_n(\mathbf{r}) = \left\{a\Psi_n + b|\Psi_n|^2\Psi_n + \frac{\hbar^2}{2m}(-i\frac{\partial}{\partial\mathbf{r}} - \frac{2e}{c}\mathbf{A})^2\Psi_n\right.$$

$$\left. - \frac{1}{2Ms^2}\left[\Psi_{n+1}(\mathbf{r})\exp[-\frac{2ie}{c}\int_{ns}^{(n+1)s}dz\,\mathbf{A_z}] - \Psi_n\right]\right\}, \tag{3a}$$

$$\hbar\gamma\left[\partial_t + i\frac{2e}{\hbar}\tilde{\Phi}\right]\Psi_{n+1}(\mathbf{r}) = \left\{a\Psi_{n+1} + b|\Psi_{n+1}|^2\Psi_{n+1} + \frac{\hbar^2}{2m}(-i\frac{\partial}{\partial\mathbf{r}} - \frac{2e}{c}\mathbf{A})^2\Psi_{n+1}\right.$$

$$\left. - \frac{1}{2Ms^2}\left[\Psi_n(\mathbf{r})\exp[-\frac{2ie}{c}\int_{(n+1)s}^{ns}dz\,\mathbf{A_z}] - \Psi_{n+1}\right]\right\}. \tag{3b}$$

Here, $\tilde{\Phi} = \Phi + \mu/2e$, we would do not consider differ[14] of $\tilde{\Phi}$ and Φ.

We also require an equation of motion for the vector potential, which is Ampère's law $\nabla\times\nabla\times\mathbf{A} = 4\pi(\mathbf{J_n} + \mathbf{J_s})$, so that $\nabla\bullet(\mathbf{J_n} + \mathbf{J_s}) = 0$. The supercurrent $\mathbf{J_s}$ is given by $\mathbf{J_s} = \frac{2e\hbar}{2mi}(\Psi^*\nabla\Psi - \Psi\nabla\Psi^*) - \frac{(2e)}{m}|\Psi|^2\mathbf{A}$, the supercurrent velocity is $\mathbf{V}_s = \mathbf{J_s}/2e|\Psi|^2$. while the normal current $\mathbf{J_n}$ is given by $\mathbf{J_n} = \boldsymbol{\sigma}^{(n)}\bullet\mathbf{E} = \boldsymbol{\sigma}^{(n)}\bullet(-\nabla\Phi - \partial_t\mathbf{A})$, with $\boldsymbol{\sigma}^{(n)}$ the normal-state conductivity tensor $\boldsymbol{\sigma}^{(n)} = \begin{bmatrix} \sigma_{xx}^{(n)} & \sigma_{xy}^{(n)} \\ \sigma_{yx}^{(n)} & \sigma_{xx}^{(n)} \end{bmatrix}$. The Onsager relations and rotational symmetry imply that $\sigma_{xy}^{(n)}(\mathbf{H}) = -\sigma_{yx}^{(n)}(\mathbf{H})$, so that the conductivity tensor may be decomposed into a diagonal piece and an anti-symmetric piece. The longitudinal normal-state conductivity $\sigma_{xx}^{(n)}$ is generally a weak function of magnetic field, $\sigma_{xy}^{(n)}$ is Hall conductivity.

1.3   Vortex Equation of Motion in the Limit $\mathbf{B} \ll \mathbf{H_{c_2}}$

We have considered a serious of approximate-evaluating, have considered true number evaluate of a time-dependant order parameter in vortex equations. We use the same method of Ref. [14,16] for equations (3a)&(3b) and obtain

$$\frac{1}{k^2}\frac{1}{r}\frac{d}{dr}[r\frac{df_{n,0}}{dr}] - Q_0^2 f_{n,0} + f_{n,0} - f_{n,0}^3 + \frac{1}{2Ms^2a}f_{n,0} - \frac{1}{2Ms^2a}f_{(n+1),0}$$

$$\times\cos[\chi_{n+1} - \chi_n - \frac{2e}{c'}\int_{ns}^{(n+1)s}dz\mathbf{A_z}] = 0 , \tag{4a}$$

$$\frac{d}{dr}\frac{1}{r}\frac{d(rQ_0)}{dr} - f_{n,0}^2 Q_0 = 0 , \tag{5a}$$

$$\frac{\sigma_{xx}^{(n)}}{k^2}\frac{d}{dr}\frac{1}{r}\frac{d(rP_1)}{dr}-\gamma_1 f_{n,0}^2 P_1 -\frac{1}{2Ms^2a}(f_{n,0}f_{n+1,0}+f_{n,0}f_{n+1,1}+f_{n,1}f_{n+1,0})$$

$$\times \sin[\chi_{n+1}-\chi_n-\frac{2e}{c'}\int\limits_{ns}^{(n+1)s}dz\mathbf{A_z}]=0 , \tag{6a}$$

$$\frac{\sigma_{xx}^{(n)}}{k^2}\frac{d}{dr}\frac{1}{r}\frac{d(rP_2)}{dr}-\gamma_1 f_{n,0}^2 P_2 -\frac{1}{2Ms^2a}(f_{n,0}f_{n+1,0}+f_{n,0}f_{n+1,1}+f_{n,1}f_{n+1,0})$$

$$\times \sin[\chi_{n+1}-\chi_n-\frac{2e}{c'}\int\limits_{ns}^{(n+1)s}dz\mathbf{A_z}]=\gamma_2 f_{n,0}\frac{\partial f_{n,0}}{\partial r}-\frac{\sigma_{xy}^{(n)}}{k}\frac{\partial h_0}{\partial r} , \tag{7a}$$

$$\frac{1}{k^2}\frac{1}{r}\frac{d}{dr}[r\frac{df_{n+1,0}}{dr}]-Q_0^2 f_{n+1,0}+f_{n+1,0}-f_{n+1,0}^3+\frac{1}{2Ms^2a}f_{n+1,0}-\frac{1}{2Ms^2a}f_{n,0}$$

$$\times \cos[-\chi_{n+1}-\chi_n-\frac{2e}{c'}\int\limits_{ns}^{(n+1)s}dz\mathbf{A_z}]=0 , \tag{4b}$$

$$\frac{d}{dr}\frac{1}{r}\frac{d(rQ_0)}{dr}-f_{n+1,0}^2 Q_0 =0 , \tag{5b}$$

$$\frac{\sigma_{xx}^{(n)}}{k^2}\frac{d}{dr}\frac{1}{r}\frac{d(rP_1)}{dr}-\gamma_1 f_{n+1,0}^2 P_1 -\frac{1}{2Ms^2a}(f_{n,0}f_{n+1,0}+f_{n,0}f_{n+1,1}+f_{n,1}f_{n+1,0})$$

$$\times \sin[\chi_n-\chi_{n+1}+\frac{2e}{c'}\int\limits_{ns}^{(n+1)s}dz\mathbf{A_z}]=0 , \tag{6b}$$

$$\frac{\sigma_{xx}^{(n)}}{k^2}\frac{d}{dr}\frac{1}{r}\frac{d(rP_2)}{dr}-\gamma_1 f_{n+1,0}^2 P_2 -\frac{1}{2Ms^2a}(f_{n,0}f_{n+1,0}+f_{n,0}f_{n+1,1}+f_{n,1}f_{n+1,0})$$

$$\times \sin[\chi_n-\chi_{n+1}+\frac{2e}{c'}\int\limits_{ns}^{(n+1)s}dz\mathbf{A_z}]=\gamma_2 f_{n+1,0}\frac{\partial f_{n+1,0}}{\partial r}-\frac{\sigma_{xy}^{(n)}}{k}\frac{\partial h_0}{\partial r} , \tag{7b}$$

$$\mathbf{h}_0 = \frac{1}{r}\frac{\partial[r\mathbf{Q}_0]}{\partial r} . \tag{8}$$

Here, c' is a constant, $P \equiv \Phi + \partial_t\chi$, $k = l/\xi$ is a parameter of Ginzburg-Landau, $l - [mb/4\pi(2e)^2|a|]^{1/2}$ is magnetic penetration depth, $\xi = \hbar/(2m|a|)^{1/2}$ is coherence length, $f_n = f_{n,0}+f_{n,1}$、 $f_{n+1}=f_{n+1,0}+f_{n+1,1}$, $f_{n,v}\equiv \mathbf{V_L}\bullet\nabla f_{n,0}$, $f_{n+1,v}\equiv \mathbf{V_L}\bullet\nabla f_{n+1,0}$, $\mathbf{V_L}$ is vortex line velocity. $\mathbf{d}$ is an infinitesimal translation vector, $f_{n,d} = \mathbf{d} \bullet \nabla f_{n,0}$, $f_{n+1,d} = \mathbf{d} \bullet \nabla f_{n+1,0}$, $\mathbf{Q} \equiv \mathbf{Q_0} + \mathbf{Q_d}$ $\mathbf{Q_d} = (\mathbf{d} \bullet \nabla)\mathbf{Q_0}$, $\mathbf{E} = -\nabla P\,k - \partial_t \mathbf{Q}$ is electric field, $\mathbf{h_0} = \nabla \times \mathbf{Q_0}$ is microscopic magnetic induction field, $\mathbf{Q} \equiv \mathbf{A} - \nabla \chi / k$, $\mathbf{P(r)} = \mathbf{V_L}[P_1(r)\cos(\theta - \theta_H) + P_2(r)\sin(\theta - \theta_H)]$.

Notice that as $r \to 0$, the scalar and vector potentials have the following behaviors: $Q_0(r) \approx -\dfrac{1}{kr} + h_0(0)r/2$, $p_1(r) \approx \dfrac{1}{r} - p_1^{(1)}r$, $p_2(r) \approx -p_2^{(1)}r$ .where $h_0(0)$ is the field at the center of the vortex, and where $p_1^{(1)}$ and $p_2^{(1)}$ are constants which are determined from the solution of Eqs.(6a) and (7a)[or Eqs.(6b) and (7b)].

From Ref. [14], we know

$$\mathbf{J_t} \times \mathbf{e_z} = \frac{\alpha_1 k}{2} \mathbf{V_L} + \frac{\alpha_2 k}{2} \mathbf{V_L} \times \mathbf{e_z} \; , \tag{9}$$

Here, α_1, α_2 would be explained at after, $\mathbf{e_z}$ is direction of z-axis.

1.4   Discussion HTSC Currents

It is just the same to evaluate equations (4a)-(7a) and (4b)-(7b), therefore, we only evaluate an equations; but it is impossible that we have completely evaluated the equations. Here we discuss a few of limited case.

Case I : $\chi_{n+1} - \chi_n - \dfrac{2e}{c'} \int\limits_{ns}^{(n+1)s} dz\, \mathbf{A_z} \to 0$

From equations (4a)-(7a) (8) (9), we obtain[14,16]

$$\alpha_1 = \gamma_1 \int\limits_0^\infty \left(f_{n,0}' \right)^2 rdr + \frac{2\sigma_{xx}^{(n)}}{k^2} P_1^{(1)} - \frac{\gamma_2}{2} \int\limits_0^\infty (f_{n,0}^2)' P_2 rdr \; , \tag{10}$$

$$\alpha_2 = -\frac{2\sigma_{xx}^{(n)}}{k^2} P_2^{(1)} + \frac{\sigma_{xy}^{(n)}}{k^2} h_0(0) - \frac{\gamma_2}{2} \int\limits_0^\infty (f_{n,0}^2)' P_1 rdr \; . \tag{11}$$

We choose [17] $$f_{n,0}(r) = Ms^2 a \frac{r}{[r^2 + \xi_v^2]^{1/2}} . \tag{12a}$$

The last integral of Equation (10) is generally quite small, is $0(\gamma_2^2, \gamma_2, \sigma_{xy}^{(n)})$, we will be dropped from now on. From equations (10) and (11), we can obtain

$$\alpha_1 = \frac{\gamma_1}{4\lambda^2} + \frac{\gamma_1 \xi K_0(\xi_v / \xi)}{\xi_v \lambda^2 K_1(\xi_v / \xi)} \; , \tag{10a}$$

$$\alpha_2 = -\frac{\gamma_2}{2\lambda^2} I(\xi_v/\xi) - \frac{\xi}{\xi_v} \frac{K_0(\xi_v/\xi)}{K_1(\xi_v/\xi)} [\gamma_2 + \frac{1}{k^2} \sigma_{xy}^{(n)}(0)] + \frac{1}{k} \sigma_{xy}^{(n)}(0) h_0(0). \quad (11a)$$

Here, ξ_v is a parameter that measures the healing length of the order parameter, it's numerically close to one. We choose $\lambda = 1/Ms^2 a$, $h_0(0) = \frac{1}{\xi_v \lambda} \frac{K_0(\xi_v/\lambda)}{K_1(\xi_v/\lambda)} = \frac{1}{\lambda^2} (\ln \lambda - 0.231)$, $\xi = (\sigma_{xx}^{(n)}/\gamma_1)^{1/2}$, $R = [r^2 + \xi_v^2]^{1/2}$. With $K_0(z)$ and $K_1(z)$ the standard modified Bessel functions[14] $I(z) = \frac{2z}{K_1(z)} \int_0^\infty \frac{K_1(x)}{x^2} dx$.

Finally, we make use of superfluid velocity $\mathbf{V}_s = \mathbf{J}_s/2e|\Psi|^2$ and $\mathbf{J}_s = kf_n^2 \mathbf{V}_s/2$、$\mathbf{J}_s' = kf_{n+1}^2 \mathbf{V}_s'/2$, we can write $f_{n,0} = 1$ and $\mathbf{V}_s = 2\mathbf{J}_t/k$ in the boundaries, $\mathbf{V}_{s1} \times \mathbf{e}_z = \alpha_1 \mathbf{V}_L + \alpha_2 \mathbf{V}_L \times \mathbf{e}_z$. Because of equations (9) (10a) (11a) and Faraday law $<\mathbf{E}> = -\mathbf{V}_L \times \mathbf{B}$, we can obtain[14,16,18]

$$\mathbf{J}_t = \frac{\alpha_1 k}{2B} < \mathbf{E} > + \frac{\alpha_2 k}{2B} < \mathbf{E} > \times \mathbf{e}_z ,$$

$$= \left[\left[\frac{\gamma_1}{4} + \frac{\gamma_1 \xi K_0(\xi_v/\xi)}{\xi_v K_1(\xi_v/\xi)} \right] \frac{[Ms^2 a_0(T - T_c)]^2}{T_c^2} \right] \frac{k}{2B} < \mathbf{E} > +$$

$$+ \left[-\frac{\gamma_2 [Ms^2 a_0(T - T_c)]^2}{2T_c^2} I(\xi_v/\xi) - \frac{\xi}{\xi_v} \frac{K_0(\xi_v/\xi)}{K_1(\xi_v/\xi)} \bullet [\gamma_2 + \frac{1}{k^2} \sigma_{xy}^{(n)}(0)] \right.$$

$$\left. + \frac{1}{k} \sigma_{xy}^{(n)}(0) h_0(0) \right] \frac{k}{2B} < \mathbf{E} > \times \mathbf{e}_z . \quad (13)$$

If $\gamma_2 < 0$, the two terms of the right of equation (13) have the same sign, $\mathbf{J}_t$ is regular transport characteristics of superconductivity current; If $\gamma_2 > 0$, may has $\alpha_2 < 0$, has $\mathbf{J}_t < 0$, superconductivity current take place reversal (anomalous transport characteristics of high temperature superconductor current). Some experiments have obtained superconductor current to reversal[19]. The superconductor current and $(T/T_C - 1)^2$ is relation of direct proportion; superconductor current $\mathbf{I}_{nl} \propto T^v$ is conquered in Ref. [19], here $v \to 2$.

If $T \to T_C$

$$\mathbf{J}_t = \left[-\frac{\xi}{\xi_v} \frac{K_0(\xi_v/\xi)}{K_1(\xi_v/\xi)} \bullet [\gamma_2 + \frac{1}{k^2} \sigma_{xy}^{(n)}(0)] + \frac{1}{k} \sigma_{xy}^{(n)}(0) h_0(0) \right] \frac{1}{2B} < \mathbf{E} > \times \mathbf{e}_z \quad (14)$$

The coefficient of $\frac{1}{2B} < \mathbf{E} > \times \mathbf{e}_z$ can be minus constant. $\mathbf{J}_t$ is minus, cause superconductor current to reversal.

Case II : $\chi_{n+1} - \chi_n - \dfrac{2e^{(n+1)s}}{c'} \displaystyle\int_{ns} dz \mathbf{A_z} \to \dfrac{\pi}{2}$

From equations (4a)-(7a) (8) (9), we obtain[14,16]

$$\alpha_1 = \gamma_1 \int_0^\infty \left(f'_{n,0}\right)^2 rdr + \gamma_1 \int_0^\infty (f_{n,0}^2)P_1 dr - \frac{\gamma_2}{2}\int_0^\infty (f_{n,0}^2)'P_2 rdr , \tag{10'}$$

$$\alpha_2 = -\gamma_1 \int_0^\infty (f_{n,0}^2)P_2 dr - \frac{\gamma_2}{2}\int_0^\infty (f_{n,0}^2)'P_1 rdr + \frac{1}{4Ms^2 a} . \tag{11'}$$

We choose[17] $\qquad\qquad f_{n,0}(r) = 2Ms^2 a \dfrac{r}{[r^2 + \xi_v^2]^{1/2}} .$ $\qquad\qquad$ (12')

Use the same method of Case I. We obtain

$$\mathbf{J_t} = \frac{\alpha_1 k}{2B} <\mathbf{E}> + \frac{\alpha_2 k}{2B} <\mathbf{E}> \times \mathbf{e_z}$$

$$= \left\{ \left[\gamma_1 + \frac{4\gamma_1 \xi K_0(\xi_v/\xi)}{\xi_v K_1(\xi_v/\xi)} \right] \bullet \left(\frac{Ms^2 a_0(T-T_c)}{T_c} \right)^2 + + \left[\frac{1}{2} - \frac{\gamma_2}{4\gamma_1}\frac{\pi}{2}\xi \right] \right.$$

$$\left. \bullet \frac{Ms^2 a_0(T-T_c)}{T_C} \right\} \frac{k}{2B} <\mathbf{E}> + \left\{ -2\gamma_2 I(\xi_v/\xi)\left(\frac{Ms^2 a_0(T-T_c)}{T_C} \right)^2 - \frac{\xi}{\xi_v}\frac{K_0(\xi_v/\xi)}{K_1(\xi_v/\xi)} \right.$$

$$\left. \bullet [\gamma_2 + \frac{1}{k^2}\sigma_{xy}^{(n)}(0)] + \frac{1}{k}\sigma_{xy}^{(n)}(0)h_0(0) - \frac{T_c}{4Ms^2 a_0(T-T_c)}[1 + \frac{\gamma_2}{2\gamma_1}\pi\xi_v] \right\} \frac{k}{2B} <\mathbf{E}> \times \mathbf{e_z} . \tag{13'}$$

If $\gamma_2 < 0$, the two terms of the right of equation (13') have the same sign, $\mathbf{J_t}$ is regular transport characteristics of superconductivity current; If $\gamma_2 > 0$, may has $\alpha_2 < 0$, has $\mathbf{J_t} < \mathbf{0}$, superconductivity current take place reversal (anomalous transport characteristics of high temperature superconductor current). Some experiments have obtained superconductor current to reversal[19].

When $\qquad\qquad\qquad\qquad \xi_v Ms^2 a_0(T/T_c - 1) \to \infty$

$$\mathbf{J} = C_{31}(T/T_c - 1)^2 \frac{1}{2B} <\mathbf{E}> + [-C_{32}(T/T_c - 1)^2 - C_{33}]\frac{1}{2B} <\mathbf{E}> \times \mathbf{e_z} . \tag{14'}$$

Here, C_{31}, C_{32}, C_{33} are constants, the superconductor current and $(T/T_C - 1)^2$ is relation of direct proportion. The equation [14'] can is a minus quantity, take superconductivity current to reversal (anomalous transport characteristics of HTSC current). Some experiments[19] have conquered superconductor current $\mathbf{I}_{nl} \propto T^\nu$, here $\nu \to 2$.

1.5 Discussed Josephen Currents

The dynamics of the moving Josephen lattice can described by coupled equations for the plane differences. Consider a layered superconductor in a magnetic field applied along the layers with transport current flowing along the c-axis (z-direction). Suppose two-layered superconductor is up and down superposition that they are parallel ample superposition, happen quasi-particle tunneling effect in between them. For simplify, we have used suffix u and L to division up and down two layered superconductors, their layer distance is even. We considered No. n layer of two layered superconductors and order parameter

for $\sum_n \Psi_{L,n}(\mathbf{r},t) = \sum_n f_{L,n}(\mathbf{r},t)\exp[\chi_{L,n}(\mathbf{r},t)]$ $\sum_n \Psi_{u,n}(\mathbf{r},t) = \sum_n f_{u,n}(\mathbf{r},t)\exp[\chi_{u,n}(\mathbf{r},t)]$

(here, $\chi_{u,n}(r,t)$ and $\chi_{L,n}(r,t)$ are equal to the angular variable only near the center of the vortex.).

$$\mathbf{J}_{u,t} = \frac{\alpha_{u,1}k}{2B} < \mathbf{E} > + \frac{\alpha_{u,2}k}{2B} < \mathbf{E} > \times \mathbf{e}_z \,, \tag{16}$$

$$\mathbf{J}_{L,t} = \frac{\alpha_{L,1}k}{2B} < \mathbf{E} > + \frac{\alpha_{L,2}k}{2B} < \mathbf{E} > \times \mathbf{e}_z \,. \tag{17}$$

The up and down two part of Josephon junctions have an angular variable ϕ_0 in z-axis[3,20,21,22]. ϕ_0 is an angular variable of quasi-particle vector, $\phi_0 \neq 0$ and $\mathbf{J}_{u,t} - \mathbf{J}_{L,t} \neq 0$, Josephon currents are not equal zero. Now we considered a special example $\phi_0 = \pi/2$.

When $\chi_{L,n+1} - \chi_{L,n} - \dfrac{2e}{c'}\int\limits_{ns}^{(n+1)s} dz\mathbf{A}_z \to \dfrac{\pi}{2}$ and $\chi_{u,n+1} - \chi_{u,n} - \dfrac{2e}{c'}\int\limits_{ns}^{(n+1)s} dz\mathbf{A}_z \to 0$

$$\mathbf{J}_{L,t} - \mathbf{J}_{u,t} = \frac{(\alpha_{L,1} - \alpha_{u,1})k}{2B} < \mathbf{E} > + \frac{(\alpha_{L,2} - \alpha_{u,2})k}{2B} < \mathbf{E} > \times \mathbf{e}_z$$

$$= \left\{ \left[\frac{3\gamma_1 \xi K_0(\xi_v/\xi)}{\xi_v K_1(\xi_v/\xi)} + \frac{3\gamma_1}{4} \right] \bullet \frac{[Ms^2 a_0(T - T_c)]^2}{T_c^2} + \left[\frac{1}{2} - \frac{\gamma_2}{4\gamma_1}\frac{\pi}{2}\xi_v \right] \right.$$

$$\left. \bullet \frac{T_c}{Ms^2 a_0(T - T_c)} \right\} \frac{k}{2B} < \mathbf{E} > + \left\{ - \frac{3\gamma_2 [Ms^2 a_0(T - T_c)]^2}{2T_c^2} I(\xi_v/\xi) \right.$$

$$\left. - \frac{T_c}{4Ms^2 a_0(T - T_c)}[1 + \frac{\gamma_2}{2\gamma_1}\pi\,\xi_v] \right\} \frac{k}{2B} < \mathbf{E} > \times \mathbf{e}_z \tag{18'}$$

When $T \to T_C$, Josephon currents $\mathbf{J} \propto (T/T_C - 1)^{-1}$; Mikael Fogelström[11] obtained $\mathbf{J} \propto T^{-1}$.

1.6   Comparison to Previous Work and Conclusion

We obtained some expressions of electric current density of superconductors and Josephon currents in theory. The currents of the expressions and electrical field are relation of direct proportion; the currents and magnetic field are relation of inverse ratio. Y.Z. Zhang etc have conquered superconductor current[19] $\mathbf{I}_{nl} \propto T^{\nu}$, here $\nu \to 2$; but we obtained $\mathbf{J} \propto (T/T_C - 1)^2$ in special condition.

G.B. Arnold etc have studied characteristics of Josephon currents with layer superconductors, and obtained expressions of $\mathbf{I}$, but different our format[3]. P. Samuelsson etc discussed anomalous characteristic of Josephon currents[4]; and obtained expressions with Josephon currents; but did not obtain expression with temperature T. J. Lsun, J. Gao and V.I. Marconi etc obtain conductibility depend on temperature $T^{4/3}$ from experiment[5,22]; but we obtained $\mathbf{J} \propto (T/T_C - 1)^{-1}$ in the special condition. Mikael Fogelström has obtained[11] $\mathbf{J} \propto T^{-1}$. E. Goldobin etc did not get expressions of superconductor currents with electrical field and magnetic field, only got about results and characterization of determine the nature from experiments[7]. Andreas Franz etc obtained expressions of electrical currents with electrical field and magnetic field[23]. Ian Affleck etc considered various reciprocity form of S-N-S, and obtained expressions of Josephon currents[24]; but did not get expressions with T/T_C.

References

[1]   Liang FangYing,Li Zuo-Hong, *Commun.Theor.Phys.* 2002 Vol.38(Beijing,China), No.3,p379-384;

[2]   L.G. Aslamazov and S.V. Lempitskii, Zh.Eksp. *Teor.Fiz.* 1982 Vol.82.1671 [*Sov.Phys.JETP* 1982 Vol.55.967]; S.V. Lempitskii.ibid. 1983 Vol.85.1072 [58.624(1983)]; A.D.Zaikim.ibid. 1983 Vol.84. 1560 [57.910(1983)];

[3]   G.B.Arnold et al, *Phys. Rev. B* 2000 Vol.62,661;

[4]   P.Samuelsson.et al, *Phys. Rev. B* 2000ll Vol.62,1319;

[5]   J.Lsun and J.Gao, *Phys. Rev. B* 2000ll Vol.62,1457;

[6]   R.Kleiner.et al, *Phys. Rev. B* 2000ll Vol.62,4086;

[7]   E.Goldobin et al, Phys. Rev. B 2000ll Vol.62,1414; E. Goldobin et al, *Phys. Rev. B* 2000ll Vol.62,1427;

[8]   Dong Z. C, *Acta Phys. Sin.* 1999 Vol.48 ,926 (in Chinese) ;

[9]   Dong Z. C, Chen G. B, *Acta Phys. Sin.* 2000ll Vol.49,2276 ;

[10] Shao B. et al, *Chinese Physics* 1999 Vol. 8,368;

[11] Mikael Fogelstr ö m, *Phys. Rev. B* 2000-I Vol.62,11812;

[12] M.S. Rzchowski, B.A. Davidson, *Phys. Rev. B* 2000-I Vol.62,11455;

[13] R.A. Klemm, A.Luther, M.R. Beasley, *Phys.Rev.B* 1975 Vol.12, 877;

[14] Alan J. Dorsey, *Phys. Rev. B* 1992 Vol.46,8376;

[15] B.I. Ivlev and N.B.Kopnin, *J.Low Temp.Phys.* 1989 Vol.77, Nos.5/6,413;

[16] Ling F. Y, Jiang W. Z, *Acta Phys. Sin.* 1997 Vol.46.,2431 (in Chinese)

[17] A.Schmid, Phys. *Kondon. Matter* 1966 Vol.5,302;

[18] Zhang Q. R et al, 1992 *High Temperature Superconductors*(Hangzhou: Zhe Jiang University Press) ;

[19] Y.Z. Zhang, R.Deltour et al, *Phys. Rev. B* 2000-I Vol.62, Number 17,11373;

[20] Q. Li, Y.N. Tsay et al, *Phys. Rev. Let.* 1999 Vol.83,4160;

[21] R.A. Klemm, A. Bille, C.T. Rieck etc, *J. Low Temp, Phys.* 1999 Vol.117,509; R.A. Klemm, G. Arnold, A. Bille, C.T. Rieck, and K. Scharnberg, *Int. J. Mod. Phys. B* 1999 Vol.13,3449; R.A. Klemm, C.T. Rieck and K. Scharnberg, *Phys. Rev. B* Vol.58, P1051;

[22] V.I. Marconi, S.Candia et al, *Phys. Rev. B* 2000II Vol.62, Number.6,4096;

[23] Andreas Franz et al, *Phys. Rev. B* 2000 I Vol.62, Number 1,119;

[24] Ian Affleck et al, *Phys. Rev. B* 2000   Vol.62. Number 2,1433.

2   Thermodynamic Properties of High Temperature Superconductor[1]

Abstract

A new systematic calculation of the specific heat contributions of vortex liquids and solids is presented. Three derivatives of the free energy with respect to the temperature of superconductor, the entropy, the specific heat, the temperature of superconductor derivative of the specific heat are continuous across the phase transition.

Keywords: thermodynamic properties, superconductor
PACS numbers: 74.25.Bt 74.25.Fy 74.60.-w

2.1   Introduction

As indicated earlier, the measurement of the specific heat played a central role in the development of conventional superconductors. The study of the specific heat of the high temperature superconductors provides valuable insight into behavior of superconductivity of these new and those usual systems. Many experiments show that the symmetry of the order parameter in high temperature superconductors (HTSC) is $d_{x^2-y^2}$, with a possible minor s-wave admixture [2,3-8]. The C is due to several individual contributions and $\partial C/\partial T$ agrees with the most recent data. The isotope effect suggests that the phonon-driven coupling of the electrons is important [10, 11]; but the magnitude of T_C seems to be above the limit that is expected for that mechanism [12]. Three derivatives of the free energy with respect to the temperature T, the entropy S, the specific heat C and the temperature derivative of the specific heat are continuous across the phase transition. Measurements of the specific heat can give information relevant to understand the mechanism of the vortex specific heat in the low magnetic fields attracted interests of the physicists in recent years[12,13-17]. In the present paper we consider a modified TDGL model[16--18] to study thermal properties in high temperature superconductors.

2.2   The Model

In the present paper we discuss the problem of the specific heat. Starting we used the modified TDGL theory to study thermal properties in high temperature superconductors. The study of upper critical field H_{c2} as a function of angle between the magnetic field H direction and the a-b plane shows that HTSC compounds the angular dependence $H_{c2}(\theta)$, near T_c can be fitted by the dependence which follows from the Ginzburg-Landau theory[19,20] superconductors: It seems reasonable, therefore, to use the Ginzburg-Landau theory to describe the static properties. The Ginzburg-Landau free energy of a superconductor is [17-19,20]

$$F = \int dV \left\{ a\left|\Psi(\mathbf{r})\right|^2 + \frac{1}{2}b\left|\Psi(\mathbf{r})\right|^4 + \frac{\hbar^2}{2m}\left|(-i\nabla - \frac{2e}{c}\mathbf{A})\Psi(\mathbf{r})\right|^2 + \frac{\mathbf{H}^2}{8\pi} \right\}, \qquad (1)$$

where a is a parameter of temperature-dependent; T denotes temperature, T_C is critical temperature; b is usual Landau parameter; $e = 1.6 \times 10^{-19}$ coulomb; $\mathbf{m}$ is the effective mass of quasi-particles moving; V is total volume of the superconductor; $\mathbf{A}$ is the vector potential, $\mathbf{H}$ is magnetic field; $\mathbf{B} = \langle \mathbf{h} \rangle$ is induction field, $\mathbf{h} = \nabla \times \mathbf{A}$ is the microscopic magnetic field; Ψ is the order parameter. We neglect the anisotropy in superconductor[17,18,19]. The order parameter is usually expressed for $\Psi = f\exp[i\delta(r,t)]$, f is amplitude.

From Eq.(1), we obtain

$$\frac{\delta F}{\delta \Psi^*} = \int dV \left\{ a\Psi + b\left|\Psi\right|^2 \Psi - \frac{\hbar^2}{2m}(\frac{\partial}{\partial \mathbf{r}} - \frac{2ie}{c}\mathbf{A})^2 \Psi \right\} \qquad (2)$$

A proper choice of f will allow all of the above equations to be solved exactly[20,21]. This is the method originally developed by Schmid who assumed an approximate order-parameter profile of the form[21] $f(r) = \dfrac{Kar}{[r^2 + \xi_v^2]^{1/2}}$. Where K is a constant, ξ_v is a healing length of the order parameter and numerically close to one[17].

2.3 Characterization of Thermodynamics

We know main quantity of thermodynamics from general principle of thermodynamics and statistics. For example free energy $F = U - TS$, inner energy $U = F - T\dfrac{\partial F}{\partial T}$, specific heat $C_V = \left(\dfrac{\partial U}{\partial T}\right)_V$ and entropy $S = -\dfrac{\partial F}{\partial T}$. Now take Eq.(1) and Eq.(2) into the some main quantity of thermodynamics, we have

Case one: $a = a_0\left(T/T_C - 1\right)$

$$U = F - T\frac{\partial F}{\partial \Psi^*}\frac{\partial \Psi^*}{\partial T}$$

$$= \int dV \left\{ a\left|\Psi\left(\mathbf{r}\right)\right|^2 + \frac{1}{2}b\left|\Psi\left(\mathbf{r}\right)\right|^4 + \frac{\hbar^2}{2m}\left|(-i\nabla - \frac{2e}{c}\mathbf{A})\Psi(\mathbf{r})\right|^2 + \frac{\mathbf{H}^2}{8\pi} \right\}$$

$$- \left(\frac{T}{T - T_C}\right) \bullet \int dV \left\{ a + b|\Psi|^2 - \frac{\hbar^2}{2m}(\frac{\partial}{\partial \mathbf{r}} - \frac{2ie}{c}\mathbf{A})^2 \right\}\Psi\Psi^*$$

$$= \int dV \left\{ -\frac{1}{2}b\left|\Psi\left(\mathbf{r}\right)\right|^4 + \frac{\mathbf{H}^2}{8\pi} \right\} - \left(\frac{T_C}{T - T_C}\right) \bullet \int dV \left\{ a + b|\Psi|^2 - \frac{\hbar^2}{2m}(\frac{\partial}{\partial \mathbf{r}} - \frac{2ie}{c}\mathbf{A})^2 \right\}\Psi\Psi^*, \quad (3)$$

$$S = \left(U - F\right)/T$$

$$= -\left(\frac{1}{T - T_C}\right) \bullet \int dV \left\{ a + b|\Psi|^2 - \frac{\hbar^2}{2m}(\frac{\partial}{\partial \mathbf{r}} - \frac{2ie}{c}\mathbf{A})^2 \right\}\Psi\Psi^*, \quad (4)$$

$$C_V = \left(\frac{\partial U}{\partial T}\right)_V. \quad (5)$$

where, a_0 is a constant, $\int dV\Psi_n\Psi_k^* = (Ka)^2 V\delta_{n,k}$, n and k are positive integers; when $n = k$, $\delta_{n,k} = 1$; when $n \neq k$, $\delta_{n,k} = 0$. V is the volume of the superconductor.

From Ref. [22], we have $-\dfrac{\hbar^2}{2m}(\dfrac{\partial}{\partial \mathbf{r}} - \dfrac{2ie}{c}\mathbf{A})^2 = -\dfrac{\hbar^2}{2m}\nabla^2 + \dfrac{2ie\hbar}{mc}\mathbf{A}\bullet\nabla + \dfrac{(2e)^2}{2mc^2}\mathbf{A}^2$. If r is very large, $\dfrac{r^2}{r^2 + \xi_v^2}$ is close to 1, $\dfrac{r}{r^2 + \xi_v^2}$ is close to 0. Taking all these into Eq.(3)、 Eq.(4) and Eq.(5) for evaluating in cylindrical coordinates system, we obtain

$$S = -\frac{V}{T_C}\left\{ a_0(Ka)^2 + +b(Ka_0)^4\left(\frac{T - T_C}{T_C}\right)^3 + \frac{2(\hbar e\mathbf{A}Ka_0)^2}{mc^2}\left(\frac{T - T_C}{T_C}\right) \right\}, \quad (6)$$

$$U = V\left\{-a_0\left(Ka_0\frac{T-T_C}{T_C}\right)^2 - b\frac{(Ka_0)^4(T-T_C)^4}{2T_C^4} + \frac{H^2}{8\pi} - b(Ka_0)^4\left(\frac{T-T_C}{T_C}\right)^3 - \frac{2(\hbar e\mathbf{A}Ka_0)^2}{mc^2}\left(\frac{T-T_C}{T_C}\right)\right\}, \quad (7)$$

$$C_V = -\frac{V(Ka_0)^2}{T_C^2}\left\{2a_0(T-T_C) + \frac{2b(Ka_0)^2(T-T_C)^3}{T_C^2} + 3b(Ka_0)^2\frac{(T-T_C)^2}{T_C} + \frac{2(\hbar e\mathbf{A})^2 T_C}{mc^2}\right\}, \quad (8)$$

$$\frac{\partial C_V}{\partial T} = -\frac{V(Ka_0)^2}{T_C^2}\left\{2a_0 + 6b\frac{(Ka_0)^2(T-T_C)^2}{T_C^2} + 6b(Ka_0)^2\frac{(T-T_C)}{T_C}\right\}, \quad (9)$$

Case two: $a = \alpha_0\left(T/T_C - 1\right) + \alpha_1\left(T/T_C - 1\right)^2$

We have considered the some reason with case one, $\Psi = \dfrac{Kare^{i\delta}}{[r^2 + \xi_v^2]^{1/2}}$. α_0, α_1 are modified

Landau parameters. If r is very large, we obtain

$$F = \int dV\left\{a|\Psi(\mathbf{r})|^2 + \frac{1}{2}b|\Psi(\mathbf{r})|^4 + \frac{\hbar^2}{2m}\left|\left(-i\nabla - \frac{2e}{c}\mathbf{A}\right)\Psi(\mathbf{r})\right|^2 + \frac{\mathbf{H}^2}{8\pi}\right\}$$

$$= V(Ka)^2\left[a + \frac{b(Ka)^2}{2} + \frac{2(e\hbar A)^2}{mc^2}\right] + \frac{H^2}{8\pi}V, \quad (10)$$

$$U = V(Ka)^2\left[a + \frac{b(Ka)^2}{2} + \frac{2(e\hbar A)^2}{mc^2}\right] + \frac{H^2}{8\pi}V$$

$$- \frac{TK^2aV}{T_C}\left\{3a + 2bK^2a^2 + \frac{4(e\hbar A)^2}{mc^2}\right\}\left[\alpha_0 + 2\alpha_1\left(\frac{T}{T_C} - 1\right)\right], \quad (11)$$

$$C_V = \left(\frac{\partial U}{\partial T}\right)_V$$

$$= -T\left\{\frac{\partial^2 F}{\partial a^2}\frac{\partial a}{\partial T} + \frac{\partial F}{\partial a}\frac{\partial^2 a}{\partial T^2}\right\}_V\frac{\partial a}{\partial T}$$

$$= -\frac{TVK^2}{T_C^2}\left[\alpha_0 + 2\alpha_1\left(\frac{T}{T_C} - 1\right)\right] * \left\{\left[6a + 6bK^2a^2 + \frac{4(e\hbar A)^2}{mc^2}\right] * \left[\alpha_0 + 2\alpha_1\left(\frac{T}{T_C} - 1\right)\right]\right.$$

$$\left. + \frac{\alpha_1}{T_C}\left[6a^2 + 4bK^2a^3 + \frac{8a(e\hbar A)^2}{mc^2}\right]\right\}, \quad (12)$$

$$S = -\frac{K^2 a V}{T_C}\left\{ 3a + 2bK^2 a^2 + \frac{4(e\hbar A)^2}{mc^2} \right\}\left[\alpha_0 + 2\alpha_1 \left(\frac{T}{T_C} - 1 \right) \right]. \qquad (13)$$

2.4   Discussion

Three derivatives of the free energy with respect to the temperature T, the entropy S, the specific heat C, the temperature derivative of the specific heat $\partial C/\partial T$ are continuous across the phase transition. We have obtained some formulae of the specific heat characteristic. The Eq.(6)-(9) agree with recent experimental data[4,13,24--27]. From Eq.(8)-(9), we know $C_V \propto (T/T_C - 1)^3, \partial C_V/\partial T \propto (T/T_C - 1)^2$,   Jeffrey   W.   Lynn   etc,   had   obtained $C(T) = A/T^{-2} + \alpha\, T + \beta\, T^3$ from experimental data[3]. Danil knapp etc had obtained some expressions[13,24,25] of the U and C_V, but did not got expression with T/T_C and $\mathbf{A}$.

When $T \ll T_C$, the algebraic symbol of C_V、$\partial C/\partial T$ may take place reversal. S increases with T in Eq.(6); but when T = 0 and T = T_C, S keeps unchanged. When T is constant, $C_V \propto \mathbf{A}^2$, $U \propto \mathbf{A}^2$, $U \propto \mathbf{H}^2$, $S \propto \mathbf{A}^2$. U increases with T in Eq.(7).

When $T \to 0K$, U、C_V、$\partial C/\partial T$ are constants; when $T \to T_C$, U、C_V are constants, The temperature derivative of the specific heat $\partial C/\partial T$ is continuous across the phase transition. The Eq.(8) agrees with experimental curve [4,23,24,25]. The expressions (6)-(9) agree with those expressions from the experimental data[4,23,24,25].

When $T \to 0K$, the specific heat C_V from Eq. (12) is a constant quantity; when $T \to T_C$, the specific heat C_V from Eq. (12) is $-\dfrac{TVK^2}{T_C^2}\dfrac{4(e\hbar A\alpha_0)^2}{mc^2}$.

S and C_V are functions of the temperature T in Eq.(6) Eq.(8) Eq.(12) Eq.(13). S and C_V is the relation with the temperature T. The identification of d-wave pairing symmetry is based on very general principles of group theory and the macroscopic quantum coherence phenomena of pair tunneling and flux quantization[28-31]. Therefore, it does not necessarily specify a mechanism for high-temperature superconductivity. It does, however, provide general constraints on possible models.

The superconducting order parameter should transform like the symmetry of the underlying crystal lattice. From a group-theoretic point of view, no admixture of s and d is allowed. Therefore, d-wave is pairing symmetry in our theory.

2.5   Comparison to Previous Work and Conclusion

Danil knapp etc had obtained some expressions[13,24,25] of the U and C_V, but did not got expression with T/T_C and $\mathbf{A}$. Previously people had done a lot of experimental work[4,24,25]. They had obtained many experimental curves about C_V (Ref.[4] and Ref.[19]). They have fitted some expressions from the experimental data, but universality of the resulting formulae is still absent. We consider a time-dependent Ginzburg-Landau modified model to study thermal properties in high temperature superconductors. A new systematic calculation of the specific heat contributions of vortex liquids and solids is presented. In the present paper, we make an attempt to calculate the specific heat characteristic in superconductors. We have obtained some expressions of the specific heat characteristic in theory. The expressions (6)-(9) approximately agree with the recent experimental data [4,19,23--27]. Three derivatives of the free energy with respect to the temperature T, the entropy S, the specific heat C, the temperature derivative of the specific heat $\partial C/\partial T$ are continuous across the phase transition. The Eq.(8) approximately agree with the experimental data[4,19,23-27]. Our Eq.(8) can be simply transformed for

$$C_V = -\frac{V(Ka_0)^2}{T_C^2}\left\{2a_0(T-T_C)+\left(\frac{2T}{T_C}+1\right)b(Ka_0)^2\frac{(T-T_C)^2}{T_C}+\frac{2(\hbar e\mathbf{A})^2 T_C}{mc^2}\right\}, \quad \text{that} \quad \text{is}$$

approximately agree with the experimental data [4,24--27]. We obtained not only the expressions of T/T_c, but also fit with experiments. Our work is all the better to describe the experimental phenomenon than the work of previously people.

References

[1]   Liang Fang-Ying, *Physica C* 2004 Vol. 402, Issues 1-2 , Pages 174-178 ;

[2]   D. Scalapino, *Phys. Rep.* 1995 Vol.250, p329;

[3]   D. Pines and P Monthoux, *J. Phys. Chem, Solids* 1995 Vol.56. p1651;

[4]   Jeffrey W.Lynn, 1990 *High Temperature Superconductivity*, by Spring-Verlag New York Inc (World publishing Corp) (p203-260);

[5]   Feng Y etc, *Chinese Phys* . 1999 Vol 8. No 5. p374(Overseas Edition);

[6]   Chen W Y, *Chinese Phys*. 2000 Vol 9.No 9. p680;

[7]   Zhou S P, *Chinese Phys*. 2001 Vol 10.No 6. p541

[8]   Lou M B, Jiao Z K etc, *Chinese Phys*. 2001 Vol 10.No 3. p229

[9]   S.L. Bud'ko, G.Lapertot, C. petrovic and C. E. Cunningham etc, *Phys. Rev. Lett.* 2001 Vol.86, p1877;

[10] D. G. Hinks, H. Claus, and J. D. Jorgensen, *Nature* 2001 Vol.411, p457;

[11] W. L. McMillan, *Phys. Rev.* 1968 Vol.167, p331;

[12] I.Vekhter, P.J. Hirschfeld, J.P. Carbotter and E.J. Novol, in physical phenomena at High magnetic Fields-III, edited by Z. Fisk. L. Gor'kov, and R Schrieffer (world scientific, singpore, 1999). P410;

[13] Danil knapp, Catherine Kallin, and A. J. Berlinsky, *Phys. Rev. B* 2001 Vol.64, 014502;

[14] Bianchi, R. Movshovich, M. Jaime etc, *Phys. Rev. B* 2001 Vol.64, 220504(R);

[15] Liang Fang-Ying, Li Zuo-Hong, *Commun. Theor. Phys.* (Beijing, China) 2002 Vol.38 No.3,p379-384;

[16] N.B.Kopnin, B.I. Ivlev, and V.A. Kalatsky, *J. Low Temp.Phys.* 1993 Vol.90. Nos.1/2, p1;

[17] Alan T. Dorsey, *Phys. Rev. B* 1992 Vol.46. p8376;

[18] Liang F Y, Jiang W. Z, *Acta Phys. Sin.* 1997 Vol 46.No 12, p2431(in Chinese);

[19] R.A. Klemm, A. Luther, M. R. Beasley, *Phys. Rev. B* 1975 Vol.12,p877;

[20] B.I. Ivlev and N.B. Kopnin, *J. Low Temp.Phys.* 1989 Vol.77.Nos.5/6, p413;

[21] Albert Schmid, Phys. Kondens. *Materie* 1966 Vol.5, p302-317;

[22] Ceng J Y, 1999 *Quantum Mechanics*, (Peking: science press) ;

[23] G.T. Furukawa, W.G. Saba, and M.L. Reilly, *Natl. Bur. Stand. Ref. Data Ser.* No.**18** (U.S.GPO,Washington.DC.1968). P1;

[24] J. S. Kim, J. Alwood, P. Kumar etc, 2002 *Phys. Rev. B.* Vol.65, p174520;

[25] Mathias J. Graf and A. V. Balatsky, *Phys. Rev. B.* 2000-II Vol.62 No.14, p9697;

[26] G. Nieva,E. N. Martinez,F. dela Cruz, D. A. Esparza and C. A. D'Ovidio, *Phys. Rev. B* 1987 Vol.36,p8780;

[27] K. Kumagai, Y. Nakamichi, I. Watanable, Y. Nakamura, H. Nakajima, N. Wada and P. Lederer, *Phys. Rev. Lett.* 1988 Vol.60,p724.

[28] E. Il'ichev, M.Grajcar, R,Hlubina etc; *Phys. Rev. Lett.* 2001 Vol.86, 5369;

[29] Yukio Tanaka and Satoshi Kashiwaya; *Phys. Rev. Lett.* 1995 Vol.74, 3451;

[30] C.C. Tsuei, J.R. Kirtley, C.C. Chi etc; *Phys. Rev. Lett.* 1994 Vol.73, 593;

[31] C.C. Tsuei and J.R. Kirtley, *Rev. Mod. Phys.* 2000 Vol.72, 969.

3 The Study of Characteristics of Superconductive Rings[1]

Abstract

The current of superconductive rings is change with jump in theory. The magnetic field of superconductive rings is quantization. If increasing magnetic field, the order parameter is gradually decreasing, leads to a decrease of the size of the jump of the flux in the vorticity. In a special condition, if the outer magnetic field is gathering, the sign of supercurrent can reversal.

PACS numbers: 74.20.-z, 74.25.Sv, 74.78.-w
Key words: Superconductive ring, Current, Superconductor

3.1 Introduction

The predecessors have made a lot of work for superconductor. The transporting characteristic of superconductor is one of the most interesting in superconductive research, but the superconductive electric current is one of the important aspects about the study of superconductor. Although quantization of the superconductive current was already proofed by scientist at 40 year ago, the quantization of electric current initiated intensive recent theoretical interest and experimental interest. The scientists take a lot of work to superconductive dish and superconductive ring[2,3-11]. They obtain the current with jump in superconductive ring, d-wave the symmetry and superconductive electric current quantization respectively. The current quantization of superconductive ring is existence in the traditional superconductivity and high temperature superconductivity.

We consider a time-dependant Ginzburg-Landau (TDGL) modified model to take into account the characteristics of superconductive rings. We make use of the models and evaluate the vortex motion equations of superconductive ring.

3.2   Model and Modified TDGL Equations

Before discuss vortex motion, we consider a modified model of time-dependant Ginzburg-Landau to take into account the characteristics of superconductive rings and suppose s, R_1 and R_2 are the superconductive rings thickness, inner radius and outer radius respectively . Our starting point is modified model of TDGL, we suppose order parameter $\Psi(\mathbf{r},t)$ is continuous complex order parameter in Ginzburg-Landau theory. We make use of the models and evaluate the vortex motion equations of superconductive ring. Our equation of vortex motion for the superconductive order parameter $\Psi(\mathbf{r},t)$ is[12,13,14]

$$\left[\partial_t + \frac{i\tilde{\mu}}{\hbar}\right]\Psi = -\Gamma\frac{\delta H}{\delta\Psi^*}. \tag{1}$$

With the Hamiltonian

$$H = \int d^3\mathbf{r}\left\{a|\Psi(\mathbf{r},t)|^2 + \frac{1}{2}b|\Psi(\mathbf{r},t)|^4 + \frac{\hbar^2}{2m}\left|\left(-i\frac{\partial}{\partial\mathbf{r}} - \frac{2e}{c}\mathbf{A}\right)\Psi(\mathbf{r},t)\right|^2 + \frac{1}{8\pi}(\nabla\times\mathbf{A})^2\right\}. \tag{2}$$

here, $\mathbf{A}$ is the vector potential with $\mathbf{h} = \nabla\times\mathbf{A}$, $\mathbf{h}$ is the microscopic magnetic induction field, $\mathbf{B} = \langle\mathbf{h}\rangle$ is induction field, m is the effective mass of quasi-particles moving, $a = a_0(T/T_C - 1)$, a_0 being a constant; T expresses temperature, T_C is critical temperature, b is a constant[11,14]. $e = 1.6\times10^{-19}$ coulomb, $\Gamma = \Gamma_1 + i\Gamma_2$ is a complex dimensionless relaxation rate, $\tilde{\mu}$ is the total chemical potential, $\tilde{\mu} = \mu + 2e\Phi_e + \delta F/\delta n_s$, Φ_e is electric potential, μ is chemical potential energy, $\delta H/\delta n_s$ is supercurrent kinetic-energy, $n_s = |\Psi|^2$ is density of the supercurrent. $\mathbf{r} = (x,y,z)$ is vector of crystal body, $\hbar$ is plank constant. The $\Psi = f\exp[i\chi(\mathbf{r},t)]$. ($f$ is amplitude. Note that a moving vortex does not possess cylindrical symmetry, so that the phase variable χ is equal to the angular variable only near the center of the vortex.).

Now, we defining a dimensionless order-parameter relaxation time $\gamma = \gamma_1 + i\gamma_2$ $= [\Gamma_1 - i(1+\Gamma_2)]/[\Gamma_1^2 + (1+\Gamma_2)^2]$. If the kinetic-energy of supercurrent $(\delta H/\delta n_s)\Psi \approx \delta H/\delta\Psi^*$. Take Eq.(2) into Eq.(1), then we can rewrite Eq.(1) as[13,15,16]

$$\hbar\gamma\left[\partial_t + i\frac{2e}{\hbar}\tilde{\Phi}\right]\Psi(\mathbf{r}) = a\Psi + b|\Psi|^2\Psi + \frac{\hbar^2}{2m}(-i\frac{\partial}{\partial\mathbf{r}} - \frac{2e}{c}\mathbf{A})^2\Psi \ , \qquad (3)$$

Where, $\tilde{\Phi} = \Phi_e + \mu/2e$, the difference between $\tilde{\Phi}$ and Φ are generally small[17], we shall neglect the difference in following.

As $\nabla \bullet \mathbf{A} = 0$, then the Eq.(3) changes to

$$\hbar\gamma\left[\partial_t + i\frac{2e}{\hbar}\tilde{\Phi}\right]\Psi(\mathbf{r}) = a\Psi + b|\Psi|^2\Psi + \frac{\hbar^2}{2m}\left(-\nabla^2 + \left(\frac{2e}{c}\mathbf{A}\right)^2\right)\Psi \ , \qquad (4)$$

3.3   Solution of Vortex Equation

We will use (ρ,θ,z) as our cylindrical coordinates, with unit vectors $\mathbf{e}_\rho$, $\mathbf{e}_\theta$ and $\mathbf{e}_z$, respectively. Origin of these coordinates locating the center of underside of the superconductive ring. The $\mathbf{e}_\rho$ and $\mathbf{e}_\theta$ parallel with underside of the superconductive ring. The $\mathbf{e}_z$ is perpendicular to the underside of superconductive ring and upward. $\nabla = \frac{\partial}{\partial\rho}\mathbf{e}_\rho + \frac{1}{\rho}\frac{\partial}{\partial\theta}\mathbf{e}_\theta + \frac{\partial}{\partial z}\mathbf{e}_z$, $\nabla^2 = \frac{1}{\rho}\frac{\partial}{\partial\rho}(\rho\frac{\partial}{\partial\rho}) + \frac{1}{\rho^2}\frac{\partial^2}{\partial\theta^2} + \frac{\partial^2}{\partial z^2}$, the parameter $\Psi(\mathbf{r},t) = \Psi(\rho,\theta,z,t)$. If $\Psi(\rho,\theta,z,t)$ is not change with $\mathbf{z}$ and t, we have $\Psi(\rho,\theta,z,t) = f_n(\rho)e^{in\theta}\chi(Z,t), 0 \le \theta \le 2\pi$, n is integer, $\nabla \bullet \mathbf{A} = 0$. The Eq.(4) change to

$$\hbar\gamma\left[\partial_t + i\frac{2e}{\hbar}\tilde{\Phi}\right]\Psi(\mathbf{r}) = a\Psi + b|\Psi|^2\Psi + \frac{\hbar^2}{2m}\left(-\frac{1}{\rho}\frac{\partial}{\partial\rho}(\rho\frac{\partial\Psi}{\partial\rho}) - \frac{1}{\rho^2}\frac{\partial^2\Psi}{\partial\theta^2} - \frac{\partial^2\Psi}{\partial z^2} + \left(\frac{2e}{c}\mathbf{A}\right)^2\Psi\right),$$

$$2ei\gamma\,\tilde{\Phi}f_n = af_n + bf_n^3 + \frac{\hbar^2}{2m}\left(-\frac{1}{\rho}\frac{\partial}{\partial\rho}(\rho\frac{\partial f_n}{\partial\rho}) + \frac{n^2}{\rho^2}f_n + \left(\frac{2e}{c}\mathbf{A}\right)^2 f_n\right), \qquad (4')$$

$$\frac{1}{\rho}\frac{\partial}{\partial\rho}(\rho\frac{\partial f_n}{\partial\rho}) + \frac{2m}{\hbar^2}\left(2e\gamma\,i\tilde{\Phi} - a - \left(\frac{2e}{c}\mathbf{A}\right)^2 - \frac{n^2}{\rho^2}\right)f_n - \frac{2m}{\hbar^2}bf_n^3 = 0 \ . \qquad (5)$$

The solution of Eq.(5) have to satisfy boundary condition $\left.\frac{df_n}{d\rho}\right|_{\rho=R_1} = 0$ and $\left.\frac{df_n}{d\rho}\right|_{\rho=R_2} = 0$. We know that solution of $\frac{1}{x}\frac{d}{dx}\left(x\frac{dy}{dx}\right) + \left(1 - \frac{\mu^2}{x^2}\right)y = 0$ is Bessel function of μ order. Suppose $\lambda^2 = \frac{2m}{\hbar^2}\left(2e\gamma\,i\tilde{\Phi} - a - \left(\frac{2e}{c}\mathbf{A}\right)^2\right)$, $u = \lambda\rho$, $\mu = \sqrt{\frac{2m}{\hbar^2}}n$, μ is not integer, the Eq.(5) change to

$$\frac{1}{u}\frac{\partial}{\partial u}(u\frac{\partial f_n}{\partial u}) + \left(1 - \frac{\mu^2}{u^2}\right)f_n - \frac{2m}{\lambda^2\hbar^2}bf_n^3 = 0 \ . \qquad (6)$$

The solution of Eq.(5) can write to

$$f_n = c_1 J_\mu(u) + c_2 J_{-\mu}(u) + f_n^* \, , \tag{7}$$

here, $c_1 J_\mu(u) + c_2 J_{-\mu}(u)$ is the general solution of Bessel equation of μ-order. $J_\mu(u)$ is Bessel function of μ order, $J_{-\mu}(u)$ is Bessel function of $-\mu$ order, f_n^* is the special solution of Eq.(6), c_1 and c_2 are constant. Take the Eq.(7) into the Eq.(6), we have

$$\frac{1}{u}\frac{\partial}{\partial u}\left(u\,\frac{\partial(c_1 J_\mu(u)+c_2 J_{-\mu}(u))}{\partial u}\right) + \left(1-\frac{\mu^2}{u^2}\right)\left(c_1 J_\mu(u)+c_2 J_{-\mu}(u)\right) - \frac{2m}{\lambda^2 h^2} b\left(c_1 J_\mu(u)+c_2 J_{-\mu}(u)\right)^3$$

$$+\frac{1}{u}\frac{\partial}{\partial u}\left(u\,\frac{\partial f_n^*}{\partial u}\right) + \left(1-\frac{\mu^2}{u^2}\right)f_n^* - \frac{2m}{\lambda^2 h^2} b f_n^{*3} - \frac{2m}{\lambda^2 h^2} 3 b f_n^{*2}\left(c_1 J_\mu(u)+c_2 J_{-\mu}(u)\right)$$

$$-\frac{2m}{\lambda^2 \hbar^2} 3 b f_n^*\left(c_1 J_\mu(u)+c_2 J_{-\mu}(u)\right)^2 = 0 \tag{8}$$

$$\left(c_1 J_\mu(u)+c_2 J_{-\mu}(u)\right)^2 + 3 f_n^{*2} + 3 f_n^*\left(c_1 J_\mu(u)+c_2 J_{-\mu}(u)\right) = 0 \tag{9}$$

$$f_n^* = \frac{-3\pm i\sqrt{3}}{6}\left(c_1 J_\mu(u)+c_2 J_{-\mu}(u)\right) \, , \tag{10}$$

$$f_n = \frac{3\pm i\sqrt{3}}{6}\left(c_1 J_\mu(u)+c_2 J_{-\mu}(u)\right). \tag{11}$$

here, $u = \lambda\rho$, $J_\mu(u)$ is Bessel function of μ order, $J_{-\mu}(u)$ is Bessel function of $-\mu$ order. The general solution f_n have to satisfy boundary condition $\left.\frac{df_n}{d\rho}\right|_{\rho=R_1} = 0$ and

$$\left.\frac{df_n}{d\rho}\right|_{\rho=R_2} = 0 .$$

We also require an equation of motion for the vector potential, the vector potential is just Ampère's law $\nabla\times\nabla\times\mathbf{A} = 4\pi(\mathbf{J_n}+\mathbf{J_s})$, so that $\nabla\bullet(\mathbf{J_n}+\mathbf{J_s})=0$. The normal current $\mathbf{J_n} = \sigma^{(n)}\bullet\mathbf{E} = \sigma^{(n)}\bullet\left(-\nabla\Phi-\partial_t\mathbf{A}\right)$. The velocity of supercurrent $\mathbf{V}_s = \mathbf{J_s}/2e|\Psi|^2$. The supercurrent $\mathbf{J_s} = \frac{2e\hbar}{2mi}\left(\Psi^*\nabla\Psi-\Psi\nabla\Psi^*\right) - \frac{(2e)}{m}|\Psi|^2\mathbf{A}$,the $\sigma^{(n)} = \begin{bmatrix}\sigma^{(n)}_{xx} & \sigma^{(n)}_{xy} \\ \sigma^{(n)}_{yx} & \sigma^{(n)}_{xx}\end{bmatrix}$ is normal-state conductivity tensor. The Onsager relations and rotational symmetry imply $\sigma^{(n)}_{xy}(\mathbf{H}) = -\sigma^{(n)}_{yx}(\mathbf{H})$, so that the conductivity tensor may be decomposed into a diagonal piece and an antisymmetric piece. The longitudinal normal-state conductivity $\sigma^{(n)}_{xx}$ is generally a weak function of the magnetic field, $\sigma^{(n)}_{xy}$ is Hall conductivity.

Suppose, the direction of outer magnetic field is the direction of $\mathbf{e}_z$, so that $\mathbf{A} = \mathbf{A}_z$. When the $\Psi(\rho,\theta,z,t)$ is not change with z and t, take the $\Psi(\rho,\theta,z,t) = f_n(\rho)e^{in\theta}\chi$ into the supercurrent

$$\mathbf{J}_s = \frac{2e\hbar}{2mi}\left(\Psi^*\nabla\Psi - \Psi\nabla\Psi^*\right) - \frac{(2e)}{m}|\Psi|^2\mathbf{A}_z$$

$$= \left[\frac{2n\hbar e}{m}\frac{1}{\rho}\mathbf{e}_\theta - \frac{(2e)^2}{m}\mathbf{A}_z^2\right]f_n^2 \tag{12}$$

The $\mathbf{J}_s$ can be equated with zero in the direction of the $\mathbf{e}_\rho$. The $\mathbf{J}_s$ cannot be equated with zero in the direction of the $\mathbf{e}_\theta$, The $\mathbf{J}_s$ is proportion with the square of order parameter. $f_n = \frac{3\pm i\sqrt{3}}{6}\left(c_1 J_\mu(u) + c_2 J_{-\mu}(u)\right)$, the f_n have to satisfy boundary condition $\left.\frac{df_n}{d\rho}\right|_{\rho=R_1} = 0$ and $\left.\frac{df_n}{d\rho}\right|_{\rho=R_2} = 0$, when $\mu \to n$ (here, n is positive number), $f_n(u)$ is not linear solution of the $J_n(u)$, therefore the supercurrent $\mathbf{J}_s$ is change with jump. From the Eq.(12), we know the $\mathbf{J}_s$ is quantization, if the outer magnetic field is gather, when $\frac{(2e)^2}{m}\mathbf{A}_z^2 > \frac{2n\hbar e}{m}\frac{1}{\rho}\mathbf{e}_\theta$, the sign of supercurrent $\mathbf{J}_s$ can reversal.

3.4   Comparison to Previous Work and Conclusion

P. Kostic and B. Veal et al. study the magnetic field of superconductive rings that is quantization[7,8-11]. Here, we know $\mathbf{J}_s$ is quantization, the magnetic field of superconductive rings is quantization. We know $f_n(u)$ is not linear solution of the $J_n(u)$ from the Eq.(11), the supercurrent $\mathbf{J}_s$ is change with jump in theory. D. Y. Vodolazo and F. M. Peeters et al. have obtained the intentional introduction of the defect in the ring, the ring has a large effect of the size and the flux with jumps[2]. D. Y. Vodolazo and F. M. Peeters et al. have obtained that with increasing magnetic field the order parameter gradually decreasing, Thus leads to a decrease of the size of the jump of the flux in the vorticity[2]. S. Pedersen and G. R. Kofod et al have experimentally investigated the magnetization of a mesoscopic aluminum loop, the magnetic field intensity periodicity observed in the magnetization measurements is expected to take integer values of the superconducting flux quanta[3] $\Phi_0 = h/2e$. C. C. Tsuei and J. R. Kirtley et al. have obtained[3] the supercurrent of superconductive ring have a form of $I_s = I_c^{ij}\sin\Delta\phi_{ij}$. They infer the inductance of the superconductive rings from the high-field asymptotic difference between the flux in a ring with junctions and the flux in the ring without junctions[3]. Here, the $\mathbf{J}_s$ can be equated with zero in the direction of the $\mathbf{e}_\rho$, the $\mathbf{J_s}$ can not be equated with zero in the direction of the $\mathbf{e}_\theta$, the $\mathbf{J_s}$ is proportional to the square of order parameter. $f_n = \frac{3 \pm i\sqrt{3}}{6}\left(c_1 J_\mu(u) + c_2 J_{-\mu}(u)\right)$, the f_n have to satisfy boundary condition $\frac{df_n}{d\rho}\Big|_{\rho=R_1} = 0$ and $\frac{df_n}{d\rho}\Big|_{\rho=R_2} = 0$, when $\mu \to n$ (here, n is positive number) , $f_n(u)$ is not linear solution of the $J_n(u)$, therefore the supercurrent $\mathbf{J_s}$ is change with jump. We obtain the $\mathbf{J_s}$ is quantization, if the outer magnetic field is gather, when $\dfrac{(2e)^2}{m}\mathbf{A}_z^2 > \dfrac{2n\hbar e}{m}\dfrac{1}{\rho}\mathbf{e}_\theta$, the sign of supercurrent $\mathbf{J_s}$ can reversal.

We consider a time-dependant Ginzburg-Landau modified model to take into account the characteristics of superconductive rings, and make use of the models and evaluate the vortex motion equations of superconductive ring. We obtain integer values of the superconducting flux quanta and obtain the flux with jumps in theory. We obtain sign reversal of the currents and obtain some expressions, the expressions are accord with the data of experiments[2,3,6-11].

Satoshi Kashiwaya and C.C.Tsuei etc have obtained d-wave symmetry in high temperature superconductors[18,19--23]. Satoshi Kashiwaya and Yukio Tanaka have obtained the d-wave pairing state in these materials has an internal phase of the pair potential, the internal phase as a function of the wavevector of the Cooper pairs has a large influence on the electric properties of tunnelling junctions. They obtained convincing evidence for d-wave symmetry in high temperature superconductors[22]. C.C. Tsuei and J.R.Kirtley etc have used the concept of flux quantization in superconducting $YBa_2Cu_3O_{7-\delta}$ ring with 0,2,and 3 grain-boundary Josephson junctions to test the pairing symmetry in the high temperature superconductors, they have obtained consistent with d-wave pairing symmetry[20].

Here, we have studied superconducting ring using TDGL eqaution. The sign reversal of the current is addressed. The superconductive ring is the paring symmetry of the pair potentials. The symmetry of superconductive ring should be considered to be d-wave symmetry in high temperature superconductors.

References

[1]   Liang Fang-ying etc al., *Physica C* 2004 Vol. 411, p89-93;

[2]   D. Y. Vodolazo and F. M. Peeters, *Phys. Rev. B* 2003Vol.67. 054506; D. Y. Vodolazo, B. J. Baelus and F. M. Peeters, *Phys. Rev. B* 2002 Vol.66. 054531; D. Y. Vodolazo, F. M. Peeters, Phys. Rev. B 2002 Vol.66. 054537;

[3]   S. Pedersen, G. R. Kofod et al, *Phys. Rev. B* 2001Vol.64. 104522;

[4]   B. J. Baelus, L. R. E. Cabral and F. M. Peeters, *Phys. Rev. B* 2004 Vol.69. 064506;

[5]   B. J. Baelus and F. M. Peeters, and V. A. Schweigert, *Phys. Rev. B* 2000 Vol.61. 9734 and Phys. Rev. B 2001 Vol.6. 144517;

[6]   C.C.Tsuei, J. R. Kirtley etc. *Phys. Rev. Lett.* 1994 Vol.73.No.4, 593;

[7]   P. Kostic, B. Veal, A. P. Paulikas etc, *Phys. Rev. B* 1996 Vol.53, 791–80,1;

[8] Terentiev, D. B. Watkins, and L. E. De Long etc, *Phys. Rev. B* 1999 Vol.60, R761–R764;

[9] W. Braunisch, N. Knauf, G. Bauer etc, *Phys. Rev. B.* 1993 Vol.48, 4030–4042;

[10] W. Braunisch, N. Knauf, V. Kataev etc, *Phys. Rev. Lett.* 1992 Vol.68, 1908–1911;

[11] K. Geim, S. V. Dubonos, J. G. S. Lok etc, *Nature* 1998 Vol.396, 144 – 146;

[12] R.A. Klemm, A.Luther, M.R. Beasley, *Phys.Rev. B.* 1975 Vol.12, 877;

[13] Alan J. Dorsey, *Phys. Rev. B.* 1992 Vol.46, 8376;

[14] B.I. Ivlev and N.B.Kopnin, *J.Low Temp.Phys.* 1989 Vol.77, Nos.5/6,413;

[15] Liang Fang-Ying, *Physica C* 2004 Vol.402, Pages 174-178; Ling F. Y, Acta Phys. Sin. 2002 Vol.51. 898(in Chinese); Fang-Ying Liang, Hong Liu etc, *Physica C* 2004 Vol.406, P115-120;

[16] Liang FangYing,Li Zuo-Hong, Commun.Theor.Phys. 2002 Vol.38(Beijing,China), No.3,p379-384, Liang Fang-Ying, Qing xin etc, *Acta Phys.* 2003 Vol.52 No.10, 2584-05 (in Chinese);

[17] A Schmid, *Phys. Kondon. Mater.* 1966 Vol.5, 302;

[18] *Rev. Mod. Phys.* 1995 Vol.67 515,

[19] *Rev. Mod. Phys.* 2000Vol.72 969,

[20] C.C. Tsuei and J.R.Kirtley etc, *Phys. Rev. Lett.* 1994 Vol.73 593,

[21] *Phys. Rev. Lett.* 1995 Vol.74 3451,

[22] Satoshi Kashiwaya and Yukio Tanaka, *Rep. Prog. Phys.* 2000 Vol.63 1641,

[23] *Phys. Rev. Lett.* 2001 Vol.86 5369,

4 Study of Thermodynamic Properties of the Type I Superconductive Film[1]

Abstract

We consider a modified model of Ginzburg-Landau to study thermal properties in type I superconductive film. Some derivatives of the free energy with respect to the temperature, the entropy, the specific heat the temperature derivative of the supercurrent et al. are studied. We have obtained some formulae of the specific heat characteristic of the type I superconductive films.

PACS: 74.25.Bt; 74.40.+k; 74.78.Db

Keywords: Thermodynamic properties, Type I superconductivity, Superconductive films, Fluctuations

4.1 Introduction

The predecessors have made a lot of research work for superconductor characteristic, but without obtained a very clear superconductive mechanism. They have the lots of the research work for the thermodynamic characteristic of type I superconductivity.[2,3--20] The recent research work[2,3,4] show fluctuation influence of order parameter of superconductivity phase change of the type I superconductive film is stronger than that of the three domain superconductivity in the zero magnetic field. Here, we have derived an effective free energy of the type I superconductive film from the modified Ginzburg-Landau model and have studied the thermodynamic characteristic of type I superconductive film. Put forward a kind of method for calculation and deduce thermodynamic function, and obtain some formulae for the specific heat characteristic of the type I superconductive film. Our results might be useful in explanation of the most recent experimental data.

4.2 The Model and Equations

We make use of an effective free energy of the type I superconductive film from the modified Ginzburg-Landau model, have studied the thermodynamic characteristic of the type I super-conductive film. Consider an effective free energy density of Ginzburg-Landau

$f(\Psi) = F(\Psi)/V$, here $V = L_1 L_2 L_0$ is volume of the superconductive film , L_1, L_2 and L_0

are the superconductive film length, breadth and thickness respectively. We have[2,3,4]

$$f(\Psi) = a|\Psi|^2 + \frac{b}{2}|\Psi|^4 + k_B T \bullet J[\rho(\Psi)] . \tag{1}$$

Here

$$J[\rho(\Psi)] = \int_0^{\wedge} \frac{dk}{2\pi} k S(k, \rho) , \tag{2}$$

$$S = \frac{1}{L_0} \sum_{k_0 = -\wedge_0}^{+\wedge_0} \ln\left[1 + \frac{\rho(\Psi)}{k^2 + k_0^2}\right] . \tag{3}$$

Here, Ψ is order parameter,

$\rho(\Psi) = \rho_0|\Psi|^2 = \rho_0 \Psi^* \Psi , \rho_0 = \left(8\pi\, e^2 / mc^2\right); a = a_0\left(T - T_{C0}\right)$ is a temperature-dependent parameter, a_0 is a constant, T denotes temperature, T_{C0} is initial critical temperature; $b > 0$ is the usual Landau parameter. They are related to zero temperature coherent length $\xi_0 = \left(h^2 / 4ma_0 T_{C0}\right)^{1/2}$. The parameter a, b and m are made sure by microscopic characteristic of the superconductive film[6,9,10,11]. $e = 1.6 \times 10^{-19}$ coulomb; m is the effective mass of quasi-particles moving; k_B is Boltzmann constant. The third part of Eq.(1) describe fluctuation effect of magnetic field[2,3]. The wave vector $\vec{q} = \left(\vec{k}, k_0\right)$ is base on integral $J(\rho)$ and summation $S(k, \rho)$ in the Eq.(2) and Eq. (3). The $\wedge$ and $\wedge_0$ are finite cut-off point of the wave vector; $\wedge$ is introduced by $k = |\vec{k}|, \vec{k} = (k_1, k_2)$. $\wedge_0$ replaces k_0. $k = [\lambda\,(T)/\xi(T)] << 1/\sqrt{2}$ is Ginzburg- Landau parameter of the type I superconductor,

$\xi\,(T)=\left(\hbar/(2m|a|)^{1/2}\right)$ is coherent length, $\lambda\,(T)=\left[mb/4\pi(2e)^2|a|\right]^{1/2}$ is London osmotic depth.

In quasi-macroapproximate treatment of Ginzburg-Landau, choose[12,13,14] $\wedge_0 \to \infty$ and integral replaces summation in the Eq.(3), we obtain

$$S = \frac{1}{L_0}\int_{-\infty}^{+\infty}\ln\left[1+\frac{\rho(\Psi)}{k^2+k_0^2}\right]dk_0 \ . \tag{4}$$

Consider $\quad \Psi = \Phi(r)e^{i\delta}, r \quad$ is restricted in L_1 and L_2, choose[6,10,11] $\Phi(r) = a\dfrac{r}{[r^2+\xi_v^2]^{1/2}}$, here $a = a_0(T-T_{C0})$, ξ_v is a parameter measuring the healing length of the order parameter[9], it's numerically close to one. We obtain

$$S = \frac{1}{L_0}\int_{-\infty}^{+\infty}\ln\left[1+\frac{\rho}{k^2+k_0^2}\right]dk_0$$

$$= \frac{1}{L_0}\left\{k_0\ln\left[1+\frac{\rho}{k^2+k_0^2}\right]\right\}\Bigg|_{-\infty}^{+\infty} + \frac{-2k^2}{L_0}\int_{-\infty}^{+\infty}\left[\frac{1}{(k^2+k_0^2)}\right]dk_0 + \frac{2}{L_0}\int_{-\infty}^{+\infty}\frac{\rho+k^2}{\rho+k^2+k_0^2}dk_0$$

$$= \frac{1}{L_0}\left\{k_0\ln\left[1+\frac{\rho}{k^2+k_0^2}\right]\right\}\Bigg|_{-\infty}^{+\infty} + \frac{1}{L_0}\left\{-2karctg\frac{k_0}{k}+2\sqrt{\rho+k^2}\,arctg\frac{k_0}{\sqrt{\rho+k^2}}\right\}\Bigg|_{-\infty}^{+\infty}$$

$$= \frac{1}{L_0}\left(-2k\pi+2\pi\sqrt{\rho+k^2}\right)$$

$$= \frac{1}{L_0}\left(-2k\pi+2\pi\sqrt{\frac{\rho_0 a^2 r^2}{[r^2+\xi_v^2]}+k^2}\right)\ . \tag{5}$$

Take Eq.(5) into Eq.(2), obtain

$$J[\rho(\Psi)] = \int_0^{\wedge}\frac{dk}{L_0}k\left(-k+\sqrt{\frac{\rho_0 a^2 r^2}{[r^2+\xi_v^2]}+k^2}\right)$$

$$= -\frac{\wedge^3}{3L_0}+\frac{1}{3L_0}\left[\left(\frac{\rho_0 a^2 r^2}{[r^2+\xi_v^2]}+\wedge^2\right)^{3/2}-\left(\frac{\rho_0 a^2 r^2}{[r^2+\xi_v^2]}\right)^{3/2}\right]\ . \tag{6}$$

Take Eq.(6) and $\Psi = \dfrac{are^{i\delta}}{[r^2+\xi_v^2]^{1/2}}$ into Eq.(1), obtain

$$f(\Psi) = a|\Psi|^2 + \frac{b}{2}|\Psi|^4 + k_B T \bullet \left\{ -\frac{\wedge^3}{3L_0} + \frac{1}{3L_0}\left[\left(\frac{\rho_0 a^2 r^2}{[r^2 + \xi_v^2]} + \wedge^2\right)^{3/2} - \left(\frac{\rho_0 a^2 r^2}{[r^2 + \xi_v^2]}\right)^{3/2}\right] \right\}$$

$$= \frac{a^3 r^2}{[r^2 + \xi_v^2]} + \frac{b}{2}\frac{a^4 r^4}{[r^2 + \xi_v^2]^2} + \frac{k_B T}{3L_0} \bullet \left\{ -\wedge^3 + \left[\left(\frac{\rho_0 a^2 r^2}{[r^2 + \xi_v^2]} + \wedge^2\right)^{3/2} - \left(\frac{\rho_0 a^2 r^2}{[r^2 + \xi_v^2]}\right)^{3/2}\right] \right\} \quad (7)$$

If $\wedge_0$ is a constant in Eq.(7), we choose[11,12] $\wedge = \wedge_0 \to \infty$ and maximum value of the r^2 as $L_1 L_2$, then the Eq.(7) changes to

$$f(\Psi) = \frac{a^3 L_1 L_2}{[L_1 L_2 + \xi_v^2]} + \frac{b}{2}\frac{a^4 (L_1 L_2)^2}{[L_1 L_2 + \xi_v^2]^2} + \frac{k_B T}{3L_0} \bullet \left\{ -\wedge^3 + \left[\left(\frac{\rho_0 a^2 L_1 L_2}{[L_1 L_2 + \xi_v^2]} + \wedge^2\right)^{3/2} - \left(\frac{\rho_0 a^2 L_1 L_2}{[L_1 L_2 + \xi_v^2]}\right)^{3/2}\right] \right\}$$

$$= a^3 \left\{ \frac{L_1 L_2}{[L_1 L_2 + \xi_v^2]} + \frac{b}{2}\frac{a(L_1 L_2)^2}{[L_1 L_2 + \xi_v^2]^2} - \frac{k_B T}{3L_0} \bullet \left(\frac{\rho_0 L_1 L_2}{[L_1 L_2 + \xi_v^2]}\right)^{3/2} \right\} \quad (8)$$

4.3   Characterization of Thermodynamics

We can know main quantities of thermodynamics from the general principle of thermodynamics and statistics. For example, free energy density $f = u - sT$, inner energy of a unit volume $u = f - T\frac{\partial f}{\partial T}$, entropy of a unit volume $s = -\frac{\partial f}{\partial T}$, and specific heat of a unit volume $C_V = \left(\frac{\partial u}{\partial T}\right)_V$. Now take Eq.(1) and (2) into the main quantities of thermodynamics, have

$$s = -3a_0^3(T - T_{C0})^2 \left\{ \frac{L_1 L_2}{[L_1 L_2 + \xi_v^2]} + \frac{b}{2}\frac{a_0(T - T_{C0})(L_1 L_2)^2}{[L_1 L_2 + \xi_v^2]^2} - \frac{k_B T}{3L_0} \bullet \left(\frac{\rho_0 L_1 L_2}{[L_1 L_2 + \xi_v^2]}\right)^{3/2} \right\}$$

$$- a_0^3(T - T_{C0})^3 \left\{ \frac{b}{2}\frac{a_0(L_1 L_2)^2}{[L_1 L_2 + \xi_v^2]^2} - \frac{k_B}{3L_0} \bullet \left(\frac{\rho_0 L_1 L_2}{[L_1 L_2 + \xi_v^2]}\right)^{3/2} \right\}$$

$$= -a_0^3(T - T_{C0})^2 \left\{ \frac{3L_1 L_2}{[L_1 L_2 + \xi_v^2]} + \frac{b}{2}\frac{4a_0(T - T_{C0})(L_1 L_2)^2}{[L_1 L_2 + \xi_v^2]^2} - \frac{k_B(4T - T_{C0})}{3L_0} \bullet \left(\frac{\rho_0 L_1 L_2}{[L_1 L_2 + \xi_v^2]}\right)^{3/2} \right\} \quad (9)$$

$$u = a_0^3(T - T_{C0})^3 \left\{ \frac{L_1 L_2}{[L_1 L_2 + \xi_v^2]} + \frac{b}{2}\frac{a_0(T - T_{C0})(L_1 L_2)^2}{[L_1 L_2 + \xi_v^2]^2} - \frac{k_B T}{3L_0} \bullet \left(\frac{\rho_0 L_1 L_2}{[L_1 L_2 + \xi_v^2]}\right)^{3/2} \right\}$$

$$+Ta_0^3(T-T_{C0})^2\left\{\frac{3L_1L_2}{[L_1L_2+\xi_v^2]}+\frac{b}{2}\frac{4a_0(T-T_{C0})(L_1L_2)^2}{[L_1L_2+\xi_v^2]^2}-\frac{k_B(4T-T_{C0})}{3L_0}\bullet\left(\frac{\rho_0L_1L_2}{[L_1L_2+\xi_v^2]}\right)^{3/2}\right\}$$

$$=a_0^3(T-T_{C0})^2\left\{\frac{L_1L_2(4T-T_{C0})}{[L_1L_2+\xi_v^2]}+\frac{b}{2}\frac{a_0(5T^2-6TT_{C0}+T_{C0}^2)(L_1L_2)^2}{[L_1L_2+\xi_v^2]^2}\right.$$
$$\left.-\frac{k_BT(5T-2T_{C0})}{3L_0}\bullet\left(\frac{\rho_0L_1L_2}{[L_1L_2+\xi_v^2]}\right)^{3/2}\right\},\tag{10}$$

$$C_V=2a_0^3(T-T_{C0})\left\{\frac{L_1L_2(4T-T_{C0})}{[L_1L_2+\xi_v^2]}+\frac{b}{2}\frac{(5T^2-6TT_{C0}+T_{C0}^2)a_0(L_1L_2)^2}{[L_1L_2+\xi_v^2]^2}-\frac{k_BT(5T-2T_{C0})}{3L_0}\bullet\left(\frac{\rho_0L_1L_2}{[L_1L_2+\xi_v^2]}\right)^{3/2}\right\}$$

$$+2a_0^3(T-T_{C0})^2\left\{\frac{2L_1L_2}{[L_1L_2+\xi_v^2]}+\frac{b}{2}\frac{(5T-3T_{C0})a_0(L_1L_2)^2}{[L_1L_2+\xi_v^2]^2}-\frac{k_B(5T-T_{C0})}{3L_0}\bullet\left(\frac{\rho_0L_1L_2}{[L_1L_2+\xi_v^2]}\right)^{3/2}\right\}$$

$$=2a_0^3(T-T_{C0})\left\{\frac{L_1L_2(6T-3T_{C0})}{[L_1L_2+\xi_v^2]}+\frac{b}{2}\frac{a_0(10T^2-14TT_{C0}+4T_{C0}^2)(L_1L_2)^2}{[L_1L_2+\xi_v^2]^2}\right.$$
$$\left.-\frac{k_B(10T^2-8TT_{C0}+T_{C0}^2)}{3L_0}\bullet\left(\frac{\rho_0L_1L_2}{[L_1L_2+\xi_v^2]}\right)^{3/2}\right\}.\tag{11}$$

4.4 Comparison with Previous Work and Conclusion

From expression (7), we choose[12,13] $\wedge=\wedge_0\to\infty$ and obtain

$$f(\Psi)=a|\Psi|^2+\frac{b}{2}|\Psi|^4+k_BT\bullet\left\{-\frac{\wedge^3}{3L_0}+\frac{1}{3L_0}\left[\left(\rho_0|\Psi|^2+\wedge^2\right)^{3/2}-\rho_0^{3/2}|\Psi|^{3/2}\right]\right\},$$

$$=a|\Psi|^2+\frac{b}{2}|\Psi|^4-k_BT\bullet\frac{1}{3L_0}\rho_0^{3/2}|\Psi|^{3/2}.\tag{7'}$$

The density of free energy $f(\Psi)$ is related to temperature and order parameter. Expression (7') is compared with the Eq.(5) of reference [2], obviously the expression (7') is more plain. When T is a constant, the coefficients of three terms in the right hand of the expression (7') will take certain values respectively, The free energy density $f(\Psi)=F(\Psi)/V$ changes with Ψ as shown in Fig.1. It can be found that Fig.1 approximately agrees with Fig.1 of reference [3].

We already to some thermodynamics measuring and deducing the calculation, respectively getting the expression (8) of the free energy density f, the expression (9) of the unit volume entropy s, the expression (10) of the unit volume inner energy u, the expression (11) of the unit volume specific heat c_V. ξ_v is a parameter measuring the healing length of the order parameter[9], it's numerically close to one. For Type I superconductive film, $L_1 L_2 \gg 1$, so that $\dfrac{L_1 L_2}{[L_1 L_2 + \xi_v^2]} \approx 1$. The expression (8) to (11) are transformed as

$$f = a_0^3 (T - T_{C0})^3 \left\{ 1 + \frac{b}{2} a_0 (T - T_{C0}) - \frac{k_B T}{3 L_0} \rho_0^{3/2} \right\}, \tag{8'}$$

$$s = -a_0^3 (T - T_{C0})^2 \left\{ 3 + 2 b a_0 (T - T_{C0}) - \frac{k_B (4T - T_{C0})}{3 L_0} \bullet \rho_0^{3/2} \right\}, \tag{9'}$$

$$u = a_0^3 (T - T_{C0})^2 \left\{ (4T - T_{C0}) + \frac{b}{2} a_0 (5T^2 - 6TT_{C0} + T_{C0}^2) - \frac{k_B T (5T - 2T_{C0})}{3 L_0} \bullet \rho_0^{3/2} \right\}, \tag{10'}$$

$$C_V = 2 a_0^3 (T - T_{C0}) \left\{ + (6T - 3T_{C0}) + \frac{b}{2} a_0 (10T^2 - 14TT_{C0} + 4T_{C0}^2) - \frac{k_B (10T^2 - 8TT_{C0} + T_{C0}^2)}{3 L_0} \bullet \rho_0^{3/2} \right\}$$

$$= 2 a_0^3 (T - T_{C0}) \left\{ 5 \left(b a_0 - \frac{2 k_B \rho_0^{3/2}}{3 L_0} \right) (T - T_{C0})^2 + \left(6 + 3 b a_0 T_{C0} - \frac{4 k_B T_{C0} \rho_0^{3/2}}{L_0} \right) (T - T_{C0}) + 3 T_{C0} - \frac{k_B T_{C0}^2 \rho_0^{3/2}}{L_0} \right\}$$

$$= 2 a_0^3 \left\{ 5 \left(b a_0 - \frac{2 k_B \rho_0^{3/2}}{3 L_0} \right) T^3 + 6 \left(1 - 2 b a_0 T_{c0} + \frac{k_B T_{C0} \rho_0^{3/2}}{L_0} \right) T^2 \right.$$

$$\left. - 9 \left(T_{C0} - b a_0 T_{C0}^2 + \frac{k_B T_{C0}^2 \rho_0^{3/2}}{3 L_0} \right) T + 3 T_{C0}^2 - 2 b a_0 T_{C0}^3 + \frac{k_B T_{C0}^3 \rho_0^{3/2}}{3 L_0} \right\}. \tag{11'}$$

From the expression (8'), obviously the unit volume free energy f is related to the four power of T. From the expression (9') to expression (11'), can find that the unit volume entropy s relates to the cube of T, the unit volume inner energy u relates to the four power of T, the unit volume specific heat C_V relates to the cube of T. The results given by the expressions (8')--(11') approximately agree with the recent experimental data. The expressions (63) and (64) of the reference [15] are the special case of our expression (11'). The experimental formula of figure (1) of the reference [16] is as a special case of our expression (11').

The coefficients and the constant in the expression (11') of C_V are taken certain values respectively, we can get Fig.2 showing the C_V / T changing with the T. The curve in Fig.2 approximately agrees with the experimental data given previously (For example, Fig. 1 of Refs. [16], Fig. 1 of Refs. [17], Fig. 3 of Refs. [18]). The expression (11') inclusion state of experimental data of the reference [15,16,17,20]. The experimental work of previously people[17,18] obtain many experimental curve of the C_V, Refs.[16,19,20] obtained some experimental curve of C_V, have obtained some mathematical expressions from the experiment curves, but did not get the ideal expression.

In the present paper, we have derived an effective free energy of the type I superconductive films from the modified Ginzburg-Landau model and obtained some expressions for the specific heat characteristic of the type I superconductive films in theory. Our work is all the better to describe the experimental phenomenon than the work of previously people. The some expressions are approximately agree with the experimental curves of Refs.[16,17--20], and might be useful in the discussions of the most recent experimental data.

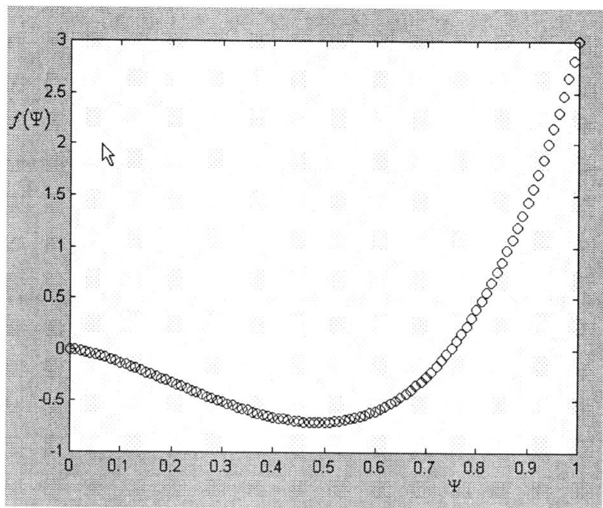

Fig.1: when $a = 3, b = 10$ and $\dfrac{k_B T}{3L_0} \rho_0^{3/2} = 5$ free energy density $f(\Psi) = F(\Psi)/V$ changes with the Ψ.

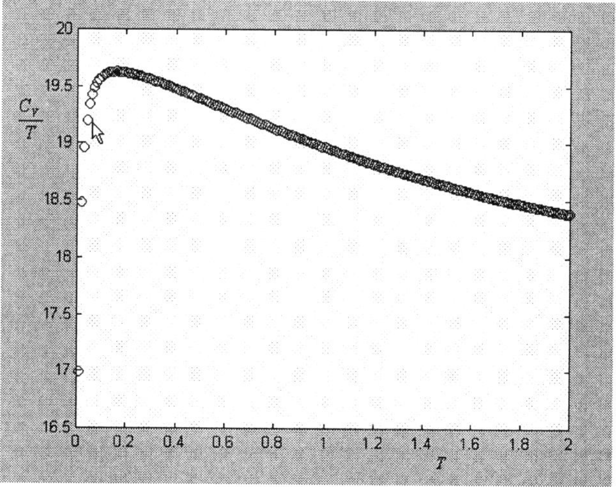

Fig. 2: the coefficients of T^3, T^2, T and constant item of the expression (11') take respectively 0.2,-1.2,20 and -0.03 ; get curve of C_V/T changes with the T.

References

[1]   Fang-Ying Liang etc al., *Physica C* 2004 Vol.406, Issues 1-2 , Pages 115-120;

[2]   R. Folk, D.V. Shopova and D. I. Uzunov, *Phys. Lett. A* 2001 Vol.281, 197;

[3]   D.V. Shopova, T. P. Todorov, D. I. Uzunov, *Mod. Phys. Lett. B* 2002 Vol. 17, 1;

[4]   B. I. Halperin, T. C. Lubensky, and S. K. Ma, *Phys. Rev. Lett.* 1974 Vol. 32, 292;

[5]   Liang F Y, *Acta Phys.* 2002 Vol.51 No.4, 0898(in Chinese)

[6]   Liang Fang-Ying Li Zuo-Hong, *Commun. Theor. Phys.* (Beijing,China) 2002 Vol. 38, p379-384;

[7]   Liang Fang-Ying, Qing xin etc, *Acta Phys.* 2003 Vol.52 No.10, 2584-05(in Chinese);

[8]   Liang Fang-Ying, *Physica C* 2004Vol.402, Issues 1-2, Pages 174-178;

[9]   N.B.Kopnin, B.I. Ivlev, and V.A. Kalatsky, J. Low Temp.Phys. 1993 Vol.90. No.1/2., p1;

[10] Alan T. Dorsey, *Phys. Rev. B.* 1992 Vol.46. 8376;

[11] Albert Schmid, *Phys. Kondens. Materie* 1966 Vol.5, 302-317, 302;

[12] L.Craco, L.De Cesare, I. Rabuffo, I. P. Takov, D. I. Uzunov, *Physica A* 1999 Vol. 270, 486;

[13] J. C. Rahola, *J. Phys. Studies* 2001 Vol.5, 304;

[14] E. M. Lifshitz and L. P. Pitaevskii, 1980 *Statistical Physics*,Part 2, [Landau and Lifshitz Course of Theoretical Physics, vol.9] (Pergamon Press, Oxford, 1980);

[15] Ralph Werner, Physical Rev.B. 2003 Vol.67, 014505;

[16] R.A.Fisher, F.Bouquet, N.E.Phillips etc, *Physical Rev.B.* 2002 Vol.65, 224509;

[17] J.S.Kim, J. Alwood, P.Kumar,and G.R.Stewart, *Physical Rev.B.* 2002 Vol.65, 174520;

[18] A.Bianchi, R. Movshovich, M.Jaime etc, *Physical Rev.B.* 2001 Vol.64, 220504(R);

[19] Yuxing Wang, Bernard Revat. Andreas Erb etc. *Physical Rev.B.* 2001 Vol.63.094508;

[20] G. T. Furukawa, W. G. Saba, and M. L. Reilly, *Natl. Bur. Stand. Ref. Data Ser.* No. **18** (U.S.GPO, Washington. DC.1968). P1;

5   Study of High Temperature Superconductor under Pressure

Abstract

When outer pressure is a constant on superconductor, the pressure intensity with the temperature is the relation of quadratic curve. The temperature is increasing with the pressure intensity.

When outer pressure on superconductor is not a constant, the external pressure intensity has a relation of partial differential equation with the temperature of superconductivity. As increasing the external pressure intensity, the temperature is rising. The critical temperature is decreasing quasi-linearly with applied hydrostatic pressure for superconductor, and observed negative pressure coefficient of the critical temperature of superconductor. In another special case, we obtain the critical temperature increases quasi-linearly with applied pressure on superconductor.

PACS: 74.62.Fj; 71.27.+a; 74.20.-z;74.25.Ld; 74.25.-q.

Keywords: Pressure; Temperature; Strongly Correlated Electrons, High temperature superconductor.

5.1   Introduction

In last a few years, there is a noticeable increase of the study of superconductivity in many elements under pressure. Recently the pressure-induced superconductivity has been found in UGe$_2$. This finding is quite interesting since the superconductivity appears in the pressure range from 1.0 to 1.6 GPa where UGe$_2$ is still in the ferromagnetic state. This is the first discovery that the same $5f$ electrons are involved with both orderings[1,2].

The discovery of unconventional superconductivity has caused an explosive growth of activities in various fields of condensed-matter research, stimulated not only studies of the basic mechanisms leading to this phenomenon, but also a widespread search for new technological applications. The different behaviors have been observed in heavy-fermions materials[3,4--15], in organic conductors[16,17], copper oxides etc al.[18,19,20]. These findings suggest that the mechanism forming Cooper pairs can be magnetic in origin. Namely, on the verge of magnetic order, the magnetically soft electron liquid can mediate spin-dependent attractive interactions between the charge carriers[14]. However, the nature of superconductivity and magnetism is still unclear when the superconductivity appears very close to the antiferromagnetism(AFM). The mechanism of the superconductivity and the symmetry of the order parameters are the main puzzles of on-going research.

Here, we consider a time-dependant Ginzburg-Landau (TDGL) modified model and calculate the modified model under pressure, study property of high temperature superconductor.

5.2   The Model of Superconductor

In the present paper   we make use of the modified TDGL theory to study the pressure effect of high temperature superconductors under outer pressure. The pressure intensity will apply work to high temperature superconductors; simultaneity the internal energy of the high temperature superconductor is increased. The modified Ginzburg-Landau free energy of a high temperature superconductor is [21,22-24]

$$F = \int dV \left\{ \bar{a}\left|\Psi\left(\mathbf{r}\right)\right|^2 + \bar{\mu}\left|\Psi\left(\mathbf{r}\right)\right|^2 + \frac{1}{2}b\left|\Psi\left(\mathbf{r}\right)\right|^4 + \frac{\hbar^2}{2m}\left|\left(-i\nabla - \frac{2e}{c}\mathbf{A}\right)\Psi(\mathbf{r})\right|^2 + \frac{\mathbf{H}^2}{8\pi} \right\}, \quad (1)$$

where, $\bar{a} = a_0\left(T/T_c - 1\right)$, $\bar{\mu} = \mu_0\left(p_0 - p\right)$, T denotes temperature, μ_0 is a constant, a_0 is a constant, p_0 is the initial pressure intensity, p is pressure intensity on high temperature superconductor; b is the usual Landau parameter; $e = 1.6 \times 10^{-19}$ coulomb; $\mathbf{m}$ is the effective mass of a quasi-particles moving; V is total volume of the high temperature superconductor; $\mathbf{A}$ is the outer vector potential, $\mathbf{H}$ is outer magnetic field; $\mathbf{B} = \langle\mathbf{h}\rangle$ is induction field, $\mathbf{h} = \nabla \times \mathbf{A}$ is the microscopic magnetic field, and Ψ is the order parameter. We neglect the anisotropy in high temperature superconductor[18,19,20]. The order parameter is usually expressed as $\Psi = f \exp[i\delta(r,t)]$, here f is an amplitude (Note that a moving vortex does not possess cylindrical symmetry, so the phase variable δ is equal to the angular variable only near the center of the vortex.)[21,22,25]. We also require an equation of motion for the vector potential, the Ampère's law $\nabla \times \nabla \times \mathbf{A} = 4\pi(\mathbf{J_n} + \mathbf{J_s})$ is to be obeyed by the vector potential , so that $\nabla \cdot (\mathbf{J_n} + \mathbf{J_s}) = 0$, therefore, the magnetic field have the vortex structure. We have

$$-\frac{\hbar^2}{2m}(\frac{\partial}{\partial \mathbf{r}} - \frac{2ie}{c}\mathbf{A})^2 = -\frac{\hbar^2}{2m}\nabla^2 + \frac{2ie\hbar}{mc}\mathbf{A} \cdot \nabla + \frac{(2e)^2}{2mc^2}\mathbf{A}^2 .$$

We consider a is a parameter of temperature and exterior pressure, $a = \bar{a} + \bar{\mu}$, have

$a = a_0\left(\frac{T}{T_C} - 1\right) + \mu_0\left(p_0 - p\right)$. From Eq.(1), we obtain

$$\frac{\delta F}{\delta \Psi^*} = \int dV \left\{ a\Psi + b|\Psi|^2 \Psi - \frac{\hbar^2}{2m}(\frac{\partial}{\partial \mathbf{r}} - \frac{2ie}{c}\mathbf{A})^2 \Psi \right\} . \qquad (2)$$

Here, a proper choice of f will allow all of the above equations to be solved exactly[24,26,27,28]. This is the method originally developed by Schmid who assumed an approximate order-parameter profile of the form[24,26] $f(r) = \dfrac{Kar}{[r^2 + \xi_v^2]^{1/2}}$, where K is a constant, ξ_v is a healing length of the order parameter and numerically close[23] to 1.

5.3   Characterization of Thermodynamics

We can obtain the main quantities of thermodynamics from the general principle of thermodynamics and statistical physics. For example, free energy $F = U - TS$, inner energy $U = F - T\frac{\partial F}{\partial T}, dF = -SdT - pdV$, expansion coefficient $\alpha = \frac{1}{V}\left(\frac{\partial V}{\partial T}\right)_p$, pressure coefficient $\beta = \frac{1}{p}\left(\frac{\partial p}{\partial T}\right)_V$, compressibility $\kappa = -\frac{1}{V}\left(\frac{\partial V}{\partial p}\right)_T$, heat capacity $C_V = \left(\frac{\partial U}{\partial T}\right)_V = T\left(\frac{\partial S}{\partial T}\right)_V$, $C_p = \left(\frac{dQ}{dT}\right)_p = T\left(\frac{\partial S}{\partial T}\right)_p$, and entropy $S = -\frac{\partial F}{\partial T}$, we have

$$C_p - C_V = \frac{TV\alpha^2}{\kappa} , \qquad (3)$$

$$\left(\frac{\partial T}{\partial p}\right)_S = \frac{TV\alpha}{C_p}$$

$$= \frac{TV\alpha}{C_V + TV\alpha^2/\kappa} \, , \tag{4}$$

In an isentropic process, from the expression (4), we have

$$\int_{T_0}^{T}\left[\frac{C_V}{TV\alpha} + \alpha/\kappa\right]dT = p - p_0. \tag{5}$$

Here, T_0 is the initial temperature. In the isentropic process, the pressure intensity is increased on high temperature superconductor; simultaneity the internal energy of the high temperature superconductor is increased.

In the isentropic process, if the coefficient $\alpha\!\big/\!\kappa$ is not function of the temperature, or the α and κ are constants, we can change the expression (5) to

$$\frac{\alpha}{\kappa}T + \frac{C_V}{\alpha V} - \frac{\Delta C_V}{\alpha V} = \Delta p + const. \tag{6}$$

When the pressure intensity is a constant, if the expansion coefficient α is not function of the temperature, or α is a constant, from $\alpha = \frac{1}{V}\left(\frac{\partial V}{\partial T}\right)_p$, we obtain $V\!\big/\!V_0 = \exp[\alpha(T - T_0)]$, here, V_0 is the initial volume.

From the following thermodynamical quantities $dF = -SdT - pdV$, $F = U - TS$ and inner energy $U = F - T\frac{\partial F}{\partial T}$, expansion coefficient $\alpha = \frac{1}{V}\left(\frac{\partial V}{\partial T}\right)_p$; and a proper choice $f(r) = \frac{Kar}{[r^2 + \xi_v^2]^{1/2}}$, We obtain

$$p = -\frac{dF}{dV} - \frac{1}{V}\left(\frac{\partial F}{\partial T}\right)\frac{1}{\alpha}$$

$$= -\left\{a + \frac{1}{2}b\frac{(Kar)^2}{[r^2 + \xi_v^2]} - \frac{\hbar^2}{2m}\frac{2\xi_v^2\left(\xi_v^2 - 3r^2\right)}{r^2[r^2 + \xi_v^2]^2} + \frac{2ieh}{mc}A\frac{2\xi_v}{r[r^2 + \xi_v^2]} + \frac{(2e)^2}{2mc^2}A^2 + \frac{\mathbf{H}^2}{8\pi}\right\}$$

$$-\frac{1}{\alpha}\frac{\partial}{\partial T}\left\{a + \frac{1}{2}b\frac{(Kar)^2}{[r^2 + \xi_v^2]} - \frac{\hbar^2}{2m}\frac{2\xi_v^2\left(\xi_v^2 - 3r^2\right)}{r^2[r^2 + \xi_v^2]^2} + \frac{2ieh}{mc}A\frac{2\xi_v}{r[r^2 + \xi_v^2]} + \frac{(2e)^2}{2mc^2}\mathbf{A}^2 + \frac{\mathbf{H}^2}{8\pi}\right\} \tag{7}$$

If $r \gg 1$, then $\dfrac{r^2}{r^2 + \xi_v^2}$ is close to 1 , and $\dfrac{r}{r^2 + \xi_v^2}$ is close to 0. From the expression (7), we have

$$p = -\left\{ a + \frac{1}{2}b(Ka)^2 + \frac{(2e)^2}{2mc^2}\mathbf{A}^2 + \frac{\mathbf{H}^2}{8\pi} \right\} - \frac{1}{\alpha}\frac{\partial}{\partial T}\left\{ a + \frac{1}{2}b(Ka)^2 + \frac{(2e)^2}{2mc^2}\mathbf{A}^2 + \frac{\mathbf{H}^2}{8\pi} \right\} \quad (8)$$

Now, we consider two special cases,

Case one:
When outer pressure on high temperature superconductor is a constant, $a = a_0\left(T/T_C - 1\right)$.

From the expression (8), we have

$$p = -\left\{ a + \frac{1}{2}b(Ka)^2 + \frac{(2e)^2}{2mc^2}\mathbf{A}^2 + \frac{\mathbf{H}^2}{8\pi} \right\} - \frac{1}{\alpha}\left(\frac{a_0}{T_C} + b(\frac{Ka_0}{T_C})^2(T - T_C) \right)$$

$$= -\left\{ \frac{1}{\alpha}\frac{a_0}{T_C} + \left(\frac{1}{\alpha}b(Ka_0)^2 + a_0 \right)\left(\frac{T}{T_C} - 1 \right) + \frac{1}{2}b(Ka_0)^2\left(\frac{T}{T_C} - 1 \right)^2 + \frac{(2e)^2}{2mc^2}\mathbf{A}^2 + \frac{\mathbf{H}^2}{8\pi} \right\} \quad (9)$$

Here, $T \leq T_C$.

Case two:
When outer pressure on high temperature superconductor is not a constant, $a = a_0\left(T/T_C - 1\right) + \mu_0(p_0 - p)$.

From the expression (8), we have

$$p = -\left\{ \frac{1}{\alpha}\frac{a_0}{T_C} + \left[a_0\left(T/T_C - 1\right) + \mu_0(p_0 - p) \right]\left[\frac{1}{\alpha}bK\left(\frac{a_0}{T_C} - \mu_0\frac{\partial p}{\partial T} \right) \right]\right.$$

$$\left. + \frac{1}{2}bK^2\left[a_0\left(T/T_C - 1\right) + \mu_0(p_0 - p) \right]^2 + \frac{(2e)^2}{2mc^2}\mathbf{A}^2 + \frac{\mathbf{H}^2}{8\pi} \right\} \quad (10)$$

Here, $T \leq T_C$.

When dT/dp is zero, the T take extremum. From the expression (9), we obtain

$$p = p_0 + \frac{bKa_0\mu_0 - \alpha T_C}{\alpha bK^2\mu_0^2 T_C}\left[1 - \exp(\alpha K(T_0 - T_C))\right] \qquad (11)$$

From the expression (11), we have

$$1 = \left\{ -\frac{a_0}{\alpha K\mu_0 T_C^2} + \left[\frac{a_0}{\alpha K\mu_0 T_C^2} + \frac{a_0}{\mu_0 T_C} - \frac{\alpha}{bK\mu_0^2}\right]\exp(\alpha K(T_0 - T_C)) \right\}\frac{dT_C}{dp} \qquad (12)$$

5.4   Discussion and Comparison to Previous Work

According to Eq.(5) and Eq.(6), in the isentropic process, as increasing the external pressure intensity on the high temperature superconductor, the internal energy of the high temperature superconductor is increased. we know that the external pressure intensity on the high temperature superconductor is proportional to the change of the heat capacity of high temperature superconductor in the isochoric process.

When outer pressure on high temperature superconductor is a constant, we obtain the relation of quadratic curve about pressure intensity with the temperature. As increasing the pressure intensity p, the temperature T is rising. Movshovic R, Graf T, Mandrus D etc. al. obtain same experimental curve of the temperature with pressure intensity[8,9,10-15], the experimental curves have a relation of quadratic curve about the pressure intensity with temperature[8,9,10-15].

When outer pressure on high temperature superconductor is not a constant, from Eq.(10), it can be found that, the external pressure intensity has a relation of partial differential equation with the temperature of superconductivity. As increasing the external pressure intensity p, the temperature T is rising.

For the ferromagnetic superconductor UGe_2, the heat-capacity and magnetization measurements under high pressure have been carried out by N. Tateiwa, T. C. Kobayashi, K. Amaya etc. Both measurements mentioned above were done using a same pressure cell in order to obtain both data for one pressure. The present results suggest the importance of the thermodynamic critical point for the appearance of the superconductivity[29]. In particular, many scientists have done a number of studies on f-electron compounds and revealed that unconventional superconductivity arises at or close to a quantum critical point, where magnetic order disappears at low temperature as a function of lattice density via application of hydrostatic pressure[7,8,9,10]. P. Modak, A. K. Verma, D. M. Gaitonde etc al., find a linear variation of the frequency with pressure up to 28 GPa without any discontinuity in the slope of the variation, pointing to the need to include the anharmonic or nonlinear terms in first-principles-based estimates of the phonon frequencies[30].

When outer pressure on high temperature superconductor is not a constant, we have $p = p_0 + \frac{bKa_0\mu_0 - \alpha T_C}{\alpha bK^2\mu_0^2 T_C}\left[1 - \exp(\alpha K(T_0 - T_C))\right]$. As increasing the external pressure, the critical temperature T_C is decreasing. If T_0 is close to T_C, from the expression (12), we have

$\left[\dfrac{a_0}{\mu_0 T_0}-\dfrac{\alpha}{bK\mu_0^2}\right]\dfrac{dT_C}{dp}=1$. When $\left[\dfrac{a_0}{\mu_0 T_0}-\dfrac{\alpha}{bK\mu_0^2}\right]<0$, the T_C decreases linearly with

applied pressure. E Saito, T Taknenobu etc al., discovered that the T_C of the high temperature superconductor MgB_2 decreases quasi-linearly with applied pressure to 1.4 GPa at a rate of - 2.0 KGPa^{-1}, their observed negative pressure coefficient of T_C in the MgB_2 high temperature superconductor assumes a very large value[31]. B. Lorenz, R. L. Meng etc al., find the transition temperature T_C decreases linearly at a large rate of 21.6 K/GPa.[32] T. Tomita, J. J. Hamlin etc al., find that the T_C decreases linearly and reversibly under pressure at the rate $dT_C/dp \approx -1.11 \pm 0.02$ K/GPa on hydrostatic pressure for superconducting MgB_2.[33]

When $\left[\dfrac{a_0}{\mu_0 T_0}-\dfrac{\alpha}{bK\mu_0^2}\right]>0$, the T_C increases linearly with increasing of the pressure. Chen

Xiao-Jia, Viktor V. Struzhkin etc al. study relation of $Bi_2Sr_2CaCu_2O_{8+\delta}$ superconductive T_C with the pressure on hydrostatic pressure 18 GPa. They find the T_C increases with increasing of the pressure originally[34], whereas the pressure attains the certain value, the T_C contrary decreases with increasing of the pressure.

5.5   Conclusion

Finally we conclude by summarizing the main points of this paper. We theoretically obtained some expressions of the temperature of superconductivity with the pressure intensity on the high temperature superconductor. We get the expression of the critical temperature with the pressure on high temperature superconductor. The critical temperature decreases quasi-linearly with applied pressure on high temperature superconductor. We believe that the results presented provide a clue to unravel the essential interplay between AFM and SC, and will to extend the universality of the understanding on the SC in strongly correlated electron systems.

References

[1] S.S. Saxena, P. Agarwal, K. Ahilan, F.M. Grosche etc, *Nature* (London) 2000 Vol. 604, 587;

[2] Huxley, I. Sheikin, E. Ressouche, N. Kernavanois etc, *Phys. Rev. B* 2001 Vol.63, 144519;

[3] F.Steglich, J.Aarts, C.D.Bredl, W.Lieke, D.Meschede et al, *Phys.Rev.Lett.* 1979 Vol. 43, 1892;

[4] H.R.Ott, H.Rudigier, Z.Fisk, and J.L.Smith, *Phys.Rev.Lett.* 1983 Vol.50, 1595;

[5] G.R.Stewart, Z.Fisk, J.O.Willis, and J.L.Smith, *Phys.Rev.Lett.* 1984 Vol.52, 679;

[6] J.L.Smith, J.O.Willis, B.Batlogg, and H.R.Ott, *J.Appl.Phys.* 1984 Vol.55, 1996;

[7] Jaccard D, Behnia K and Sierro J, *Phys. Lett. A* 1992 Vol.63 475;

[8] Movshovic R, Graf T, Mandrus D, Thompson J D, Smith J L and Fisk Z, *Phys. Rev. B* 1996 Vol.53 8241;

[9] Mathur N D, Grosche F M, Julian S R,Walker I R, Freye D M et al, *Nature* 1998 Vol.394, 39;

[10] Hegger H, Petrovic C, Moshopoulou E G, Hundley M F et al, Phys. Rev. Lett. 2000 Vol.84, 4986; Y. Taguchi, M. Hisakabe, Y. Ohishi, S. Yamanaka, and Y. Iwasa; *Phys. Rev. B* 2004 Vol.70, 104506;

[11] Bellarbi B, Benoit A, Jaccard D, Mignot J M and Braun H F, *Phys. Rev. B* 1984 Vol.30 1182;

[12] Thomas F, Thomasson J, Ayache C, Geibel C and Steglich F, *Physica B* 1993 Vol.186-188 303;

[13] Kawasaki Y, Ishida K, Mito T, Thessieu C, Zheng G –q et al, *Phys. Rev. B* 2001 Vol.63 R140501;

[14] Kawasaki Y, Ishida K, Kawasaki S, Mito T, Zheng G –q et al, *J. Phys. Soc. Jpn.* 2004 Vol.73 194;

[15] Muramatsu T, Tateiwa N, Kobayashi T C, Shimizu K et al, *J. Phys. Soc. Jpn.* 2001 Vol.70 3362;

[16] D.Jerome, D.Mazaud, M.Ribault, K.Bechgaard, J.Physique Lett. 1980 Vol.41, L95;

[17] D.Jerome, and H.J.Schulz, *Adv.Phys.* 2002 Vol.51 293[This article is originally published in *Adv.Phys.* **31** 299 (1982)];

[18] J.G.Bednorz, and K.A.M'uller, *Z.Phys.,B* 1986 Vol.64; *Rev.Mod.Phys.* 1988Vol.60 585;

[19] M.K.Wu, et al., *Phys.Rev.Lett.* 1987 Vol.58, 908;

[20] R.J.Cava, B.Batlogg, K.Kiyano, H.Takagi, J.Krajewski et al, Phys.Rev. 1994 Vol.49, 11890;

[21] Alan T. Dorsey, *Phys. Rev. B* 1992 Vol.46. p8376;

[22] Liang F Y, Jiang W. Z, *Acta Phys. Sin.* 1997 Vol.46.No 12, p2431(in Chinese); Liang Fang-Ying, Li Zuo-Hong, *Commun. Theor. Phys.* (Beijing, China) 2002Vol.38 No.3,p379-384;

[23] R.A. Klemm, A. Luther, M. R. Beasley, *Phys. Rev. B* 1975 Vol.12,p877;

[24] B.I. Ivlev and N.B. Kopnin, J. *Low Temp.Phys.* 1989 Vol.77.Nos.5/6, p413;

[25] N.B.Kopnin, B.I. Ivlev, and V.A. Kalatsky, *J Low Temp.Phys.* 1993 Vol.90.Nos.1/2, p1;

[26] Albert Schmid, *Phys. Kondens. Materie* 1966 Vol.5, p302-317;

[27] Fang-Ying Liang etc, Physica C 2004 Vol. /Issue 406/1-2 ,P115-120; Liang Fang-Ying etc, *Physica C* 2004 Vol.411,P89-93;

[28] Liang Fang-Ying, *Physica C* 2004Vol.402, Issues 1-2, P174-178;

[29] N. Tateiwa, T. C. Kobayashi, K. Amaya, etc, *Phys. Rev. B* 2004 Vol.69, 180513(R);

[30] P. Modak, A. K. Verma, D. M. Gaitonde etc al., *Phys. Rev. B* 2004 Vol.70, 184506;

[31] E Saito, T Taknenobu, T Ito etc al., *J. Phys.: Condens. Matter* 2001 Vol.13 L267–L270;

[32] Lorenz, R. L. Meng, and C. W. Chu etc al., *Phys. Rev. B* 2001 Vol. 64, 012507;

[33] T. Tomita, J. J. Hamlin, and J. S. Schilling etc al., *Phys. Rev. B* 2001 Vol.64, 092505.

[34] Chen Xiao-Jia, Viktor V. Struzhkin et al., 2004 *Phys. Rev. B* **70**, 214502.

In: New Topics in Superconductivity Research
Editor: Barry P. Martins, pp. 195-222

ISBN: 1-59454-985-0
© 2006 Nova Science Publishers, Inc.

Chapter 6

STUDIES OF CU-BASED HIGH TEMPERATURE SUPERCONDUCTORS BY USING COINCIDENCE DOPPLER BROADENING OF THE ELECTRON POSITRON ANNIHILATION RADIATION MEASUREMENT TECHNIQUE

Mahuya Chakrabarti, D. Sanyal[1]
Variable Energy Cyclotron Centre,
1/AF, Bidhannagar, Kolkata 700064, India
A. Sarkar, S. Chattopadhyay
Department of Physics, University of Calcutta,
92 A. P. C. Road, Kolkata 700009, India

Abstract

In the present work an attempt has been taken to study the variation of positron annihilation parameters, specially those which are probing the electron momentum distributions, due to superconducting transition in three different high T_c superconducting oxides (single crystalline $Bi_2Sr_2CaCu_2O_{8+\delta}$, single crystalline $SmBa_2Cu_3O_{7+x}$ and polycrystalline $La_{0.7}Y_{0.3}Ca_{0.5}Ba_{1.5}Cu_3O_z$) and also to identify the core electrons with which positrons are annihilating in these cuprate HTSC systems. This will help to understand the reasons of the variation of positron annihilation parameters due to superconducting transition in these HTSC systems in a better way. The anisotropy of the EMD in different crystallographic orientations in the layered structured HTSC system has also been studied by using the positron annihilation technique. The two detector coincidence Doppler broadening of the electron positron annihilation radiation (CDBEPAR) measurement, having peak to background ratio better than 14000 : 1, have been used to study the temperature dependent (300 K to 30 K) electron momentum distributions in these high T_c superconducting oxides. The CDBEPAR data are analysed both by conventional lineshape analysis and the ratio curve analysis.

---

[1] E-mail address: dirtha@veccal.ernet.in

1 Introduction

After the discovery of high temperature superconducting (HTSC) oxides [1-2], different experimental techniques have been employed [3] to understand the mechanism of the high temperature superconductivity. Positron annihilation technique [4-6] which is a very efficient nuclear solid state technique to study the variations of the electron number density and the electron momentum distributions (EMD) due to phase transition, have also been employed to probe any kind of variations of the electron number density and the electron momentum distributions due to the superconducting transition. Since 1987 there are large number of reports of positron annihilation studies on different HTSC oxides [7-44], some are reviewed in Section 3. Employing positron annihilation techniques it may not be possible to probe directly the "superconducting electrons" i.e., the electrons or holes which are forming the Cooper pair. But from the variations of the temperature dependent positron annihilation parameters [7-44] it has been concluded that at or near the superconducting transition temperature there occurs some structural changes which may be linked with the mechanism of high temperature superconductivity.

The layered structured HTSC are highly structurally anisotropic and hence positrons are not uniformly probing all the sites in these HTSC. Thus for better understanding of the positron annihilation results in the HTSC system it is very important to identify the core electrons with which positrons are annihilating.

Another widely discussed phenomenon regarding high T_c superconducting oxides is the anisotropy of its properties in different crystallographic directions. Hence it is also interesting to study the anisotropy of the EMD in different crystallographic orientations in the HTSC by employing positron annihilation techniques.

In the present work an attempt has been taken to study the variation of positron annihilation parameters, specially those which are probing the electron momentum distributions, due to superconducting transition in three different high T_c superconducting oxides (single crystalline $Bi_2Sr_2CaCu_2O_{8+\delta}$, single crystalline $SmBa_2Cu_3O_{7+x}$ and polycrystalline $La_{0.7}Y_{0.3}Ca_{0.5}Ba_{1.5}Cu_3O_z$) and also to identify the core electrons with which positrons are annihilating in these cuprate HTSC systems [45,46]. This will helps to understand the reasons of the variation of positron annihilation parameters due to superconducting transition in these HTSC systems in a better way. The anisotropy of the EMD in different crystallographic orientations in the layered structured HTSC system has also been studied by using the positron annihilation technique [47,48]. The two detector coincidence [49] Doppler broadening of the electron positron annihilation radiation (CDBEPAR) measurement, having peak to background ratio better than 14000 : 1, have been used to study the temperature dependent (300 K to 30 K) electron momentum distributions in these high T_c superconducting oxides. The CDBEPAR data are analysed both by conventional lineshape analysis and the ratio curve analysis [50,51].

The CDBEPAR S-parameter vs. temperature graph for these three different HTSC samples shows a step like increase in the value of the S-parameter at their respective superconducting transition temperature region which clearly indicate superconductivity induced redistribution of the electron momentum distributions at the superconducting transition region for these HTSC.

Comparison between the ratio-curves (constructed with reference to the CDBEPAR spectra of pure Al and Cu metals) for single crystalline $Bi_2Sr_2CaCu_2O_{8+\delta}$ and polycrystalline $La_{0.7}Y_{0.3}Ca_{0.5}Ba_{1.5}Cu_3O_z$ HTSC samples shows that positrons are relatively more probing the Cu site in the $La_{0.7}Y_{0.3}Ca_{0.5}Ba_{1.5}Cu_3O_z$ HTSC than the $Bi_2Sr_2CaCu_2O_{8+\delta}$ HTSC. The analyses of the CDBEPAR spectra at different sample temperatures in case of $Bi_2Sr_2CaCu_2O_{8+\delta}$ HTSC have been done by constructing ratio-curves with respect to room temperature (298 K) CDBEPAR spectrum. The results indicates less annihilation of the positrons with the 3d electrons of Cu ions and more annihilation with the 2p electrons of the O ions, which suggest a shift of the apical oxygen ion towards the Bi-O plane. In case of $La_{0.7}Y_{0.3}Ca_{0.5}Ba_{1.5}Cu_3O_z$ HTSC ratio-curves for different sample temperatures have been constructed with respect to the CDBEPAR spectra of pure Cu metal. Here also a less annihilation of positrons with the 3d electrons of Cu ions just above T_c strongly suggest an increase of effective positive charge at the superconducting Cu-O plane due to onset of superconductivity.

2   Probing Solids by Positron Annihilation Techniques

2.1   Basics of the Positron Annihilation Technique

Positron annihilation technique is a nuclear solid state technique [4-6] to study the electron number density, characterization of defects and the electron momentum distributions in a material. Entering a solid, energetic positrons (from a radioactive source e.g., ^{22}Na, ^{64}Cu, ^{58}Co, etc.) become thermalized within 1 to 10 ps by producing electron-hole pairs and phonons and then diffuse ($\sim$ 100 nm) inside the material. The eventual annihilation of the thermalized positron with an electron in the studied material is in general ($\sim$ 99.7 %) a two 511 keV γ-annihilation process.

There are three basic techniques using positron annihilation in a material.

* *Positron annihilation lifetime measurement technique*: to study the electron number density and to characterize the possible defect sites in a material.
* *Doppler broadening of the electron positron annihilation γ-radiation measurement technique* to study the electron momentum distributions in a material.
* *Angular correlation of annihilation radiation (ACAR) spectroscopy* to study the electron momentum distributions in a material.

Figure 1 shows the schematic representation of the main positron annihilation techniques.

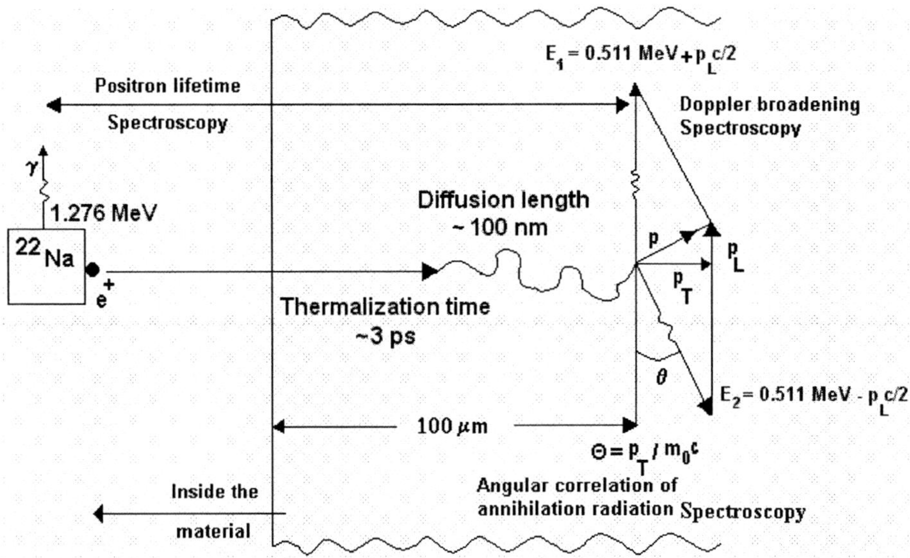

Figure 1: Schematic representations of the three basic positron annihilation techniques:

(i)   Positron annihilation lifetime spectroscopy,
(ii)   Doppler broadening spectroscopy and
(iii)   Angular correlation of the annihilation radiation spectroscopy.

In the non-relativistic limit, positron annihilation rate, λ, is given by the overlap integral of the electron density $n_-(\mathbf{r})$ at the annihilation site and the positron density $n_+(\mathbf{r}) = |\Psi^+(\mathbf{r})|^2$ [6],

$$\lambda = 1/\tau = \pi r_0^2 c \int |\Psi^+(\mathbf{r})|^2 \, \gamma \, n_-(\mathbf{r}) \, d\mathbf{r}$$

where r_0 is the classical electron radius, c the speed of light, and $\mathbf{r}$ the position vector. γ ($\sim$ $1+\Delta n_-/n_-$) is the correlation function and it describes the increase of the electron density (Δn_-) due to the Coulomb attraction between electron and positron. Thus the positron annihilation lifetime, τ, (which is the reciprocal of the positron annihilation rate, λ) is inversely proportional to the electron number density. Therefore, by measuring the positron annihilation lifetime one can obtain directly the information about the electron density at the site of positron annihilation [5].

In the center of mass frame (in case of two photon annihilation process), the energy of the annihilating photon is exactly $m_0 c^2 = 511$ keV (m_0 is the rest mass of the electron or the positron) and the two photons are moving exactly in the opposite direction, i.e., the emission angle between the two 511 keV γ-rays are 180°. But the electron-positron pair has some momentum, p, which is entirely due to the momentum of the electron, as before annihilation, the positron is thermalized and hence its momentum is almost negligible ($\sim$ meV). During the annihilation process, the momentum of the electron-positron pair (p) is transferred to the photon pair to conserve the momentum. As a result of which the 511 keV annihilation γ- rays are Doppler shifted [6] by an amount $\pm \Delta E$ in the laboratory frame. Where

$$\pm \Delta E = p_L c/2$$

p_L ($p\cos\theta$) is the component of the electron momentum, p, along the direction of the detection of the annihilating γ- rays. So by measuring the Doppler broadening of the electron positron annihilation 511 keV γ-radiation spectrum one can study the momentum distributions of the electrons at the positron annihilation site. Figure 2 represents the Doppler shift of the electron positron pair along the detector direction due to non zero momentum of the electron positron pair.

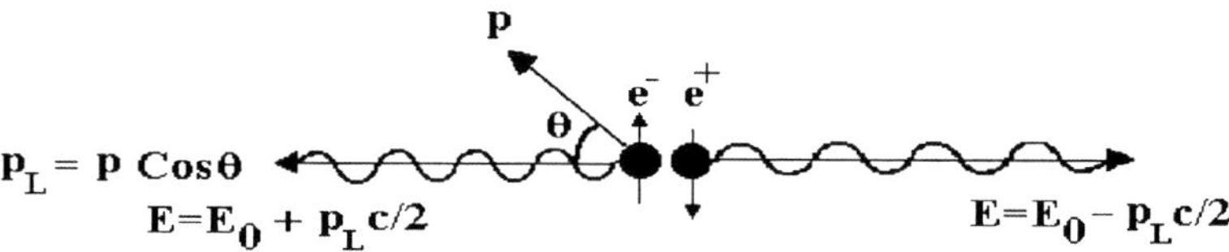

Figure 2: Schematic representation of Doppler shift of the annihilating γ- rays along the detector direction.

2.2   Experimental Setup for the Coincidence Doppler Broadening of the Electron Positron Annihilation Radiation Measurement

Using a high resolution HPGe detector one can measure the Doppler broadening of the electron positron annihilation γ-radiation (DBEPAR) spectrum. The central portion of the DBEPAR spectrum (as shown in Figure 3) represents those 511 keV γ-rays, which are less Doppler shifted, i.e., coming from the annihilation of positrons with the lower momentum electrons. Similarly the wing portion of the DBEPAR spectrum represents those 511 keV γ-rays, which are more Doppler shifted, i.e., coming from the annihilation of positrons with the higher momentum electrons, e.g., core electrons. Now it is very important to study the annihilation of positrons with the core electrons in a particular material. Hence it is very important to increase the statistics of the counts in the DBEPAR spectrum, particularly in the wing portion. Unfortunately, the Compton part of the 1.274 MeV γ-ray is always present in the photo-peak of the 511 keV γ-rays and is more prominent as a background in the wing portion of the DBEPAR spectrum. The typical peak to background ratio of a DBEPAR spectrum is $\sim$ 50 : 1. This peak to background ratio can be improved more than 10000 : 1 by using another NaI(Tl) detector in the opposite direction, i.e., the 511 keV – 511 keV coincidence technique [47,49] by using two oppositely directed (angle between the two detector is 180°) detectors. Using two HPGe detectors in opposite direction one can increase the peak to background ratio better than 10^5 : 1.

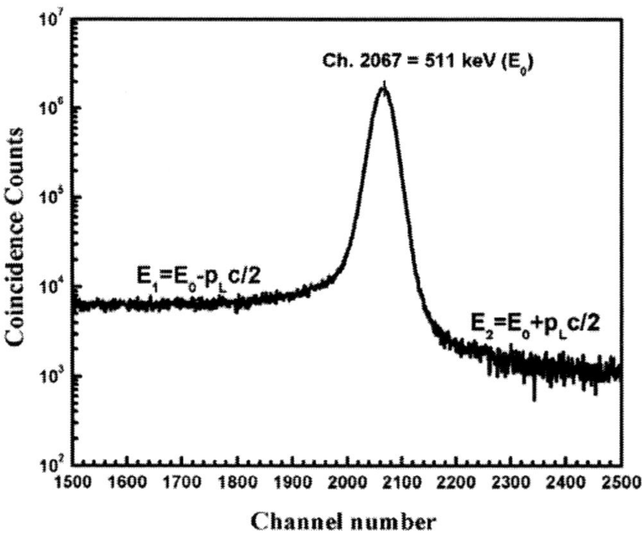

Figure 3: A typical Doppler broadening spectrum with two detectors in coincidence [one HPGe and another NaI(Tl) detectors].

The block diagram of such a Coincidence DBEPAR (CDBEPAR) spectrometer, used in the present experiments, is shown in Figure 4. In the present case, one n-type HPGe detector (OXFORD GC 13117 Detector) of efficiency 13 % having an energy resolution of 1.10 keV for the 514-keV γ-ray line of ^{85}Sr with 6 μs shaping time constant in the spectroscopy amplifier is used as a primary detector to measure the Doppler broadening of the electron positron annihilation γ-radiation spectrum. The detector crystal has an active volume of 40 cm^3 (47.6 mm diameter and 23.5 mm length). The detector is always placed in cryostat containing liquid nitrogen. The bias voltage (+ 2400 V) is given to the HPGe by TENELEC TC 950 High Voltage Supply. A $3^{//} \times 3^{//}$ NaI(Tl) crystal optically coupled to a RCA 8850 photomultiplier tube has been placed at 180° with the HPGe detector for the purpose of coincidence measurement. The detection of the oppositely directed 511 keV γ-rays by the NaI(Tl) detector reduces the background of the annihilation γ-ray spectrum recorded in the HPGe channel under the 511 keV photo-peak and adds to the precision of the measurement.

A total of ~ 6 $\times$ 10^6 to 10^7 coincidence counts have been recorded under the photo-peak of the 511 keV γ-ray coincidence-DBEPAR spectrum at a rate of 110 counts per second. The CDBEPAR spectrum is recorded in a PC based 8k multi-channel analyzer.

Background has been calculated from 607 keV to 615 keV energy range of the spectrum. The achieved peak to background ratio in the present case is ~ 14000 : 1. The system stability has been checked frequently during the progress of the experiment.

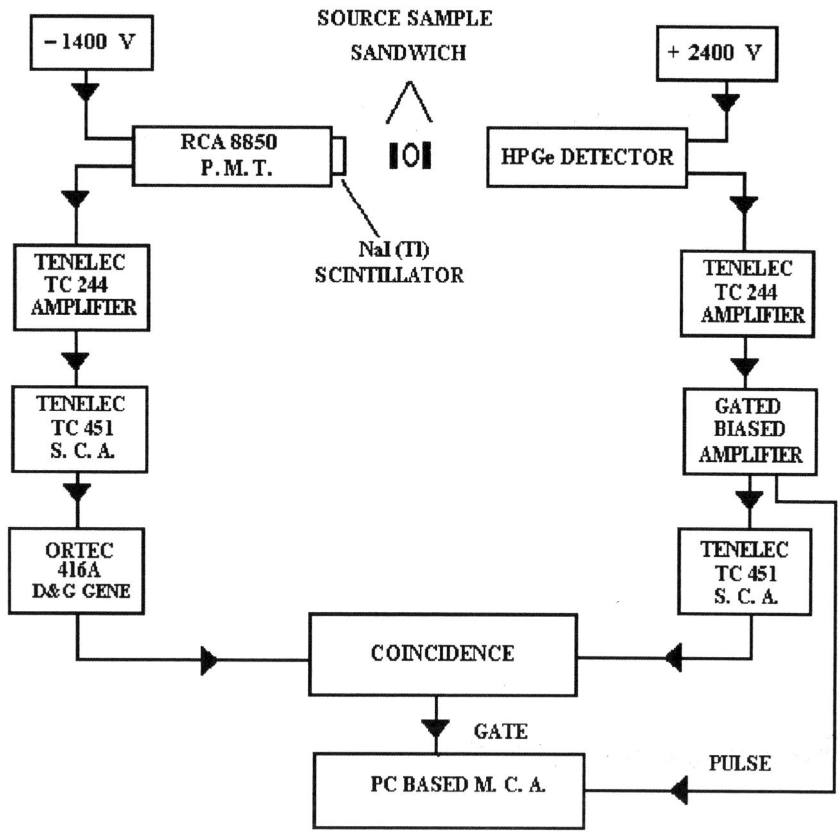

Figure 4: Block diagram of coincidence DBEPAR spectrometer.

2.3   The CDBEPAR Data Analysis

A.   Line Shape Analysis

The coincidence doppler Broadening of the electron positron annihilation 511 keV γ-ray spectrum has been analyzed by evaluating the so called line-shape parameters [5] (S-parameter and W-parameter). The S-parameter is calculated as the ratio of the counts in the central area of the 511 keV photo peak ($|$ 511 keV - E_γ $|$ $\leq$ 0.85 keV) and the total area of the photo peak ($|$ 511 keV - E_γ $|$ $\leq$ 4.25 keV). The S-parameter represents the fraction of positron annihilating with the lower momentum electrons with respect to the total electrons annihilated. During all measurements the value of the S parameter is kept fixed around 0.45 to 0.5 by suitable selecting the energy range. The W-parameter represents the relative fraction of the counts in the wings region (1.6 keV $\leq$ $|E_\gamma$ -511 keV$|$ $\leq$ 4 keV) of the annihilation line with that under the whole photo peak ($|$ 511 keV - E_γ $|$ $\leq$ 4.25 keV). The W-parameter corresponds to the positrons annihilating with the higher momentum electrons. The statistical error is 0.2 % on the measured line-shape parameters.

B.   Ratio-Curve Analysis

To identify the contributions of the valence and the core electron momentum involved in the annihilation process ratio curve analysis [50,51] have been followed. Ratio-curve is defined as point to point ratio of area normalized CDBEPAR spectrum of the material under study with an area normalized CDBEPAR spectrum of reference sample. Reference sample should be a highly pure defects free sample. In the present studies defects free 99.9999 % pure Al single crystal and 99.9999 % pure Cu single crystal have been taken as reference samples.

3   Earlier Positron Annihilation Studies on HTSC Oxides

3.1   Y-Based high T_c Superconducting Oxide

Temperature dependence of positron annihilation parameters on Y-123 high T_c superconductors was first studied by Ishibashi et al. in 1986 [12]. In 1987 Jean et al. [13] measured the positron annihilation parameters for $YBa_2Cu_3O_{6+\delta}$ (δ=0.8 for the superconducting sample and δ=0 for the non-superconducting sample). The superconducting critical temperature, T_c, of the sample was 88 K. They observed a significant change in the positron annihilation intermediate lifetime component, τ_2, and the Doppler broadened positron annihilation shape parameter, S parameter, with temperature for the superconducting and non-superconducting sample. The lifetime components τ_1 and τ_3 remain almost constant with temperature. According to them the lifetime component τ_2 and its intensity I_2 is due to the positron annihilation at the oxygen vacancies present in the sample. In non-superconducting sample the intensity I_2 was 8 % $\pm$ 2 %, larger than that in the superconducting sample. The lifetime component τ_2 and the S parameter decrease abruptly below T_c and the nature of the slope dτ/dT and dS/dT are different for the temperature above T_c and below T_c. From their measurement they conclude that in this high temperature superconductors the electronic structure is different for the normal state and the superconducting state, and the oxygen vacancies present in the sample changes the electron density (higher in the superconducting state than the normal state) as well as the electron momentum distributions at T_c. In 1988, Smedskjaer et al. [14] measured the Doppler broadened positron annihilation parameters for polycrystalline$YBa_2Cu_3O_{7-\delta}$ (δ = 0.1) HTSC as a function of temperature. They observed that the value of the S parameter changes abruptly near T_c and between 11 K to T_c the line-shape parameter increases by an amount $\Delta S/S \approx 6 \times 10^{-3}$. According to them, at T>T_c the change in the S parameter may be due to the positron trapping in extended defects with low binding energy, but at T<T_c the change in the S parameter is due to the changes in the electronic structures. Harshman et al. [15] measured the positron annihilation lifetime parameters for single crystalline $YBa_2Cu_3O_7$ (Y1:2:3) HTSC. Above T_c they observed that the single lifetime component τ is independent of temperature but below T_c it increases with decreasing temperature. According to them this peculiar behavior of τ may be linked with the superconducting gap, redistribution of electronic charge or may be due to lattice relaxations.

Usmar et al. [16] measured the positron lifetime component for $YBa_2Cu_3O_7$ and the Doppler broadened S parameter for $YBa_2Cu_3O_{7-x}$ (x = 0.7, 0.4 and 0.0) samples. At T>T_c, the longer lived component τ_2, is independent of temperature but it changes from 210 $\pm$ 2 ps to 200 $\pm$ 2 ps across T_c and continues to decrease up to the lowest temperature is reached (12 K).

In case of $YBa_2Cu_3O_7$ and $YBa_2Cu_3O_{6.6}$ superconductors, at $T>T_c$ S parameter remains almost constant, but at the superconducting transition region S parameter decreases about 0.2% and 0.4% respectively. No such change is observed for the non-superconducting $YBa_2Cu_3O_{6.3}$ sample. According to them as the oxygen concentration increases from sample to sample, such change (decreasing nature) in S parameter is observed.

Wang et al. [17] studied the temperature dependence of the positron annihilation mean lifetime component, τ, and the S parameter for $YBa_2Cu_3O_{7-\delta}$ At 100 K, τ has a value of 257 $\pm$ 1 ps and it increases to 268 $\pm$ 1 ps at 150 K. Then it decreases. At superconducting phase the value of S parameter is lower than the value at normal phase. They concluded that such changes in the annihilation parameters are due to positron localization in a region of lower electron density. Similar types of studies have been reported by large number of groups [18-23].

Jean et al. [24] measured the temperature dependence of the positron lifetime parameters across T_c in undoped, Zn doped and Ga doped Y-1:2:3 HTSC. According to them, in case of undoped and Ga doped Y-1:2:3 HTSC, at $T<T_c$ the bulk lifetime decreases but it increases for Zn doped Y-1:2:3 HTSC. In Zn doped Y-1:2:3 HTSC, the positron density distribution (PDD) shifts from the CuO chain towards the CuO_2 layer. Therefore the electron density at CuO_2 layer decreases which effectively increase the positron lifetime value. But in case of Ga doped Y-1:2:3 HTSC, PDD is in the CuO chain and therefore below T_c the lifetime value decreases. From these they concluded that in this Y-1:2:3 system if the positrons probe the CuO_2 layer, then below T_c the positron lifetime component τ and the S parameter increases but when the positrons probe the CuO chains both of the positron annihilation parameter (τ and S parameter) decrease below T_c.

Y-1:2:3 superconductors have one CuO chains but in Y-1:2:4 superconductors $(YBa_2Cu_4O_8)$ there are two CuO chains. Sundar et al. [25] measured the temperature dependence of the positron annihilation parameters in this Y-1:2:4 system. They observed that the bulk lifetime τ and the S parameter decrease below T_c. The long lived component τ_2 is independent of temperature and the lifetime component τ_1 decreases at $T<T_c$. From this they have concluded that in Y-1:2:4 system also the positrons probe the region between the CuO double chains and there is a decrease in the annihilation parameters below T_c.

Depending upon the oxygen concentration, $La_{1.85}Sr_{0.15}CuO_4$ may be superconductor or non-superconductor. Jean et al. [19] measured the positron annihilation parameters for $La_{1.85}Sr_{0.15}Cu_4$ superconductor (T_c = 33 K) in the temperature range 10 K to 297 K. They observed only a single lifetime component of the order of $170 - 190$ ps present for this sample. Both τ and S parameter increase below T_c which is similar to the results obtained for single crystalline $YBa_2Cu_3O_7$ but different from $YBa_2Cu_3O_{6.8}$.

3.2   Tl-based high T_c Superconducting Oxide

The $Tl_2Ba_2Ca_{n-1}Cu_nO_{2n+4}$ system is another interesting class of HTSC (maximum T_c = 125 K with n = 3). In this HTSC system Tl-O layer plays an important role in the mechanism of superconductivity. This system grows some special interest as their T_c value increases with increasing number of CuO_2 layers. Since few years there has been large number of attempts to study this Tl-based HTSC system by employing several experimental techniques [26-33]. Jean et al. [22] studied the temperature dependence of the positron annihilation parameters for

$Tl_{2.2}Ca_2Ba_2Cu_3O_{10.3+\delta}$ HTSC. They observed that the shortest lifetime component, τ_1, decreases monotonically from 138 ps (at $2T_c$) to 100 ps (near T_c). But its intensity remains constant throughout the temperature range $2T_c$ to 298 K. They also observed that the temperature dependence of the bulk lifetime τ_B for this $Tl_{2.2}Ca_2Ba_2Cu_3O_{10.3+\delta}$ HTSC is similar to those observed in $YBa_2Cu_3O_7$ [15] and $La_{1.85}Sr_{0.15}CuO_4$ [19] HTSC samples. In 1990, Sundar et al. [34] measured the positron annihilation lifetime and the Doppler broadening parameters in Tl-Ba-Ca-Cu-O system at different temperatures. For Tl-2201 and Tl-2212 systems T_c were 20 K and 104 K respectively. In Tl-2212, the lifetime parameters are $\tau_1=160 \pm 5$ ps, $\tau_2=380 \pm 10$ ps and $I_2 = (15 \pm 2)$ % respectively whereas for Tl-2201, these are $\tau_1 = 215 \pm 2$ ps, $\tau_2 = 500 \pm 30$ ps and $I_2 = (5 \pm 1)$% respectively. With deceasing temperature τ_1 decreases but τ_2 and I_2 remains constant with temperature.

Wang et al. [35] studied the variation of the Doppler broadening parameters for the Tl-system in 1223 phase in the temperature range 77 K to room temperature. They observed that both S-parameter and W-parameter have some wavy nature and the decrease in the S parameter between the temperatures 130 K to 150 K is about 1% which indicates that there is a change in the electron momentum distribution in this temperature region. According to them there is a transformation from free electrons to localized electrons in this temperature range.

3.3 Bi-based high T_c Superconducting Oxide

$Bi_2Sr_2Ca_{n-1}Cu_nO_{2n+4}$ or Bi-22(n-1)n high temperature superconductors are another class of interesting HTSC system. After the discovery of this Bi-Sr-Ca-Cu-O system, large number of efforts has been made to study the structural and physical property of this system [36,37]. The temperature dependence of the positron annihilation parameters across T_c in this HTSC system have been studied by several groups. Singh et al. [38] calculate the positron wave function for $Bi_2Sr_2CaCu_2O_{8+\delta}$ HTSC by using the general potential linearized augmented plane wave method. According to them in this $Bi_2Sr_2CaCu_2O_{8+\delta}$ HTSC system, positron wave function is quite delocalized in the unit cells, and therefore is expected to probe the CuO layer derived electronic states. In 1991, Sundar et al., [39] measured the positron annihilation lifetime values in Bi-Sr-Ca-Cu-O system. They observed that the lifetime parameters are independent of temperature. They also measured the variation of the positron annihilation lifetime parameters as a function of annealing temperature. In this case, the lifetime component τ_1 decreases from 210 ps (at 800^0 C) to 195 ps (at 100^0 C). Same trend has been observed for τ_B also. According to them the decrease in lifetime on lowering the annealing temperature is correlated with the increase in oxygen content in the system. From Positron Density Distribution (PDD) calculation they also concluded that in this HTSC system positron density is confined in the Bi-O plane. The temperature independence in the lifetime parameters is thus due to the lack of positrons at the superconducting CuO_2 layers across T_c in this HTSC system. Similarly Zhang et. al. [40] studied the temperature dependence of the positron annihilation lifetime parameters in the two mixed phase Bi-Sr-Ca-Cu-O HTSC system. Above 120 K, these parameters are independent of temperature, but below this temperature there are some strong irregular oscillations in these parameters and after 83 K these parameters increase again. According to them the strange behavior in the lifetime parameters in this temperature region is due to the phase transition occurring in the Bi-Sr-Ca-Cu-O (2:2:2:3) phase.

In 1994, Pujari et al. [41] studied the temperature dependence of the positron annihilation parameters in $Bi_{2-x}Pb_xCa_2Sr_2Cu_3O_y$ and $Bi_2Sr_2CaCu_2O_y$ superconductors. At room temperature the lifetime values for Bi-2:2:1:2 sample are $\tau_1 = 220$ ps, $\tau_2 = 377$ ps and $\tau_B = 245$ ps, whereas for Bi-2:2:2:3 sample these are 223 ps, 396 ps and 230 ps respectively. In the superconducting state the values of τ_B are 226 ps and 238 ps for Bi-2:2:2:3 and Bi-2:2:1:2 samples respectively. The S parameter also decreases at the superconducting transition temperature region from its value at the non superconducting temperature region. Similar type of change in the bulk lifetime values for the Bi-2223 samples have been also observed by Tang et al., [42]. In the year 1994, Tang et al., [43] studied the variation of positron annihilation parameters with the annealing temperature in $Bi_{2-x}Pb_xSr_2Ca_2Cu_3O_{10}$ HTSC. They observed that the mean lifetime component rises slightly below 300 K and then increases rapidly with temperature. According to them the annealing treatment in vacuum creates oxygen vacancies in the HTSC system which causes the electron density to decrease. The increase in the mean lifetime value is thus due to these oxygen vacancies present in the samples. Sanyal et al. [9] measured the positron annihilation lifetime values in the temperature range 30 K to 300 K for $(Bi_{0.92}Pb_{0.17})_2Sr_{1.91}Ca_{2.03}Cu_{3.06}O_{10+\delta}$ HTSC. The T_c value of the sample was 104 K. The value of the S parameter decreases from its normal value (at room temperature) at the T_c region and τ_2 shows a broad minimum at around 175 K. They observed a decreasing nature of the mean positron lifetime component due to superconducting transition, which indicate an increase of the electron density at the positron annihilation sites. According to them such type of change in the annihilation parameters may be linked with the charge transfer model effects or other lattice instabilities.

3.4 Other High T_c Superconducting Oxide

Another interesting class of superconductors is $RE_{1+x}Ba_{2-x}Cu_3O_{7-\delta}$ or RE-123 where RE = La, Nd, Sm, and Eu. These superconductors have lower critical temperature (T_c) than that of other class of superconductors like Hg-Ba-Ca-Cu-O, Tl- Ba-Ca-Cu-O, and Bi-Sr- Ca-Cu-O. Li et al. [44] measured the positron annihilation parameters for $Eu_{1+x}Ba_{2-x}Cu_3O_{7-\delta}$ superconductors with $x < 0.06$ (orthorhombic phase with low level of Eu substitution), $0.06 < x < 0.10$ (beginning of orthorhombic phase transition) and $x > 0.10$ (strong orthorhombic phase transition). From the variation of the mean positron annihilation lifetime component τ_m and the Doppler broadening of the annihilation radiation lineshape parameter (S parameter) with x it is clear that for $x < 0.06$, τ_m increases and the S parameter decreases but in the region $0.06 < x < 0.10$, τ_m decreases and S parameter increases again in the region $x > 0.10$, τ_m increases and S parameter decreases. According to them, such type of change in the annihilation parameter is due to structural phase transition (O-T phase transition) in this material.

4   Present *CDBEPAR* Studies on High T_c Superconducting Oxides

Presently we have carried out CDBEPAR measurement on three different high temperature superconductors, namely, single crystalline $Bi_2Sr_2CaCu_2O_{8+\delta}$ (T_c = 91 K) [52]single crystalline $SmBa_2Cu_3O_{7+x}$ (T_c = 94 K) and polycrystalline $La_{0.7}Y_{0.3}Ca_{0.5}Ba_{1.5}Cu_3O_z$ (T_c = 78.5 K) [46,53] across their superconducting critical temperature. The conventional line-shape analysis of the CDBEPAR spectrum indicates a superconductivity induced redistribution of the electron momentum distributions around the superconducting transition temperature region. To identify the core electrons with which positrons are annihilating in these complex structured materials, ration-curve analysis have also been done. In the first section we will discuss the conventional line-shape analysis of the CDBEPAR spectra for these three different HTSCs. Ratio-curve analyses of the CDBEPAR spectra will be discussed in the next section. Anisotropy of the EMD in the different crystallographic orientations of the single crystalline $Bi_2Sr_2CaCu_2O_{8+\delta}$ HTSC will be discussed in the last section.

4.1 Conventional Line-Shape (S, W and S/W - Parameter) Analysis of the CDBEPAR Spectra

A   In case of Single Crystalline $Bi_2Sr_2CaCu_2O_{8+\delta}$ HTSC

The temperature dependent (30 K to 300 K) coincidence Doppler broadening of the electron positron annihilated γ-radiation (CDBEPAR) measurement has been carried out on single crystalline $Bi_2Sr_2CaCu_2O_{8+\delta}$ (Bi-2212) high T_c superconducting sample along the crystallographic c –axis of the crystal.

The distribution of positrons in these layered-structured HTSC oxides is not uniform. The annihilation characteristics of the positrons from such a structurally complex material bear the information related to the region where positron density distribution is maximum. The positron density distribution calculations for $Bi_2Sr_2CaCu_2O_{8+\delta}$ HTSC by Sundar et al., [39] show that the positron density is maximum in the Bi-O planes. It is attributed to the fact that a fraction of positrons are mainly annihilating at the oxygen site [9] of the Bi-O plane. These annihilating 511 keV γ-rays are less Doppler broadened and contribute to the central part of the Doppler broadened 511 keV γ-ray spectrum. Thus the variation of S-parameter with temperature may be correlated with the variation of the momentum of electron at the oxygen site of Bi-O plane.

Figure 5 shows the variation of the S-parameter with sample temperature for single crystalline $Bi_2Sr_2CaCu_2O_{8+\delta}$ HTSC. As mentioned earlier S-parameter reflects the contribution of the lower momentum electrons in the Doppler broadened 511 keV γ-ray spectrum whereas the W-parameter is associated with the higher momentum electrons. A change in the S-parameter is associated with the redistribution of electron momentum inside the material. In the present experiment an increase of S-parameter in the temperature region from 92 K to 116K has been observed. The observed step like increase of S-parameter at 116 K has a magnitude of ~ 0.8 %, which is in agreement with the earlier results [7,9]. Just at the superconducting transition temperature T_c (91 K) S-parameter suddenly comes back nearly to its original value.

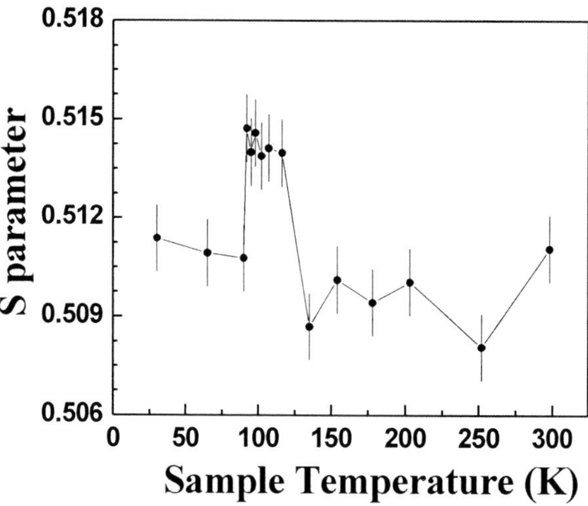

Figure 5: Variation of S-parameter as a function of sample temperature for single crystalline Bi-2212 HTSC.

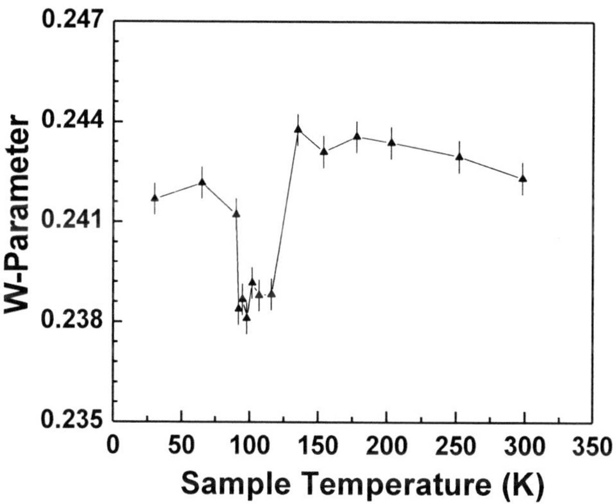

Figure 6: Variation of W-parameter as a function of sample temperature for single crystalline Bi-2212 HTSC.

Figure 6 represents the variation of the W – parameter with sample temperature for single crystalline Bi-2212 HTSC. From the figure a decreasing nature in the W- parameter in the temperature region from 92 K to 116K has been observed. Just at the superconducting transition temperature, T_c (91 K) W-parameter suddenly comes back nearly to its original value which also confirms the redistribution of electron momentum at and around the superconducting transition region.

The S/W-parameter represents the fraction of lower momentum electron over their higher momentum counterparts. The variation of S/W parameter with temperature is shown in Figure 7. The same behavior of S/W-parameter and S-parameter with temperature is visible from Figures 7 & 5. It represents that transfer of electrons [13,16,54] between higher and lower momentum states is associated in the process of superconducting transition which starts far above T_c.

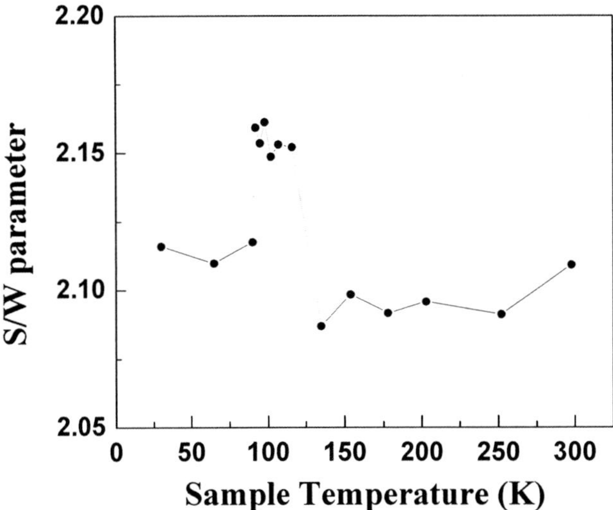

Figure 7: Variation of S/W-parameter as a function of sample temperature for single crystalline Bi-2212 HTSC.

The increase of the S-parameter suggests either the positrons are less annihilating with the core electrons or an increase of the number of lower momentum electrons at the positron annihilation site. This step-like increase of S-parameter above T_c and its coming back to the original value at T_c may be linked with the possibility of the local structural changes [11] which in a way favour the "charge transfer model" valid for these cuprate superconductor [42]. According to the charge transfer model due to the onset of superconductivity, charge is transferred from the Bi-O plane to Cu-O plane (superconducting plane). This could be possible by considering a structural change which helps to increase the coupling of the p-type state in the Bi-O band in such a way that the number of the d-type electrons in Cu-O band decreases. In this way the effective hole density in the Cu-O layer and the electron density at the Bi-O layer (positron annihilation site) increases. In earlier experiments of temperature dependent positron lifetime studies on Bi-based polycrystalline HTSC [9] a step-like decrease of mean positron lifetime around the superconducting transition region has been observed, which supports the charge transfer model. The increased number of electrons in the Bi-O band increases the probability of positrons to be annihilated with lower momentum electrons of the oxygen site and therefore increases the value of S-parameter. Thus the structural changes favour electron momentum redistribution at and around the superconducting transition region.

B   In Case of Single Crystalline SmBa$_2$Cu$_3$O$_{7+x}$ HTSC

The structure of a unit cell of RBa$_2$Cu$_3$O$_{7+x}$ (where R=Sm, Y, Nd) HTSC is also inhomogeneous and layered. The positrons are also not uniformly distributed inside the whole unit cell of these layered structured superconductors rather they populate in certain layers. In these layered structured cuprate superconductor certain layers are important in the light of superconductivity. In case of R-based HTSC, there are two Cu sites, CuO chain and CuO_2 plane. It is observed that CuO_2 plane is the most important subband with respect to the superconductivity. But positron density distribution calculations [24] show that positrons prefer to go to the CuO chains in these $RBa_2Cu_3O_{7+x}$ or R-123, R = Y, Sm, Nd etc. Thus by employing positron annihilation technique one can probe the CuO chain. From positron density distribution calculations [24] by Jean et al., it is established that positron will prefer to go to the CuO chain when the $YBa_2Cu_3O_{7+x}$ is undoped, but for the samples with Zn substitution, positrons will prefer to go to the CuO_2 plane. In the present work an effort has been made to probe the electron momentum distribution for pure R-123 (R= Sm) single crystals by employing coincidence Doppler broadening of the electron positron annihilation γ-ray line-shape measurement technique and to observe whether the variation of the lineshape parameters with sample temperature in this HTSC are also similar in nature with the previously discussed HTSC.

Figure 8 represents the variation of S parameter with sample temperature for Sm-123 HTSC. It is seen from the Figure 8 that the S parameter increases just at the superconducting transition temperature from its value at the non-superconducting zone. Then again it decreases and comes back to its original value at 75 K. The variation of S - parameter with sample temperature for Sm-123 single crystalline HTSC is similar to the variation of the S parameter for single crystalline Bi-2212 HTSC. Thus from Figure 8 it is clear that in this HTSC also there occur some structural changes as a result of which a redistribution of the electron momentum distribution has been observed around the superconducting transition region.

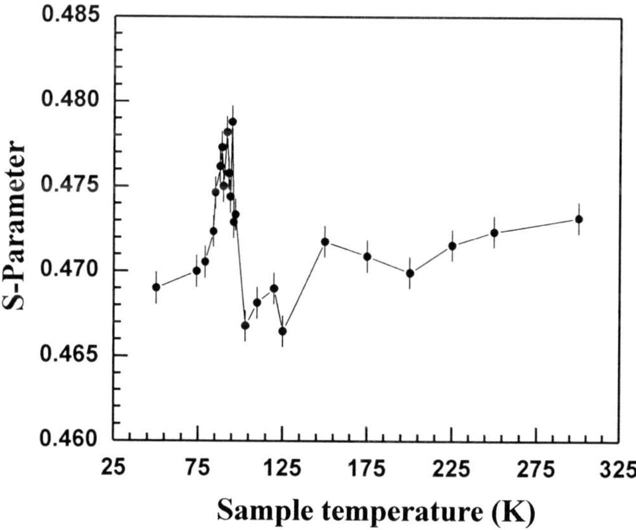

Figure 8: Variation of S parameter with sample temperature for single crystalline Sm-123 HTSC.

Figure 9 represents the variation of the S/W parameter with sample temperature for single crystalline Sm-123 HTSC. It is clear from Figure 9 that the nature of the variation of the S/W parameter with sample temperature is similar to that of S parameter with sample temperature. The S/W-parameter represents the fraction of lower momentum electron over their higher momentum counterparts.

Figure 10 represents the variation of the W parameter with sample temperature for single crystalline Sm-123 HTSC. The W parameter decreases at T_c from its value at non-superconducting region. Then it goes back to its original value at 75 K. From Figure 10 it can be concluded that at the superconducting transition there occurs some structural changes.

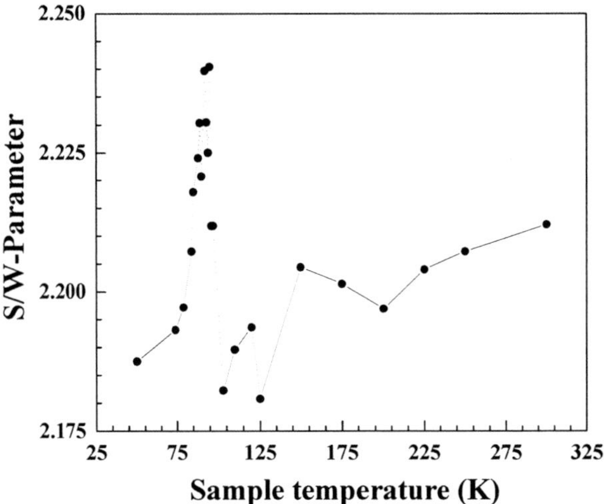

Figure 9: Variation of the S/W parameter with sample temperature for single crystalline Sm-123 HTSC

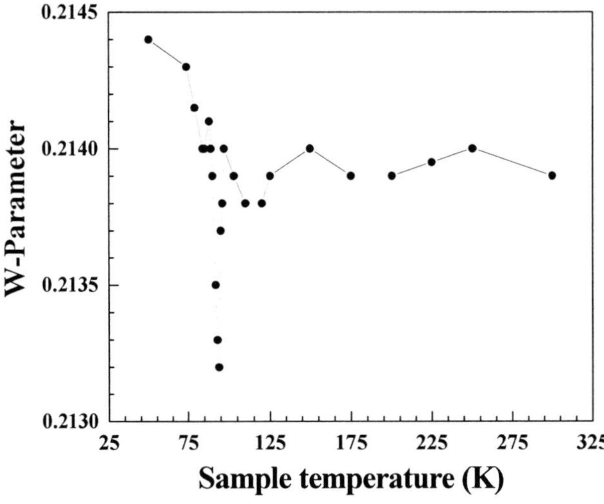

Figure 10: Variation of the W parameter with sample temperature for single crystalline Sm-123 HTSC.

C   In Case of Polycrystalline $La_{0.7}Y_{0.3}Ca_{0.5}Ba_{1.5}Cu_3O_z$ HTSC

The temperature dependent S-parameter for the tetragonal LYCBCO type polycrystalline $La_{0.7}Y_{0.3}Ca_{0.5}Ba_{1.5}Cu_3O_z$ high T_c superconductor is shown in Figure 11.

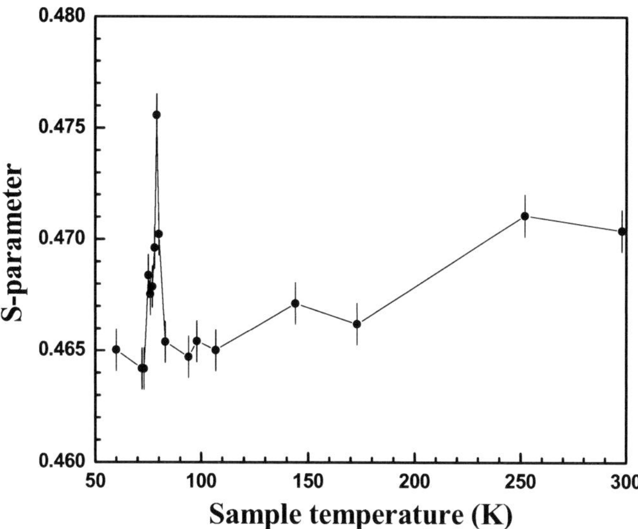

Figure 11: Variation of S-parameter as a function of sample temperature for the $La_{0.7}Y_{0.3}Ca_{0.5}Ba_{1.5}Cu_3O_z$ polycrystalline high T_c superconductor.

The S vs. T graph (Figure 11) shows a sharp peak at the superconducting transition temperature region. The results are also in agreement with the previous results for the single crystalline $Bi_2Sr_2CaCu_2O_{8+\delta}$ HTSC and Sm-123 HTSC where an increase of S-parameter with sample temperature has been observed just above the superconducting transition temperature. The increase of the S-parameter, in general, suggests either the positrons are less annihilating with the core electrons or the annihilation increases with the lower momentum electrons at the positron annihilation site.

Summary

Temperature dependent CDBEPAR measurement technique followed by the lineshape analysis have been carried out on three different type of HTSC oxides e.g., single crystalline $Bi_2Sr_2CaCu_2O_{8+\delta}$ (T_c = 91 K), single crystalline $SmBa_2Cu_3O_{7+x}$ (T_c = 94 K) and polycrystalline $La_{0.7}Y_{0.3}Ca_{0.5}Ba_{1.5}Cu_3O_z$ (T_c = 78.5 K). S-parameter vs. temperature graph of the single crystalline $Bi_2Sr_2CaCu_2O_{8+\delta}$ (Bi-2212) high T_c superconducting sample shows a step like increase in the value of the S-parameter ($\sim$ 0.8 %) at the temperature region of 116 K (far above T_c) to 92 K. The temperature dependent S-parameter for single crystalline Sm-123 and polycrystalline $La_{0.7}Y_{0.3}Ca_{0.5}Ba_{1.5}Cu_3O_z$ HTSC also shows a peak like change of S-parameter at their respective superconducting transition temperature region. Thus for all these three HTSC (irrespective of single crystalline or polycrystalline) a superconductivity induced redistribution of the electron momentum has been observed around the superconducting transition region.

4.2 Ratio Curve Analysis of the CDBEPAR Spectra

B In case of Single Crystalline $Bi_2Sr_2CaCu_2O_{8+\delta}$ HTSC

In the previous section it has been observed that the superconductivity induced changes in the positron annihilation lineshape parameters are very small < 1%, which is in agreement with the previously reported results [7]. To observe such a small change in the ratio curve the room temperature (298 K) CDBEPAR spectrum of the $Bi_2Sr_2CaCu_2O_{8+\delta}$ HTSC has been chosen as a reference spectrum for the construction of the ratio-curves [50,51] at the other sample temperature CDBEPAR spectrums. Ratio-curves at 14 different temperatures (252, 203, 178, 154, 135, 116, 107, 102, 98, 95, 92, 90, 65 and 30 K) have been constructed for the single crystalline Bi-2212 HTSC sample. Figure 12 represents the ratio-curves at some selected temperature points (T = 203, 135, 116, 98, 92, 90, and 30 K). In between these temperature points, ratio-curves follow the same trend.

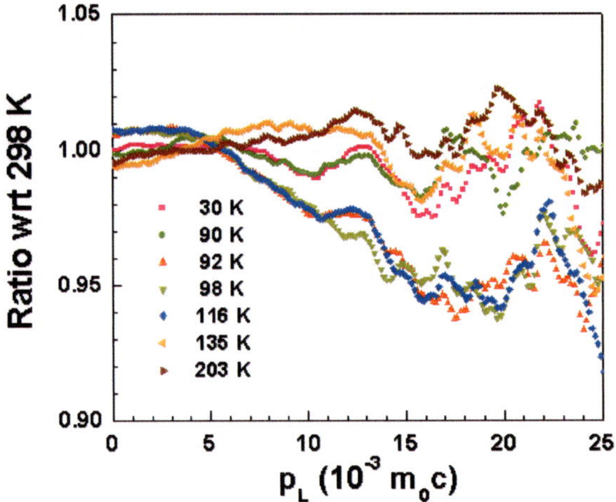

Figure 12: Ratios of the experimental electron-positron momentum distributions at different sample temperatures to the electron-positron momentum distributions at the room temperature (298 K) for the single crystalline Bi-2212 HTSC.

It is clear from Figure 12 that the ratio-curves for the temperature region 116 K to 92 K have shown a dip in the momentum range $(10 < p_L < 25) \times 10^{-3}$ m_0c. But below T_c (91 K) the dip in the ratio-curves disappears. Now the annihilation of positrons with the 3d electrons of Cu atom is predominant in the momentum region $(15 < p_L < 40) \times 10^{-3}$ m_0c [55]. Thus the dip observed in the ratio-curves for the Bi-2212 HTSC at temperatures between 116 K and 92 K (see Figure 12) can be interpreted as showing that less positrons annihilate with 3d electrons of the Cu ions [45].

To observe it clearly we define two area-parameters, R_O and R_{Cu}. One of the area parameter, R_O, is defined as the total area under the ratio-curve (Figure 12) in the momentum range 0 to 5×10^{-3} m_0c, which is a good measure of the fraction of positrons annihilating with the 2p electrons of the oxygen ions. Similarly, R_{Cu} is the total area under the ratio-curve from 12×10^{-3} m_0c to 25×10^{-3} m_0c, which is a good measure of the fraction of positrons annihilating with the 3d electrons of the Cu ions. Figures 13 (a) and (b) represents the variation of R_O and R_{Cu} respectively with sample temperatures. It is clear from Figures 13 (a) & (b) that just above the superconducting transition temperature (116 K to 92 K), positrons

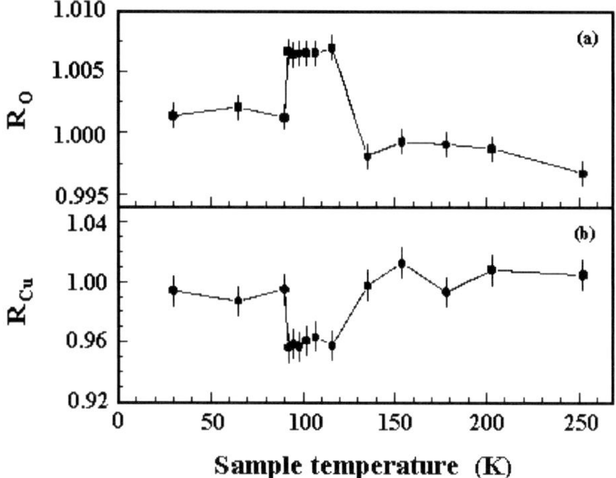

Figure 13: Variation of the area-parameters (a) R_O and (b) R_{Cu} with sample temperature for the single crystalline Bi-2212 HTSC.

are more likely annihilating with the 2p electrons of the oxygen ions than with the 3d electrons of Cu ions. From the positron density distribution calculations for Bi-2212 system [39] it is expected that positrons are mainly probing the Bi-O plane and a small fraction is expected to probe the Cu-O plane. To explain the above observations the possibility of an effective shift of the "apical oxygen" ions towards the Bi-O plane is considered, so that the probability of positrons to be annihilated with the 2p electrons of the oxygen ions increases. This type of structural changes also results a decrease in the number of the 3d electrons in the Cu-O band, which in a way support the "charge transfer model" valid for these cuprate superconductors [42].

B.   In Case of Polycrystalline $La_{0.7}Y_{0.3}Ca_{0.5}Ba_{1.5}Cu_3O_z$ HTSC

The S vs. T graph (Figure 11) for tetragonal $La_{0.7}Y_{0.3}Ca_{0.5}Ba_{1.5}Cu_3O_z$ polycrystalline high T_c superconductor shows a sharp peak at the superconducting transition temperature region. Analyzing the CDBEPAR spectra by constructing the ratio-curves with the reference samples like pure Al or Cu, the contributions of the core electron momentum of different elements in the Doppler broadening spectra can be identified. A comparison between the ratio-curves of polycrystalline LYCBCO HTSC sample at 79 K (just above T_c = 78.5 K) and single crystalline $Bi_2Sr_2CaCu_2O_{8+\delta}$ HTSC sample at 92 K (just above T_c = 91 K) constructed with defect free Al and Cu metals has been shown in Figure 14. From Figure 14 it is observed that the ratio-curves with respect to defect free Al single crystal has a peak around the momentum value 11×10^{-3} m_0c for all cuprate superconductors although the peak height is different. This peak is in general observed in oxide materials, probably due to the annihilation of positrons with the electrons of the oxygen ions [56].   The wide portion of the ratio-curve in the momentum range $(15 < p_L < 30) \times 10^{-3}$ m_0c, specially in case of the LYCBCO oxide HTSC

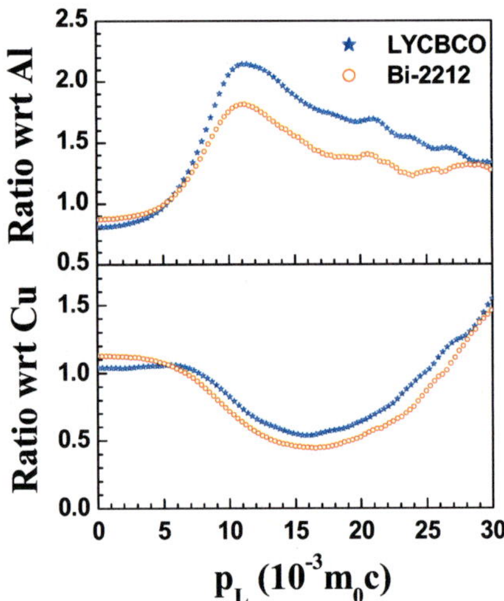

Figure 14: Ratios of the experimental electron-positron momentum distributions for polycrystalline LYCBCO HTSC at 79 K and single crystalline $Bi_2Sr_2CaCu_2O_{8+\delta}$ or (Bi-2212) HTSC at 92 K to the electron-positron momentum distributions for the defects free Al and Cu metals respectively.

sample, indicates that a fraction of positrons are annihilating with the core electrons (3d electrons) of Cu ions, as the annihilation of positrons with core electrons of Cu ions is mainly represented by the momentum range $(15 < p_L < 40) \times 10^{-3}$ $m_0 c$ [50,55]. Ratio-curves with respect to defect free Cu metal results a dip at the momentum value $\sim 16 \times 10^{-3}$ $m_0 c$ (Figure 14) for all the three HTSCs. The dip at the momentum value $\sim 16 \times 10^{-3}$ $m_0 c$ indicates that with respect to Cu metal positrons are less annihilating with the 3d electrons of Cu ions in these cuprate superconductors. It is also observed that the depth of the dip at $\sim 16 \times 10^{-3}$ $m_0 c$ is less in case of LYCBCO HTSCs, compared to $Bi_2Sr_2CaCu_2O_{8+\delta}$ HTSC which indicate that positrons are relatively more annihilating with the 3d electrons of Cu ions in case of LYCBCO HTSC.

In the temperature dependent CDBEPAR studies on $Bi_2Sr_2CaCu_2O_{8+\delta}$ it is observed that just above the superconducting transition temperature positrons are relatively less annihilating with the 3d electrons of Cu ions. Thus for LYCBCO sample it is also important to study the temperature dependence of the fraction of positrons annihilating with the 3d electrons of the Cu ions, particularly at the superconducting transition region. Ratio-curves at different sample temperatures (252, 173, 144, 107, 98, 94, 83, 80, 79, 78, 77, 76, 75, 73, 72, and 60 K) for the LYCBCO HTSC with the defects free Cu metal have been constructed. Ratio-curves at some important temperatures (T = 107, 83, 79, 75 and 72 K) have been shown in Figure 15. In between these temperatures, ratio-curves follow the same trend. It is clear from Figure 15 that the depth of the dip at $\sim 16 \times 10^{-3}$ $m_0 c$ is more when the sample is at 79 K compared with the non-superconducting temperature points (107 K, 83 K) or below the critical temperature points (75 K, 72 K). This implies that the annihilation of positrons with the 3d electrons of Cu ions in LYCBCO cuprate superconductor has temperature dependence.

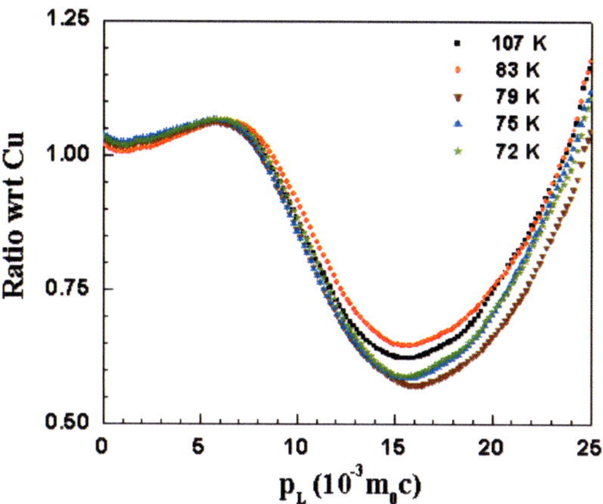

Figure 15: Ratios of the experimental electron-positron momentum distributions at different sample temperatures to the electron-positron momentum distributions of the defects free Cu metal for the LYCBCO polycrystalline HTSC.

To observe it clearly, an area-parameter R_{Cu}, which is calculated as the total (normalized) area under the ratio-curve (Figure 15) from 12×10^{-3} m_0c to 25×10^{-3} m_0c momentum range, and is a good measure of the fraction of positrons annihilating with the 3d electrons of the Cu ions have been constructed. Figure 16 represents the variation of R_{Cu} with the sample temperature. From Figure 16 it is clear that the annihilation of positrons with the 3d electrons of Cu ions decreases sharply at 79 K, i.e., just above the superconducting critical temperature, T_c. In hole doped tetragonal $La_{0.7}Y_{0.3}Ca_{0.5}Ba_{1.5}Cu_3O_z$ cuprate superconductor Cu-O plane is considered as the superconducting plane. Less annihilation of positrons with the Cu 3d electrons at the superconducting transition temperature may be an indication of sudden increase of effective positive charge at the Cu-O plane which repels the positrons to go to the Cu-O plane. The sudden increase of the positron annihilation lineshape parameter (S-parameter) around the superconducting critical temperature (Figure 11) is also due to the less annihilation of positrons with the 3d electrons of Cu ions.

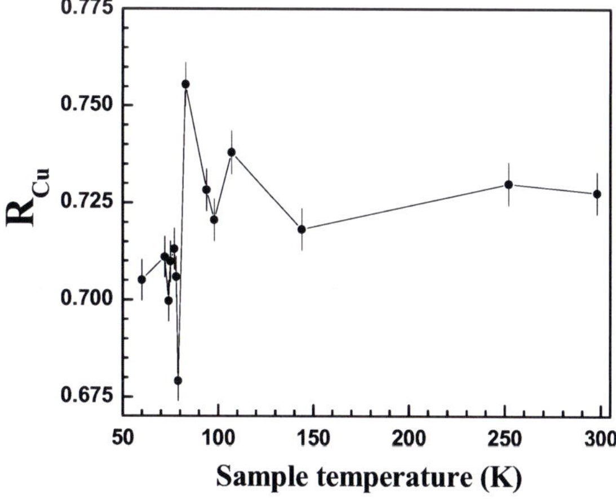

Figure 16: Variation of the area-parameter (R_{Cu}) as a function of sample temperature for the LYCBCO polycrystalline high T_c superconductor.

Summary

Present analysis of the CDBEPAR spectrum by constructing the ratio-curves helps to understand with higher certainty the variation of the CDBEPAR line-shape parameter with temperature at or around the superconducting transition temperature for the high T_c cuprate superconductors. Comparison between the ratio-curves (constructed with reference to the CDBEPAR spectra of pure Al and Cu metals) for single crystalline $Bi_2Sr_2CaCu_2O_{8+\delta}$ and polycrystalline $La_{0.7}Y_{0.3}Ca_{0.5}Ba_{1.5}Cu_3O_z$ HTSC samples shows that positrons are relatively more probing the Cu site in the $La_{0.7}Y_{0.3}Ca_{0.5}Ba_{1.5}Cu_3O_z$ HTSC than the $Bi_2Sr_2CaCu_2O_{8+\delta}$ HTSC. The analyses of the CDBEPAR spectra at different sample temperatures in case of $Bi_2Sr_2CaCu_2O_{8+\delta}$ HTSC have been done by constructing ratio-curves with respect to room temperature (298 K) CDBEPAR spectrum. Less annihilation of the positrons with the 3d electrons of Cu ions and more annihilation with the 2p electrons of the O ions suggest a shift of the apical oxygen ion towards the Bi-O plane. In case of $La_{0.7}Y_{0.3}Ca_{0.5}Ba_{1.5}Cu_3O_z$ HTSC ratio-curves for different sample temperatures have been constructed with respect to the CDBEPAR spectra of pure Cu metal. Here also a less annihilation of positrons with the 3d electrons of Cu ions just above T_c strongly suggest an increase of effective positive charge at the superconducting Cu-O plane due to onset of superconductivity.

4.3　Studies of the Direction Oriented CDBEPAR Measurement in the Single Crystalline $Bi_2Sr_2CaCu_2O_{8+\delta}$ HTSC : to Probe the Anisotropy of the EMD Along the Two Different Crystallographic Directions (Along the c-axis and Along the a-b Plane) in These Highly Anisotropic Crystal Structured System

The widely discussed phenomenon regarding high T_c superconducting oxides is the anisotropy of its properties in different crystallographic direction [57]. The structure of the unit cell of such HTSC materials consists of a stack of conducting (Cu-O plane) and non conducting planes (Figure 17) [57]. Crystallographic "c-axis" is the axis perpendicular to these planes. Among the HTSC compounds Bi-2212 is the most anisotropic in nature. In the polycrystalline HTSC sample presence of large number of defects as well as grain boundaries may affect the positron annihilation parameters. Considering these facts good quality single crystalline Bi-2212 HTSC has been chosen in the present experiment to explore more reliable results. In the present work an effort has been made to observe the manifestation of such anisotropy in the electron momentum distribution by using CDBEPARL measurement technique.

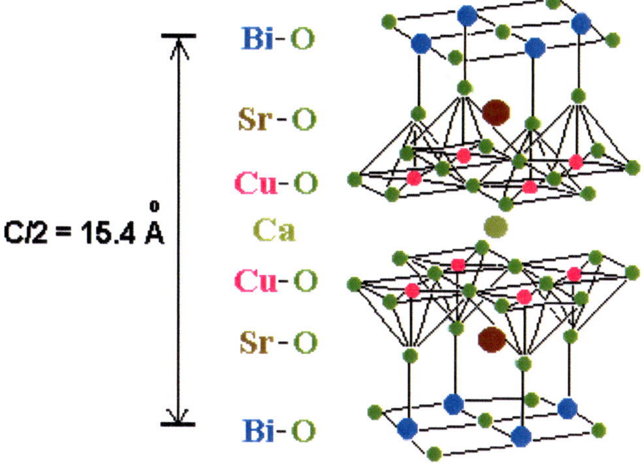

Figure 17: Structure of $Bi_2Sr_2CaCu_2O_{8+\delta}$ high temperature superconductor

To study the anisotropy of the Electron Momentum Distribution (EMD) of the single crystalline Bi-2212 HTSC sample CDBEPAR experiments have been carried out in two different crystallographic orientations. The first orientation is such that the crystallographic c-axis of the "ordered single crystal" makes an angle 0° (position A) with the joining axis of the HPGe and NaI(Tl) detectors. In this orientation one can probe the c-axial component of the electron momentum as p_L is directed along the c-axis of the single crystal (Figure 2). In the other orientation the angle is 90° (position B) to probe the component of the electron momentum along the a-b plane of the single crystal.

Figure 18 shows the variation of S-parameter with sample temperature for the two different orientations of the single crystalline Bi-2212. S-parameter reflects the contribution of low momentum electrons in the Doppler broadened 511 keV spectrum. A change in the S-parameter is associated with the redistribution of electron momentum inside the material. In the present experiment an increase of S-parameter in the temperature region from 92 K to 116K has been observed. The typical feature of S vs. T variation is similar in the two different orientations of the crystal (for position A and position B) indicating a common mechanism involved with the superconductivity in all directions. It represents that transfer of electrons [16,24,54] between higher and lower momentum states is associated in the process of superconducting transition which starts far above T_c. The most important result is the difference of S-parameter value for the two different orientations of the sample (Figure 18). The value of S-parameter in the entire temperature range (30K to 300K) is higher for position B than position A. The difference between the magnitudes of the S-parameter in these two orientations [ΔS = (S at position B) – (S at position A)] is a measure of the anisotropy of the electron momentum distributions. Figure 18 shows that ΔS is almost constant with temperature. Such type of temperature independent anisotropy in the electron momentum distributions (EMD) has been observed by probing high T_c superconductors by two dimensional Angular correlation of annihilation radiation (2D-ACAR) [58] and Compton scattering experiment [59]. The authors of Ref. 59 have observed a significant amount (~ 1 %) of anisotropy in the EMD in the Bi-2212 superconductor.

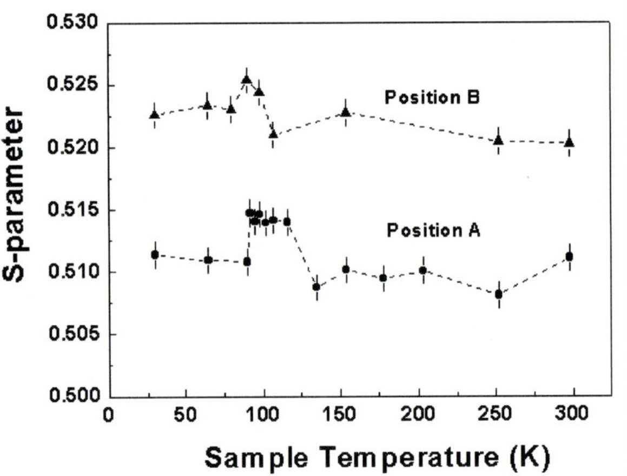

Figure 18: Variation of S-parameter as a function of sample temperature for the two different orientations of the single crystalline Bi-2212 HTSC. For "Position A" crystallographic c-axis of the "ordered single crystal" makes an angle $0°$ with the joining axis of the HPGe and NaI(Tl) detectors. For "position B" the angle is $90°$.

Presently observed anisotropy in the EMD by the CDBEPAR measurement technique is more than 2 % which is twice as observed by the Compton scattering technique. Present results indicate an increased value of electron momentum at the Bi-O plane along the c-axis than a-b plane of the Bi-2212 crystal. Therefore from the above measurements anisotropy in the electron momentum distribution between the a-b plane and the c-axis of the single crystalline Bi-2212 HTSC has successfully been probed. This anisotropy has been found to be of temperature independent.

Now to identify the contributions of the core electron momentum in the observed anisotropy of EMD we have analyzed the CDBEPAR spectra by ratio – curve method.

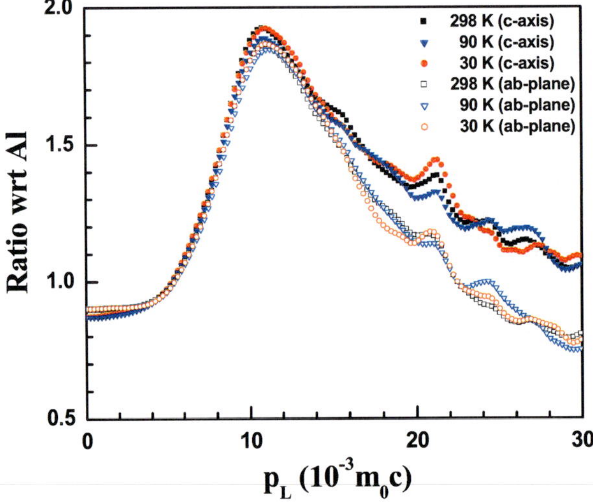

Figure 19: Ratios of the experimental electron-positron momentum distributions for the single crystalline $Bi_2Sr_2CaCu_2O_{8+\delta}$ along the two different crystallographic directions (c-axis and a-b plane) to the electron-positron momentum distributions for the defects free pure Al metal.

The coincidence DBEPAR spectra at three different sample temperatures (298 K, 90 K and 30 K) for both the orientations have been analyzed by constructing the "ratio-curves" [50,51] with defects free 99.9999 % pure Al and Cu metals respectively.

Figure 19 represents the ratio-curves for three important temperature points, 298 K (room temperature, non-superconducting point), 90 K (just below the superconducting transition temperature) and 30 K (superconducting region) for both the crystal orientations, i.e., along the crystallographic c-axis and along the a-b plane, with respect to the reference CDBEPAR spectrum of 99.9999 % pure Al single crystal. Figure 19 shows a peak at momentum value $\sim$ 11×10^{-3} m_0c. This peak is in general observed in oxide materials, probably due to the annihilation of positrons with the 2p electrons of the oxygen ion [56]. The peak height at the momentum value $\sim 11 \times 10^{-3}$ m_0c is more for the ratio-curves along the c-axis compared to the ratio-curves along the a-b plane of the single crystal. Another important feature of Figure 19 is that the ratio-curves along the c-axis shows higher value at the momentum range $(15 < p_L < 30) \times 10^{-3}$ m_0c compared to the ratio-curves along the a-b plane. Next ratio-curves (Figure 20) with respect to defect free Cu metal for the both directional cases have been constructed. It shows a dip at the momentum value $\sim 16 \times 10^{-3}$ m_0c for both the directions. The ratio-curves for the a-b plane CDBEPAR spectra for all the sample temperatures shows a lower value in the momentum range $(15 < p_L < 30) \times 10^{-3}$ m_0c compared to the c-axial CDBEPAR spectra. The annihilation of positrons with the Cu 3d electrons is mainly represented by the momentum range $(15 < p_L < 30) \times 10^{-3}$ m_0c in the ratio-curve.

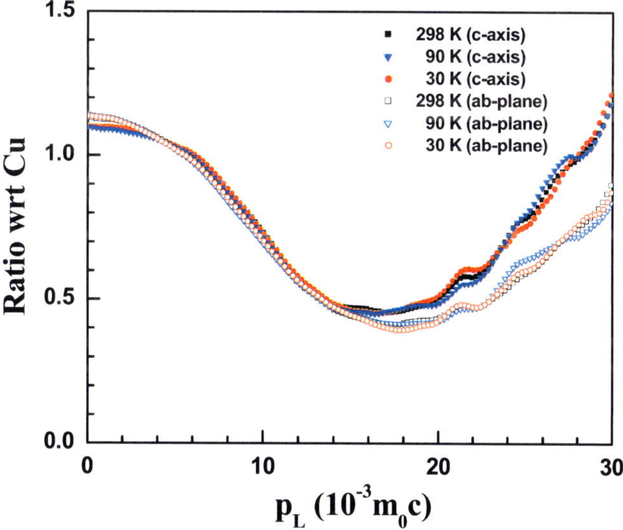

Figure 20: Ratios of the experimental electron-positron momentum distributions for the single crystalline $Bi_2Sr_2CaCu_2O_{8+\delta}$ along the two different crystallographic directions (c-axis and a-b plane) to the electron-positron momentum distributions for the defects free pure Cu metal.

Thus from Figures 19 and 20 it is clear that the coincidence DBEPAR spectra for the two different orientations (c-axis and a-b plane) of the single crystalline $Bi_2Sr_2CaCu_2O_{8+\delta}$ HTSC are not same, rather compared to a-b plane, c-directional ratio-curves shows a higher contributions from the momentum of 2p electrons of oxygen ions and 3d electrons of the Cu ions. This suggests the possibility that the momentum component (p_L) of the 2p electrons of oxygen ions and 3d electrons of the Cu ions are more along the crystallographic c-axis than the a-b plane. It is also clear from Figures 19 and 20 that the nature of the ratio-curves are almost same for all the three sample temperatures (298 K, 90 K and 30 K), which indicates a temperature independent nature of the momentum components along the c-axis and along the a-b plane of the single crystal. Thus, the observed anisotropy of the EMD is due to the higher contributions of momentum of 2p electrons of oxygen ions and 3d electrons of the Cu ions along the crystallographic c-axis of the $Bi_2Sr_2CaCu_2O_{8+\delta}$ HTSC.

Summary

Anisotropy of the electron momentum distributions (EMD) in the highly anisotropic crystal structured cuprate high T_c superconducting system is a widely discussed phenomenon. The anisotropy in the EMD in the two different crystallographic orientations in a HTSC system has been probed successfully by employing CDBEPAR spectroscopy followed by ratio-curve analysis. Present ratio-curve analysis of the coincidence DBEPAR spectra for the two different crystallographic orientations of the single crystalline $Bi_2Sr_2CaCu_2O_{8+\delta}$ HTSC samples indicate that the contributions of the momentum of Cu 3d electrons and oxygen 2p electrons are relatively more towards the crystallographic c-axis than the a-b plane. Moreover, it has been observed that this anisotropy is almost temperature independent over the temperature range 300 K to 30 K.

Acknowledgement

M. Chakrabarti and S. Chattopadhyay gratefully acknowledge the CSIR, New Delhi for providing financial assistance.

References

[1] J. G. Bednorz and K. A. Mullar, Z. *Phys. B* **64** (1986) 189; *Rev. Mod. Phys.* **60** (1988) 588.

[2] M. K. Wu, J. R. Ashburn, C. J. Torng, P. H. Hor, R. L. Meng, L. Gao, Z. J. Huang, Y. Q. Wang and C. W. Chu, *Phys. Rev. Lett.* **58** (1987) 908.

[3] Donald M. Ginsberg (Eds.), *Physical Properties of High Temperature Superconductors* V, World Scientific, Singapore, 1996; A. A. Manuel, *J. of Phys. C* **1** (1989) SA107.

[4] P. Hautojarvi (Eds.), *Positron in Solids*, Springer-Verlag, Berlin, 1979, p. 145.

[5] P. Hautojarvi and C. Corbel, in: A. Dupasquier and A. P. Mills Jr. (Eds.), *Positron Spectroscopy of Solids*, IOS Press, Ohmsha, Amsterdam, 1995, p. 491

[6] R. Krause-Rehberg and H. S. Leipner (Eds.), *Positron Annihilation in Semiconductors*, Springer Verlag, Berlin, 1999, p.5 -126.

[7] Udayan De and D. Sanyal, in: A. V. Narlikar (Eds.), *Studies of High Temperature Superconductors*, **29**, Nova Science Publishers Inc., New York, 1999.

[8] Udayan De, P. M. G. Nambissan, D. Sanyal and D. Banerjee, *Phys. Lett. A* **222** (1996) 119.

[9] D. Sanyal, D. Banerjee and Udayan De, *Phys. Rev. B* **58** (1998) 15226; and references therein.

[10] Udayan De, D. Sanyal, S. Chaudhuri, P. M. G. Nambissan, Th. Wolf and H. Wuhl, *Phys. Rev. B* **62** (2000) 14519.

[11] P. M. Horn, D. T. Keane, G. A. Held, J. L. Jordan-Sweet, D. L. Kaiser, F. Holtzberg and T. M. Rise, *Phys. Rev. Lett.* **59** (1987) 2772.

[12] S. Ishibashi, A. Yamaguchi, Y. Suzuki, M. Doyama, H. Kumakura and K. Togano, *Japan J. Appl. Phys.* **26** (1987) L688.

[13] Y. C. Jean, S. J. Wang, H. Nakanishi, W. N. Hardy, M. Y. Hayden, R. F. Kiefl, R. L. Meng, P. H. Hor, Z. J. Huang and C. W. Chu, *Phys. Rev. B* **36** (1987) 3994.

[14] L. C. Smedskjaer, B. W. Veal, D. G. Leginni, A. P. Paulikas and L. J. Nowicki, *Phys. Rev. B* **37** (1988) 2330.

[15] D. R. Harshman, L. F. Schneemeyer, J. V. Waszczak, Y. C. Jean, M. J. Fluss, R. H. Howell and A. L. Wachs, *Phys. Rev. B* **38** (1988) 848.

[16] S. G. Usmar, P. Sferlazzo, K. G. Lynn and A. R. Moodenbaugh, *Phys. Rev. B* **36** (1987) 8854.

[17] S. J. Wang, S. V. Naidu, S. C. Sharma, D. K. De, D. Y. Jeong, T. D. Black, S. Krichene, J. R. Reynolds and J. M. Owens, *Phys. Rev. B* **37** (1988) 603.

[18] G. Balogh, W. Puff, L. Liszkay and B. Molnar, *Phys. Rev. B* **38** (1988) 2883.

[19] Y. C. Jean, J. Kyle, H. Nakanishi, P. E. A. Turchi, R. H. Howell, A. L. Wachs, M. J. Fluss, R. L. Meng, P. H. Hor, Z. J. Huang and C. W. Chu, *Phys. Rev. Lett.* **60** (1988) 1069.

[20] V. L. Sedov, M. A. Hafiz, I. E. Graboy, A. R. Kaul and V. P. Sabatin, *Mater. Sci. Forum* **105-110** (1992) 1217.

[21] S. G. Usmar, K. G. Lynn A. R. Moodenbaugh, M. Suenaga and R. L. Sabatini, *Phys. Rev. B* **38** (1988) 5126.

[22] Y. C. Jean, H. Nakanishi, M. J. Fluss, A. L. Wachs, P. E. A. Turchi, R. H. Howell, Z. Z. Wang, R. L. Meng, P. H. Hor, Z. J. Huang and C. W. Chu, *J. Phys.: Condens. Matter* **1** (1989) 2989.

[23] E. C. von Stetten, S. Berko, S. S. Li, R. R. Lee, J. Brynestad, D. Singh, H. Krakauer, W. E. Pickett and R. E. Cohen, *Phys. Rev. Lett.* **60** (1988) 2198.

[24] Y. C. Jean, C. S. Sundar, A. Bharathi, J. Kyle, H. Nakanishi, P. K. Tseng, P. H. Hor, R. L. Meng, Z. J. Huang, C. W. Chu, Z. Z. Wang, P. E. A. Turchi, R. H. Howell, A. L. Wachs and M. J. Fluss, *Phys. Rev. Lett.* **64** (1990) 1593.

[25] S. Sundar, A. Bharathi, Y. C. Jean, P. H. Hor, R. L. Meng, Y. Y. Xue, Z. J. Huang and C. W. Chu, *Phys. Rev. B* **41** (1990) 11685.

[26] S. S. P. Parkin, V. Y. Lee, A. I. Nazaal, R. Savoy, T. C. Huang, G. Gorman and R. Beyers, *Phys. Rev. B* **38** (1988) 6531.

[27] P. Halder, K. Chen, B. Maheswaeran, A. Roig-Janicki, N. K. Jaggi, R. S. Markiewicz and B. C. Giessen, *Science* **241** (1989) 1198.

[28] Jung, M. H. Whangbo, N. Herron and C. C. Toradi, *Physica C* **160** (1989) 381.

[29] R. Bishop, R. L. Martin, K. A. Muller and Z. Tesanovic, *Z. Phys. B* **76** (1989) 76.

[30] T. Suzuki, M. Nagoshi, Y. Fukuda, Y. Syono, M. Kikuchi, N. Kobayashi and M. Tachiki, *Phys. Rev.* **40** (1989) 5184.

[31] R. V. Kasowski, W. Y. Hsu and F. Herman, *Phys. Rev. B* **38** (1988) 6470.

[32] Junod, D. Eckert, G. Triscone, V. Y. Lee and J. Muller, *Physica C* **159** (1989) 215.

[33] Y. Shimakawa, Y. Kubo, T. Manako and H. Igrashi, *Phys. Rev. B* **40** (1989) 11400.

[34] S. Sundar, A. Bharti, W. Y. Ching, Y. C. Jean, P. H. Hor, R. L. Meng, Z. J. Huang and C. W. Chu, *Phys. Rev. B* **42** (1990) 2193.

[35] Yun-bo Wang, Guo-hui Cao, Yang Li, Qing-zhu Ma, Yinghuan Guo and Ru-zhang Ma, *Solid State Commun.* **96** (1995) 717.

[36] H. Maeda, Y. Tanaka, M. Fukutomi and T. Asano, *Jpn. J. Appl. Phys. Lett.* **27** (1988) L209.

[37] W. Chu, J. Bechtold, L. Gao, P. H. Hor, Z. J. Huang, R. L. Meng, Y. Y. Sunn, Y. Q. Wang and Y. Y. Xue, *Phys. Rev. Lett.* **60** (1988) 941.

[38] Singh, W. E. Pickett, R. E. Cohen, H. Krakauer and S. Berko, *Phys. Rev. B* **39** (1989) 9667.

[39] S. Sundar, A. Bharti, W. Y. Ching, Y. C. Jean, P. H. Hor, R. L. Meng, Z. J. Huang and C. W. Chu, *Phys. Rev. B* **43** (1991) 13019.

[40] M. Zhang, C. Q. Tang, T. Gen and G. Y. Li, *Phys. Rev. B* **47** (1993) 3435.

[41] P. K. Pujari, T. Datta, Udayan De and B. Ghosh, *Phys. Rev. B* **50** (1994) 3438.

[42] Z. Tang, S. J. Wang, X. H. Gao, G. C. Ce and Z. X. Zhao, *Phys. Letts. A* **178** (1993) 320.

[43] Z. Tang, S. J. Wang and X. H. Gao, *Phys. Rev. B* **50** (1994) 3209.

[44] Yang Li, Yang Liu, Ruifei Duan, Xiaotao Xiong, Baoyi Wang, Guohui Cao, Long Wei, D. N. Zheng, Z. X. Zhao and Joseph H. Ross Jr., *Physica C* **402** (2004) 179.

[45] Mahuya Chakrabarti, A. Sarkar, D. Sanyal, G.P. Karwasz and A. Zecca, *Phys. Lett. A* **321** (2004) 376.

[46] Mahuya Chakrabarti, K. R. Mavani, S. Chattopadhyay, A. Sarkar and D. Sanyal, *Physics Letters A* **329** (2004) 231.

[47] Mahuya Chakrabarti, A. Sarkar, S. Chattapadhayay, D. Sanyal, A. K. Pradhan, R. Bhattacharya and D. Banerjee, *Solid State Commun.* **128** (2003) 321.

[48] Mahuya Chakrabarti, A. Sarkar, S. Chattapadhayay, D. Sanyal and A. Chakrabarti *Physica C* **416** (2004) 25.

[49] K. G. Lynn and A. N. Goland, *Solid State Commun.* **18** (1976) 1549.

[50] R.S. Brusa, W. Deng, G. P. Karwasz and A. Zecca, *Nucl. Instr. & Meth. B* **194** (2002) 519.

[51] P. Asoka-Kumar, M. Alatalo, V. J. Ghosh, A. C. Kruseman, B. Nielsen and K. G. Lynn, *Phys. Rev. Lett.* **77** (1996) 2097.

[52] K. Pradhan, G. D. Gu, K. Nakao, N. Koshizuka and D. Kanjilal, *Phys. Rev. B* **61** (2000) 14374.

[53] K. R. Mavani, D. S. Rana, R. Nagarajan and D. G. Kuberkar, *Physica C* **403** (2004) 304.

[54] R. S. Brusa, R. Grisenti, S. Liu, S. Oss, O. Pilla, A. Zecca, A. Dupasquier and F.A. Matacotta, *Physica C* **156** (1988) 65.

[55] V. J. Ghosh, M. Alatalo, P. Asoka-Kumar, B. Nielsen, K. G. Lynn, A. C. Kruseman and P. E. Nijnarends, *Phys. Rev. B* **61** (2000) 10092.

[56] U. Myler and P. J. Simpson, *Phys. Rev. B* **56** (1997) 14303.

[57] N. M. Plakida, *High-Temperature Superconductivity*, Springer Verlag, Berlin, 1995.

[58] L. P. Chan, D. R. Harshman, K. G. Lynn, S. Massidda and D. B. Mitzi, *Phys. Rev. Lett.* **67** (1991) 1350.

[59] S. Manninen, K. Hamalainen, M. A. G. Dixon, M. J. Cooper, D. A. Cardwell and T. Buslaps, *Physica C* **314** (1999) 19.

In: New Topics in Superconductivity Research
Editor: Barry P. Martins pp. 223-237

ISBN 1-59454-985-0
© 2006 Nova Science Publishers, Inc.

Chapter 7

COMPARATIVE STUDY OF STATISTICS OF COOPER'S ELECTRON PAIRS IN LOW-TEMPERATURE SUPERCONDUCTORS AND COUPLED HOLES PAIRS IN HIGH T_c CERAMICS

I.G. Kaplan and O. Navarro*
Instituto de Investigaciones en Materiales, Universidad Nacional Aut
ónoma de México, Apartado Postal 70-360, 04510, México, D. F

Abstract

It is well known that the system of Cooper's pair is described by boson symmetric wave functions, but Cooper's pair operators are bosons only when the moments **k** are different and they are fermions for equal **k**. The analysis of trilinear commutation relations for the Cooper pair (pairon) operators reveals that they correspond to the modi ed parafermi statistics of rank $p = 1$. Two different expressions for the Cooper pair number operator are presented. We demonstrate that the calculations with a Hamiltonian expressed via pairon operators is more convenient using the commutation properties of these operators without presenting them as a product of fermion operators. This allows to study problems in which the interactions between Cooper's pairs are also included. The problem with two interacting Cooper's pairs is resolved and its generalization in the case of large systems is discussed.

It is shown that in site representation, the hole-pair operators obey the same commutation relations (paulion) as the Cooper pair operators in impulse representation, although the latter describe delocalized quasiparticles. In quasi-impulse representation, the hole-pair operators are also delocalized and their exact commutation relations correspond to a modi ed parafermi statistics of rank M (M is the number of sites in a "superlattice" formed by the centers of mass of each hole pair). From this follows that one state can be occupied by up to M pairs. Even in the absence of dynamic interaction, the system of hole pairs is characterized by some immanent interaction, named after Dyson as kinematic interaction. This interaction appears because of the deviation of the quasiparticle statistics from the Bose (Fermi) statistics and its magnitude

---

*E-mail address: kaplan at iim.unam.mx(Corresponding author)

depends on the concentration of hole pairs. In spite of the non-bosonic behavior, there is no statistical prohibition on the Bose-Einstein condensation of coupled hole pairs.

Keywords: Cooper's pair commutation relations, Pairing interactions, Strongly correlated electron systems.

1   Introduction

The theory of the low temperature superconductivity was created by Bardeen, Cooper, and Schrieffer (BCS) [1] only after Cooper [2] had shown that two electrons interacting above the Fermi sea of non-interacting electrons can couple in a stable pair, if the interaction resulting from virtual exchange of phonons is attractive near the Fermi surface. As was demonstrated in a more sophisticated study [3], in a full agreement with the Cooper assumption, the largest binding energy of the Cooper pair corresponds to electrons with the opposite momenta and spins. In the second quantization formalism, the operators of creation, b_k^+, and annihilation, b_k, of Cooper's pair in a state $(k\alpha, -k\beta)$, are de ned as simple products of the electron creation and annihilation operators, $c_{k\sigma}^+$ and $c_{k\sigma}$, satisfying the fermion commutation relations,

$$b_k^+ = c_{k\alpha}^+ c_{-k\beta}^+; \qquad b_k = c_{-k\beta}c_{k\alpha}. \tag{1}$$

Let us call these operators, following Schrieffer [3], as "pairon" operators.

The Cooper pair has the total spin $S = 0$. Hence, in accordance with the Pauli principle, the wave functions describing the Cooper pair system have the boson permutation symmetry, that is, they are symmetric under permutations of pairs. But the pairon operators (1) do not obey the boson commutation relations. It is easy to show by direct calculation. Namely,

$$\begin{aligned}
\left[b_k, b_{k'}^+\right]_- &= \left[b_k^+, b_{k'}^+\right]_- = \left[b_k, b_{k'}\right]_- = 0 && \text{for } k \neq k', \tag{2} \\
\left[b_k, b_k^+\right]_- &= 1 - \hat{n}_{k\alpha} - \hat{n}_{-k\beta}, \tag{3} \\
\left(b_k^+\right)^2 &= \left(b_k\right)^2 = 0, \tag{4}
\end{aligned}$$

where $\hat{n}_{k\alpha} = c_{k\sigma}^+ c_{k\sigma}$ is the electron number operator. As follows from Eqs. (2)–(4), for $k \neq k'$ the Cooper pairs are bosons, while for $k = k'$ they do not obey the boson commutation relations, although they obey the Pauli principle and have the fermion occupation numbers for one-particle states.

Thus, the pairon operators may not be considered neither as the Bose operators, nor as the Fermi operators. This is the reason that the problem with the model Hamiltonian of the BCS theory

$$H = \sum_k \varepsilon_k b_k^+ b_k + \sum_{k',k} V_{kk'} b_{k'}^+ b_k \tag{5}$$

cannot be directly solved by transforming Hamiltonian (5) to the diagonalized form

$$H = \sum_n \varepsilon_k' B_k^+ B_k \tag{6}$$

by means of some unitary transformation

$$B_n = \sum_k u_{nk} b_k, \qquad\qquad B_n^+ = \sum_k u_{nk}^* b_k^+.$$ (7)

The unitary transformation is canonical only for the Bose or Fermi operators. In general case, it is not canonical; it does not preserve the commutation properties of the operators transformed. Therefore, practically all calculations in the BCS approach were performed using the fermion properties of electron operators forming the Cooper pair.

In 1953 Green [4] and, independently, Volkov [5] have shown that the boson and fermion commutation relations do not exhaust all physically possible commutation relations for eld operators in the second quantization formalism. The eld operators satisfying the requirements of causality, relativistic invariance and positivity of energy can obey more general commutation relations than the boson and fermion ones. This new statistics was called parastatistics, it is shortly described in next section.

In section 3 we analyze the commutation relations for the pairon operators and show that the trilinear commutation relations correspond to the modi ed parafermi statistics of rank $p = 1$. In section 4, we demonstrate that the calculations with a Hamiltonian expressed via the pairon operators is more convenient to perform using the commutation properties of these operators without presenting them as a product of fermion operators. This allows to study problems in which the interactions between Cooper's pairs are also included [6]. The solution of the simplest problem with two interacting Cooper's pairs is presented.

Although, all elementary particles known at present are bosons or fermions, the parastatistics can be realized for quasiparticles. As was shown by one of the authors [7], the quasiparticles in a periodical lattice (the Frenkel excitons and magnons) obey a modi ed parafermi statistics of rank M, where M is the number of equivalent lattice sites within the delocalization region of collective excitations. Later on, the results [7] for the Frenkel excitons and magnons where extended to polaritons [8, 9], defectons in quantum crystals [10] and to the Wannier-Mott excitons [11]. The statistics and some properties of a system of noninteracting holes was studied in ref. [12]. Below, in section 5 and 6, we discuss in detail the statistics and properties of the coupled hole pairs [13].

2   Short Account of Parastatistics

Green [4] introduced in 1953 a generalized method of eld quantization with trilinear commutation relations, which include the boson and fermion commutation relations as particular cases,

$$\left[\left[a_k^+, a_{k'} \right]_\pm, a_{k''} \right]_- = -2\delta_{kk''} a_{k'},$$

$$\left[\left[a_k^+, a_{k'} \right]_\pm, a_{k''}^+ \right]_- = 2\delta_{k'k''} a_k^+,$$ (8)

$$\left[\left[a_k^+, a_{k'}^+ \right]_\pm, a_{k''}^+ \right]_- = \left[\left[a_k, a_{k'} \right]_\pm, a_{k''} \right]_- = 0.$$

The relations (8) with upper sign at the inner brackets are called the paraboson commutation relations and with lower sign are called the parafermion commutation relations.

These names are connected with the fact that the ordinary boson and fermion operators also obey the relations (8). But the above relations are ful lled by a more general type of operators given by the so-called Green Ansatz [4]:

$$a_k = \sum_{\rho=1}^{p} d_{\rho k},$$ (9)

having an in nite set of solutions, labeled by the integers $p(p = 1, 2, ..., \infty)$.

For parabosons, the operators $d_{\rho k}$ obey the following commutation relations:

(a) $\rho = \rho'$, as for bosons

$$\left[d_{\rho k}, d_{\rho k'}^{+}\right]_{-} = \delta_{kk'} \;\; ; \;\; \left[d_{\rho k}, d_{\rho k'}\right]_{-} = \left[d_{\rho k}^{+}, d_{\rho k'}^{+}\right]_{-} = 0$$ (10)

(b) $\rho \neq \rho'$, as for fermions

$$\left[d_{\rho k}, d_{\rho' k'}^{+}\right]_{+} = \left[d_{\rho k}, d_{\rho' k'}\right]_{+} = \left[d_{\rho k}^{+}, d_{\rho' k'}^{+}\right]_{+} = 0.$$ (11)

For parafermions:

(a) $\rho = \rho'$, as for fermions

$$\left[d_{\rho k}, d_{\rho k'}^{+}\right]_{+} = \delta_{kk'} \;\; ; \;\; \left[d_{\rho k}, d_{\rho k'}\right]_{+} = \left[d_{\rho k}^{+}, d_{\rho k'}^{+}\right]_{+} = 0$$ (12)

(b) $\rho \neq \rho'$, as for bosons

$$\left[d_{\rho k}, d_{\rho' k'}^{+}\right]_{-} = \left[d_{\rho k}, d_{\rho' k'}\right]_{-} = \left[d_{\rho k}^{+}, d_{\rho' k'}^{+}\right]_{-} = 0.$$ (13)

As follows from Eqs. (10) and (12), the solution of the Green Ansatz (9) for $p = 1$, reduces to the usual boson and fermion operators in paraboson and parafermion cases, respectively. It is worth-while to note that such mixed boson-fermion behavior of the operators $d_{\rho k}$ is well known in solid state physics and called as *paulion*, see the book by Davydov [14]. The value of p in the Green Ansatz is called the rank of parastatistics. The application of operators (9) to the vacuum state is easily found using Eqs. (10)-(13). So,

$$a_k|0\rangle = 0, \qquad a_k a_k^{+}|0\rangle = p\delta_{kl}|0|\rangle \qquad \text{for all } k, l.$$ (14)

For the parafermi statistics, the maximum occupation numbers of one state is equal to the rank of parastatistics p

$$\left(a_k^{+}\right)^{N}|0\rangle \; \neq \; 0 \qquad \text{for } N \leq p,$$ (15)

$$\left(a_k^{+}\right)^{p+1}|0\rangle \; = \; 0.$$ (16)

At $p = 1$, the parafermi statistics is reduced to the Fermi-Dirac statistics. For parabose statistics there is no restriction in the occupation numbers, as in the Bose-Einstein statistics.

The particle number operator in parastatistics is de ned as

$$\hat{N}_k = \frac{1}{2}\left[a_k^{+}, a_k\right]_{\pm} \mp \frac{1}{2}p,$$ (17)

where the upper sign for parabosons and the lower sign for parafermions. It satis es the general property of the particle number operators

$$\left[\hat{N}_k, a_l^+\right]_- = \delta_{kl} a_k^+ \tag{18}$$

and in the Bose (Fermi) case turns to

$$\hat{N}_k = a_k^+ a_k. \tag{19}$$

As was shown in Refs. [15, 16], the parafermion and paraboson algebras are the Lie algebras of the orthogonal and symplectic groups, respectively. For example, the algebra for ν parafermion operators a_r and their adjoints $a_r^+ (r = 1, 2, ..., \nu)$, which constitutes a parafermi ring, is the Lie algebra of the orthogonal group $SO_{2\nu+1}$ in $2\nu + 1$ dimensions; more details about the parastatistics is presented in the book by Ohnuki and Kamefuchi [17].

In spite of numerous studies of all known elementary particles, see Refs. [18, 19], the elementary particles obeying the parastatistics were not revealed. On the other hand, as discussed in Refs. [20, 21], the ordinary fermions, which differ by some internal quantum numbers but are similar dynamically, can be described by the parafermi statistics. In this case, fermions with different internal quantum numbers are considered as non-identical distinguishable particles. The parafermi statistics of rank p describes systems with p different types of fermions. As a result, *quarks* with 3 colors obey the parafermi statistics of rank $p = 3$; *nucleons* in nuclei (isotope spin $1/2$) obey the parafermi statistics of rank $p = 2$.

Let us stress, the fermions characterized by the same value of internal quantum number obey the Fermi-Dirac statistics. It is the system of p fermion groups with different internal quantum numbers that obey the parafermi statistics of rank p.

3   Statistics of Cooper's Pairs

As was discussed in the Introduction, the pairon operators, Eq. (1), obey boson commutation relations only in the case of different momenta. For equal momenta, the right-hand part of commutation relation (3) contains the products of fermion operators that re ects the fermion structure of pairon operators.

$$\left[b_k, b_k^+\right]_- = 1 - c_{k\alpha}^+ c_{k\alpha} - c_{-k\beta}^+ c_{-k\beta}. \tag{20}$$

To operate with the pairon operators, the commutation relations for these operators do not have to include other kinds of operators. One of the ways to achieve this goal is to calculate trilinear commutation relations, as it is formulated in the parastatistics [4, 22].

The direct calculation leads to the following trilinear commutation relations

$$\left[\left[b_k^+, b_{k'}\right]_-, b_{k''}^+\right]_- = 2\delta_{kk'}\delta_{kk''} b_k^+, \tag{21}$$

$$\left[\left[b_k^+, b_{k'}\right]_-, b_{k''}\right]_- = -2\delta_{kk'}\delta_{kk''} b_k. \tag{22}$$

These relations coincide with the trilinear commutation relations (8) for the parafermi statistics for $k = k' = k''$. For different k, k', and k'' the relations are different. In the parafermi statistics in relations (8) corresponding to Eq. (21) instead of the two presented Kronecker symbols there is one, namely, $\delta_{k'k''}$; and in relation corresponding to Eq. (22), $\delta_{kk'}$ is absent. Thus, the pairon operators satisfy some modi ed parafermi statistics of the rank $p = 1$. The latter follows from Eq. (4), since the parastatistics of rank p satisfy Eqs. (15) and (16).

As follows from the de nition of the particle number operator $\hat{N}_k$

$$\hat{N}_k|N_k\rangle = N_k|N_k\rangle. \tag{23}$$

For the boson and fermion number operators the well-known expression $\hat{N} = a_k^+ a_k$ is valid. But it is quite not evident that the same expression is valid for the pairon number operator. In the parafermi statistics, the particle number operator is de ned as

$$\hat{N}_k = \frac{1}{2}\left([a_k^+, a_k]_- + p\right), \tag{24}$$

for pairons $p = 1$ and

$$\hat{N}_k = \frac{1}{2}\left([b_k^+, b_k]_- + 1\right). \tag{25}$$

As follows from the trilinear commutation relations (21) and (22), the operator (25) satis es the commutation relations for the particle number operator that were established earlier for bosons and fermions, see Ref. [23],

$$\left[\hat{N}_k, b_k^+\right]_- = b_k^+, \tag{26}$$

$$\left[\hat{N}_k, b_k\right]_- = -b_k. \tag{27}$$

It is easy to check that for fermions, Eq. (25) is equivalent to the standard expression $\hat{n}_k = c_k^+ c_k$. Let us study it in the pairon case.

Using Eq. (3), the expression for the pairon number operator (25) can be written as

$$\hat{N}_k = \frac{1}{2}\left(\hat{n}_{k\alpha} + \hat{n}_{-k\beta}\right). \tag{28}$$

This is quite natural that the number of Cooper's pairs is two times less than the number of electrons forming pairs. It can be proved that from Eq. (28) follows that the expression $\hat{N}_k = b_k^+ b_k$ may be also used for the pairon number operators. Let us do it.

The product $b_k^+ b_k$ is equal to the product of the fermion number operators $\hat{n}_{k\alpha}\hat{n}_{-k\beta}$, but in the general case

$$\hat{n}_{k\alpha}\hat{n}_{-k\beta} \neq \frac{1}{2}\left(\hat{n}_{k\alpha} + \hat{n}_{-k\beta}\right). \tag{29}$$

Pairons operators possess the fermions occupation numbers, $n_{k\alpha}$ and $n_{-k\beta}$ equal to 0 or 1, in this case, and only in this case, the left-hand part of Eq. (29) is equal to its right-hand part. Thus,

$$\frac{1}{2}\left(\hat{n}_{k\alpha} + \hat{n}_{-k\beta}\right) = b_k^+ b_k \tag{30}$$

and from Eqs. (28) and (30) follows that although Cooper's pairs are neither bosons nor fermions, for the operators of their number, the traditional form $\hat{N}_k = b_k^+ b_k$ can be used.

If one substitutes the equality

$$\hat{n}_{k\alpha} + \hat{n}_{-k\beta} = 2b_k^+ b_k, \tag{31}$$

into the commutation relation (3), it transforms into

$$\left[b_k, b_k^+\right]_- = \left(1 - 2b_k^+ b_k\right) \tag{32}$$

or

$$\left[b_k, b_k^+\right]_+ = 1. \tag{33}$$

Thus, for equal **k**, the pairon operators obey the fermion commutation relations, while for different **k**, they obey the boson commutation relations.

Despite the fact that each Cooper's pair has the total spin $S = 0$, the pairons are not bosons, because for equal momenta **k** they behave as fermions. However, for different **k**, the pairons obey the Bose-Einstein statistics and can occupy one energy level, that is, they can undergo the phenomenon of the Bose-Einstein condensation. However, in this case all electrons composed into the condensed Cooper pairs must have different momenta **k**.

The Eqs. (2) and (32) can be combined into one commutation relation

$$\left[b_k, b_{k'}^+\right]_- = \delta_{kk'} \left(1 - 2b_k^+ b_k\right). \tag{34}$$

The application of pairon operators to the vacuum state follows from their de nition, Eq. (1),

$$b_k|0\rangle = 0, \qquad b_k b_{k'}^+|0\rangle = \delta_{kk'}|0\rangle. \tag{35}$$

The relations (34) and (35) are suf cient for performing calculations using only the pairon operators.

4 Generalized Model Hamiltonian and the Problem of Two Interacting Cooper's Pairs

Let us add to the BCS Hamiltonian the term describing the interaction among Cooper's pairs. The generalized model Hamiltonian is

$$H = 2\sum_k \epsilon_k b_k^\dagger b_k - {\sum_{k,k'}}' V_{k,k'} b_{k'}^\dagger b_k + \frac{1}{2} {\sum_{k_1',k_2'}}' {\sum_{k_1'',k_2''}}' V_{k_1'k_2',k_1''k_2''} b_{k_1''}^\dagger b_{k_2''}^\dagger b_{k_1'} b_{k_2'}, \tag{36}$$

where primes in sums denote that $k \neq k'$, $k_1' \neq k_2'$ and $k_1'' \neq k_2''$. The other restriction is concern with the potential energy of interpair interaction in which $k_1'' \neq k_1'$ and $k_2'' \neq k_2'$. While according to the Cooper model, $V_{k,k'} > 0$, the sign of $V_{k_1'k_2',k_1''k_2''}$ is not restricted, it can be both positive for a repulsive interpair interaction and negative for an attractive interaction.

In general, the variational wave function of the system with N pairs can be presented as

$$|\Psi(1, 2, ..., N)\rangle = {\sum_{k_1,k_2,...,k_N}}' \alpha(k_1, k_2, ..., k_N) b_{k_1}^+ b_{k_2}^+ ... b_{k_N}^+ |0\rangle, \tag{37}$$

where $k_1 \neq k_2 \neq ... \neq k_N$ because of the fermion condition (4), or it can be presented in the BCS form, which is not as precise as the variational function (37) but easier for calculations

$$|\Psi(1, 2, ..., N)\rangle = \prod_{k=1}^{N} (u_k + v_k b_k^+)|0\rangle. \tag{38}$$

As an illustration of operations with the Cooper pair operators using their properties (34) and (35), we consider the model of two interacting pairs described by the BCS wave function

$$\begin{aligned} |\Psi(1, 2)\rangle &= \prod_{k_j=1}^{2} \left(u_{k_j} + v_{k_j} b_{k_j}^{\dagger} \right) |0\rangle \\ &= \left(u_{k_1} u_{k_2} + u_{k_1} v_{k_2} b_{k_2}^{\dagger} + u_{k_2} v_{k_1} b_{k_1}^{\dagger} + v_{k_1} v_{k_2} b_{k_1}^{\dagger} b_{k_2}^{\dagger} \right) |0\rangle. \end{aligned} \tag{39}$$

Following the BCS theory, we assume in the general Hamiltonian (36) that the interaction energies do not depend upon the value of moments

$$H = 2\sum_{k} \epsilon_k b_k^{\dagger} b_k - V_0 \sum_{k,k'}{}' b_{k'}^{\dagger} b_k + \frac{V_1}{2} \sum_{k_1',k_2'}{}' \sum_{k_1'',k_2''}{}' b_{k_1''}^{\dagger} b_{k_2''}^{\dagger} b_{k_1'} b_{k_2'}. \tag{40}$$

Using the properties of Cooper's pair operators, Eqs. (34) and (35), we calculate the expectation value for the energy $W = \langle \Psi(1, 2) | H | \Psi(1, 2) \rangle$ with the Hamiltonian (40) and wave function (39). The results is

$$W = 2v_{k_1}^2 \epsilon_{k_1} + 2v_{k_2}^2 \epsilon_{k_2} - 2V_0 u_{k_1} v_{k_1} u_{k_2} v_{k_2} + V_1 v_{k_1}^2 v_{k_2}^2. \tag{41}$$

By a minimization procedure respect to v_{k_1} and v_{k_2} using the Lagrange multiplicators, the following de nitions

$$\Delta_{k_1} \equiv V_0 u_{k_2} v_{k_2}; \qquad \Delta_{k_1}' \equiv V_1 v_{k_2}^2, \tag{42}$$

$$\Delta_{k_2} \equiv V_0 u_{k_1} v_{k_1}; \qquad \Delta_{k_2}' \equiv V_1 v_{k_1}^2, \tag{43}$$

and introducing

$$E_{k_1}' = \sqrt{\left(\epsilon_{k_1} + \frac{\Delta_{k_1}'}{2} \right)^2 + \Delta_{k_1}^2}, \qquad E_{k_2}' = \sqrt{\left(\epsilon_{k_2} + \frac{\Delta_{k_2}'}{2} \right)^2 + \Delta_{k_2}^2}; \tag{44}$$

we obtain that

$$v_{k_1}^2 = \frac{1}{2} \left(1 - \frac{\epsilon_{k_1} + \frac{\Delta_{k_1}'}{2}}{E_{k_1}'} \right), \qquad u_{k_1}^2 = \frac{1}{2} \left(1 + \frac{\epsilon_{k_1} + \frac{\Delta_{k_1}'}{2}}{E_{k_1}'} \right), \tag{45}$$

$$v_{k_2}^2 = \frac{1}{2} \left(1 - \frac{\epsilon_{k_2} + \frac{\Delta_{k_2}'}{2}}{E_{k_2}'} \right), \qquad u_{k_2}^2 = \frac{1}{2} \left(1 + \frac{\epsilon_{k_2} + \frac{\Delta_{k_2}'}{2}}{E_{k_2}'} \right), \tag{46}$$

and

$$u_{k_1} v_{k_1} = \frac{1}{2} \left\{ 1 + \frac{\left(\epsilon_{k_1} + \frac{\Delta'_{k_1}}{2}\right)}{E'^2_{k_1}} \right\}^{1/2} = \frac{\Delta_{k_1}}{2E'_{k_1}}. \tag{47}$$

The calculation of the quasiparticle excitation energy results in

$$E'_k = \sqrt{\left(\epsilon_k + \frac{\Delta'_k}{2}\right)^2 + \Delta_k^2}. \tag{48}$$

Thus, the quantities introduced by Eq. (44) and entered into Eqs. (45)-(47) have the physical sense as the quasiparticle excitation energies.

Using, as in the BCS theory, the approximations

$$\begin{aligned} \Delta_{k_1} &= \Delta_{k_2} = \Delta \\ \Delta'_{k_1} &= \Delta'_{k_2} = \Delta' \\ \epsilon_{k_1} &\simeq \epsilon_{k_2} = \epsilon \end{aligned} \tag{49}$$

we obtain the explicit expressions for the parameters Δ' and Δ

$$\Delta' = V_1 \frac{V_0 - 2\epsilon}{2V_0 + V_1}, \tag{50}$$

$$\Delta = \sqrt{\left(\frac{V_0}{2}\right)^2 - \left\{\epsilon + \frac{V_1}{2}\left(\frac{V_0 - 2\epsilon}{2V_0 + V_1}\right)\right\}^2}. \tag{51}$$

In the expression for E'_k and u_k^2 and v_k^2, as well, the electron energy ϵ_k enters with the additive term $\Delta'_k/2$. According to Eq. (50), this term depends upon the interaction energies V_1 and V_0, and it disappears when $V_1 = 0$. Thus, one can say that the interpair interaction leads to the renormalization of the electron energy. This renormalization is proportional to the interaction potential V_1 between pairs. On the other hand, the interpair interaction leads to an augmentation of the gap, which is according to Eq. (48) equal to

$$\sqrt{\Delta_k^2 + (\Delta'_k/2)^2}. \tag{52}$$

The interpair interaction increases the energy of the quasiparticle excitation.

For comparison with the BCS theory, we have to neglect the interaction between pairs, $V_1 = 0$. With this condition, expressions for u_k^2, v_k^2 and E_k are reduced to the BCS expressions. But in the case of Δ, (Eq. 51), it is not so, it is reduced to

$$\Delta = \sqrt{\left(\frac{V_0}{2}\right)^2 - \epsilon^2}. \tag{53}$$

This formula differs from the exponential expression for the energy gap in the BCS theory. The difference is connected with the fact that in a system with a nite number of particles, the distances between the energy levels are also nite, there is a discrete set of energy levels.

The approximation $\epsilon_{k_1} = \epsilon_{k_2} = \epsilon$, Eq. (49), is valid only for systems with $N \gg 1$, for which the energy spectrum is continuous and after integration one obtains the exponential dependence as in the BCS theory.

Finally, it is important to mention that the two-pairon system was considered as an illustration of applying the commutation relation for pairons, Eq. (34). The employment of pairon operators b_k^+, b_k in place of presenting them as product of fermion operators reduces twice the number of operators in the Hamiltonian. For instance, for the two-pairon system, the interaction term contains four pairon operators instead of eight fermion operators.

Let us note that the problem of two interacting pairons has been solved exactly. The real physical problem for N interacting pairons with $N \to \infty$ is very complicated and can be solved only approximately. For its solution one should, similar to the BCS theory, consider the interacting pairons in the nearest neighborhood of the Fermi level and take into account an approximation similar to that given by Eq. (49), so that the summation over k can be replaced by an integration.

5 Statistics and Properties of Coupled Hole Pairs

At present, it is well established that the conductivity in high-T_c ceramics has a hole origin with charge of carriers equal to $+2e$. Here, we present the results of our study of statistics and some physical properties of the hole-pair system [13]. But before we shortly consider the properties of isolated holes in superconducting ceramics.

Usually, holes are described as fermions. It came from atomic physics: the closed electronic shell after one electron is knocked out (the "hole" formation) has the same angular and spin momentum properties as the electronic shell with one electron. But in general case, the holes can have different values of spin S. For example, in the CuO_2 planes in high-T_c ceramics where the hole conductivity is revealed, all spins are paired, the so-called Zhang-Rice singlet [24, 25] is realized. The holes in high-T_c ceramics (at least in the CuO_2 planes) can be considered as spinless positive charged quasiparticles. On the CuO_2 plane, the hole is delocalized among Cu and four O coupled by covalent bonding [25].

In second quantization formalism in the site representation, the model Hamiltonian for one type of spinless holes is

$$H = \epsilon_0 \sum_n b_n^\dagger b_n + \sum_{nn'} M_{nn'} b_n^\dagger b_{n'} + \sum_{nn'} V_{nn'} b_n^\dagger b_{n'}^\dagger b_{n'} b_n, \tag{54}$$

where ϵ_0 is the energy for the hole creation in a lattice, $M_{nn'}$ is the so-called hopping integral and $V_{nn'}$ is the hole-hole interaction term. As we showed in [12], the hole creation, $b_n^\dagger$, and hole annihilation, b_n, operators are characterized by the paulion properties:

$$[b_n, b_{n'}^\dagger]_- = [b_n, b_{n'}]_- = [b_n^\dagger, b_{n'}^\dagger]_- = 0 \qquad \text{for } n \neq n', \tag{55}$$

$$[b_n, b_n^\dagger]_+ = 1; \quad [b_n, b_n]_+ = [b_n^\dagger, b_n^\dagger]_+ = 0, \tag{56}$$

the operators acting on different sites obey the boson commutation relations, while the operators acting on one site obey the fermion commutation relations.

Suppose that the hole-hole interaction term in (54) is attractive. In this case under some conditions, the coupled hole pairs can be formed.

The operators of creation and annihilation of the hole pair are defined as usual:

$$
\begin{aligned}
a_t^\dagger &= b_n^\dagger b_m^\dagger, \\
a_t &= b_m b_n,
\end{aligned}
\tag{57}
$$

where t denotes the localization point of the center of mass of the coupled hole pair. In high-T_c ceramics, the hole-pair localization region is not large: the correlation length in the CuO_2 planes is of the order of $(10-12)$A.

It is easy to verify that the hole-pair operators $a_t^\dagger$ and a_t obey the same paulion commutation relations (55) and (56) as the hole operators. Let us note that the Cooper pair operators also obey the paulion commutation relations, see Eqs. (2) and (33). However, there is an essential difference: the Cooper pair operators are defined in the impulse space and so they are completely delocalized, on the other hand, the hole-pair operators $a_t^\dagger$ and a_t are defined in the site representation and are localized at some regions of the lattice. As we show below, in the quasi-impulse representation, the statistics of hole pairs radically changes.

Let us assume that all hole pairs have the same size and the region of the hole pair localization can be repeated in crystal so that the points t form a "superlattice" with M sites. The model Hamiltonian for hole pairs can be presented as

$$
H = \sum_t \epsilon_p a_t^\dagger a_t + \sum_{tt'} M_{tt'} a_t^\dagger a_{t'},
\tag{58}
$$

where $\epsilon_p = 2\epsilon_0 + V_0$ is the energy of the coupled hole pair, V_0 is the attractive potential between holes which we assume to be the same for all pairs, as in the BCS approach, $M_{tt'}$ is the hopping integral for a hole pair moving as a whole entity. The Hamiltonian (58) can be transformed by some unitary transformation:

$$
A_\mathbf{q} = \frac{1}{\sqrt{M}} \sum_{t=1}^{M} u_{\mathbf{q}t} a_t, \quad A_\mathbf{q}^\dagger = \frac{1}{\sqrt{M}} \sum_{t=1}^{M} u_{\mathbf{q}t}^* a_t^\dagger
\tag{59}
$$

to the diagonalized form in the quasi-impulse space,

$$
H = \sum_\mathbf{q} \epsilon_\mathbf{q} A_\mathbf{q}^\dagger A_\mathbf{q}.
\tag{60}
$$

For simple lattices with one site per cell, the unitary transformation (59) is completely determined by the translation symmetry of the lattice and the coefficients $u_{\mathbf{q}n} = \exp(-i\mathbf{q}r_n)$. The self-energy of the diagonalized Hamiltonian (60) is equal to

$$
\epsilon_\mathbf{q} = \epsilon_p + \sum_{t'(\neq t)} M_{tt'} \exp[i\mathbf{q}\cdot(\mathbf{r}_t - \mathbf{r}_{t'})].
\tag{61}
$$

Since the operators (57) obey neither the boson nor the fermion commutation relations, the unitary transformation in general case is not canonical; this means that it does not preserve the commutation properties of the operators transformed. In particular, the operators (59) do not describe the paulion quasiparticles. As we showed in [12] for holes in quasi-momentum space, such operators obey the modified parafermi statistics of rank M (where

M is the number of lattice sites at which the hole can be created) with the values of quasi-momentum de ned from the quasi-momentum conservation law. This statistics has been introduced by Kaplan [7] in 1976 for the Frenkel excitons and magnons.

For lattices, diagonalized by an exponential unitary transformation, the operators (59) obey the following trilinear commutation relations:

$$[[A_{\mathbf{q}}^{\dagger}, A_{\mathbf{q}'}]_{-}, A_{\mathbf{q}''}]_{-} = -2M^{-1}A_{\overline{\mathbf{q}}}, \qquad \overline{\mathbf{q}} = \mathbf{q}' + \mathbf{q}'' - \mathbf{q} \tag{62}$$

$$[[A_{\mathbf{q}}^{\dagger}, A_{\mathbf{q}'}]_{-}, A_{\mathbf{q}''}^{\dagger}]_{-} = 2M^{-1}A_{\overline{\mathbf{q}}}^{\dagger}, \qquad \overline{\mathbf{q}} = \mathbf{q} - \mathbf{q}' + \mathbf{q}''. \tag{63}$$

Eqs. (62) and (63) have not the Kronecker symbols, as the parafermi trilinear commutation relations (8), the value of $\overline{\mathbf{q}}$ is determined by the quasi-momentum conservation law. Thus, the commutation relations (62) and (63) correspond to the modi ed parafermi statistics [7]. It can be proved that the rank of parastatistics is equal to the number of sites, M, in the superlattice. This means that one state can be occupied by up to M hole pairs:

$$(A_{\mathbf{q}}^{+})^{N} |0\rangle \neq 0, \quad N \leq M \tag{64}$$

$$(A_{\mathbf{q}}^{+})^{M+1} |0\rangle = 0. \tag{65}$$

The state with N noninteracting pairs, each with the same $\mathbf{q}$, is de ned by the usual expression

$$|N_{\mathbf{q}}\rangle = C_N (A_{\mathbf{q}}^{\dagger})^N |0\rangle, \tag{66}$$

where the normalization factor C_N can be found by the induction method using the operator equation obtained from the commutation relation (63)

$$\begin{aligned} A_{\mathbf{q}} A_{\mathbf{q}'}^{\dagger} A_{\mathbf{q}''}^{\dagger} &= A_{\mathbf{q}'}^{\dagger} A_{\mathbf{q}} A_{\mathbf{q}''}^{\dagger} + A_{\mathbf{q}''}^{\dagger} A_{\mathbf{q}} A_{\mathbf{q}'}^{\dagger} - A_{\mathbf{q}''}^{\dagger} A_{\mathbf{q}'}^{\dagger} A_{\mathbf{q}} - \frac{2}{M} A_{\overline{\mathbf{q}}}^{\dagger}, \\ \overline{\mathbf{q}} &= \mathbf{q}' + \mathbf{q}'' - \mathbf{q}. \end{aligned} \tag{67}$$

The expression for C_N differs from that for a Bose system and is given by:

$$C_N = \left[N! \left(1 - \frac{1}{M} \right) \left(1 - \frac{2}{M} \right) \cdots \left(1 - \frac{N-1}{M} \right) \right]^{-\frac{1}{2}}. \tag{68}$$

Now, it is easy to nd

$$A_{\mathbf{q}}^{\dagger} |N_{\mathbf{q}}\rangle = \sqrt{(N_{\mathbf{q}} + 1) \left(1 - \frac{N_{\mathbf{q}}}{M} \right)} |N_{\mathbf{q}} + 1\rangle \tag{69}$$

$$A_{\mathbf{q}} |N_{\mathbf{q}}\rangle = \sqrt{N_{\mathbf{q}} \left(1 - \frac{N_{\mathbf{q}} - 1}{M} \right)} |N_{\mathbf{q}} - 1\rangle. \tag{70}$$

As $M \longrightarrow \infty$, relations (69) and (70) turn into the well known relations for bosons. From Eqs. (69) and (70) it follows that

$$A_{\mathbf{q}}^{\dagger} A_{\mathbf{q}} |N_{\mathbf{q}}\rangle = N_{\mathbf{q}} \left(1 - \frac{N_{\mathbf{q}} - 1}{M} \right) |N_{\mathbf{q}}\rangle. \tag{71}$$

Thus, the operator $A_{\mathbf{q}}^{\dagger} A_{\mathbf{q}}$ is not a particle number operator in a state $\mathbf{q}$, as in the case of boson, fermion and paulion operators. It can be proved that for the modi ed parafermi statistics, the operator of particle number in a state $\mathbf{q}$ does not exist, see the Refs. [7, 12]. This is the consequence of the absence of the Kronecker symbols in the commutation relations (62) and (63). What can be de ned is the operator of the total number of hole pairs, $\hat{N}$. For the commutator, the following relation is valid [7, 12]:

$$[A_{\mathbf{q}}, A_{\mathbf{q}}^{\dagger}]_{-} = 1 - \frac{2\hat{N}}{M}.$$

(72)

Only for small concentrations, $\langle \hat{N} \rangle / M \ll 1$, the hole pairs satisfy the Bose statistics.

6   Some Properties of the Hole-Pair System

As we showed above, the operator $A_{\mathbf{q}}^{\dagger} A_{\mathbf{q}}$ is not the hole-pair number operator; so, the diagonalized Hamiltonian (60) does not describe the ideal gas of the hole pairs. The latter does not exist in principle. Even in the absence of dynamical interactions, some immanent interaction in the hole-pair system is always present. The origin of this interaction, which called after Dyson [26] the kinematic interaction, is in the deviation of the hole-pair statistics from the Bose (Fermi) statistics.

Let us estimate the magnitude of the kinematics interaction in the state (66) with N noninteracting hole pairs, each pair with energy $\epsilon_{\mathbf{q}}$ (61). Using the equation (67) for shifting the operator $A_{\mathbf{q}}$ to the right, after straightforward although cumbersome calculations, we obtain [7, 12]

$$
\begin{aligned}
E(N_{\mathbf{q}}) &= \left\langle N_{\mathbf{q}} \left| \sum_{\mathbf{q}'} \epsilon_{\mathbf{q}'} A_{\mathbf{q}'}^{\dagger} A_{\mathbf{q}'} \right| N_{\mathbf{q}} \right\rangle \\
&= N \epsilon_{\mathbf{q}} \left(1 - \frac{N-1}{M} \right) + \frac{N}{M-1} \frac{N-1}{M} \sum_{\mathbf{q}'(\neq \mathbf{q})} \epsilon_{\mathbf{q}'} \\
&= N \left[\epsilon_{\mathbf{q}} + \frac{N-1}{M} (\bar{\epsilon} - \epsilon_{\mathbf{q}}) \right],
\end{aligned}
$$

(73)

where

$$\bar{\epsilon} = \frac{1}{M-1} \sum_{\mathbf{q}'(\neq \mathbf{q})} \epsilon_{\mathbf{q}'}$$

(74)

is the mean energy of the hole-pair band. The second term in the last line of Eq. (73) is the kinematic interaction. It is proportional to the concentration of hole pairs and its magnitude is larger the larger is the difference between the $\epsilon_{\mathbf{q}}$ and the mean energy of the hole pair band. According to Eq. (73), there is an immanent coupling among all states of the hole pair band. Therefore, we cannot de ne the independent quasi-particles in some particular state. As we mentioned above, the ideal gas of the hole pairs does not exist fundamentally. It can exist only in the low concentration limit in which the kinematic energy becomes small and we get the case of the Bose statistics, cf. Eq. (72).

In high T_c superconducting ceramics, the maximum T_c is achieved for a hole concentration in CuO_2 planes equal to $0.2 - 0.25$ per CuO_2 unit [27, 28]. The same order of magnitude has to be for the hole-pair concentration because the latter is counted not per CuO_2 units, but per the number of sites M in the superlattice. Thus, the deviations from the Bose statistics for the hole-pair system are not negligible and have to be taken into account.

As we showed above, the hole pairs obey the modi ed parafermi statistics of rank M, so, one state can be occupied by up to M hole pairs. The number of hole pairs cannot exceed the number of sites M in the superlattice. This means that, in spite of the non-boson behavior of the hole-pair system, there is no statistical prohibition for the Bose-Einstein condensation. On the other hand, the hole-pair system is always non-ideal (because of the kinematic interaction). For a rigorous study of the Bose-Einstein condensation phenomenon, one has to include also a dynamic interaction and consider an interplay between kinematic and dynamic interactions to study the stability of the Bose condensate, as was done for the molecular exciton system in Ref. [29].

Acknowledgement

This work was partially supported by grants from CONACYT (México) 41226-F, 46770-F and from UNAM by IN102203, IN107305.

References

[1] J. Bardeen, L.N. Cooper, and J.R. Schriefer, Phys. Rev. **106**, 162 (1957); Ibid.**108**,1175 (1957).

[2] L.N. Cooper, *Phys. Rev.* **104**, 1189 (1956).

[3] J.R. Schrieffer, *Theory of Superconductivity*, Addison-Wesley, Redwood City, California, 1988.

[4] H.S. Green, *Phys. Rev.* **90**, 270 (1953).

[5] D.V. Volkov, *Sov. Phys. JETP* **9** 1107 (1959); Ibid. **11** 375 (1960).

[6] I.G. Kaplan, O. Navarro and J.A. Sanchez, *Physica C* **419**, 13 (2005).

[7] I.G. Kaplan, *Theor. Math. Phys.* **27** , 466 (1976).

[8] A.N. Avdyugin, Yu. D. Zavorotnev and L.N. Ovander, *Sov. Phys. Solid State* **25** , 1437 (1983).

[9] B.A. Nguen, *J. Phys.: Condensed Matter* **1**, 9843 (1989).

[10] D.I. Pushkarov, *Phys. Status Solidi* (**b**) **133** , 525 (1986).

[11] B.A. Nguen and N.C. Hoang, *J. Phys.: Condens. Matter* **2**, 4127 (1990).

[12] I.G. Kaplan and O. Navarro, *J. Phys.: Condens. Matter* **11**, 6187 (1999).

[13] I.G. Kaplan and O. Navarro, *Physica C* **341-348**, 217 (2000).

[14] A.S. Davydov, *Theory of Molecular Excitons* McGraw-Hill, New York, 1962.

[15] S. Kamefuchi and Y. Takahachi, *Nucl. Phys.* **36**, 177 (1962).

[16] C. Rayan and E.C.G. Sudarshan, *Nucl. Phys.* **47**, 207 (1963).

[17] Y. Ohnuki and S. Kamefuchi, *Quantum Field Theory and Parastatistics* Springer-Verlag, Berlin, 1982.

[18] O.W. Greenberg and A.M. Messiah, *Phys. Rev.* **138**, B1155 (1965).

[19] R.C. Hilborn and G.M. Tino, Eds., *Spin-Statistics Connection and Commutation Relations*, *AIP Conf. Proc.* No. **545**, AIP, Melville, New York, 2000.

[20] N.A. Chernikov, *Acta Physica Polonica* **21**, 52 (1962).

[21] A.B. Govorkov, *Sov. J. Part. Nucl.* **14**, 520 (1983).

[22] I.G. Kaplan, in *Fundamental World of Quantum Chemistry. A Tribute Volume to the Memory of Per-Olov Lowdin*, Eds. E.J. Brandas and E.S. Kryachko, Kluwer Academic Publ., Dordrecht, 2003, pp.183-220.

[23] S.S. Schweber, *An Introduction to Relativistic Quantum Field Theory*, Row Petersen, New York, 1961.

[24] F.S. Zhang and T.M. Rice, *Phys. Rev. B* **37**, 3759 (1988).

[25] I.G. Kaplan, J. Soullard, J. Hernández-Cobos, and R. Pandey, *J. Phys.: Condens. Matter* **11**, 1049 (1999).

[26] F. Dyson, *Phys. Rev.* **102**, 1217 (1956).

[27] M.W. Shafer et al., *Phys. Rev. B* **39**, 2914 (1989).

[28] H. Zhang and H. Sato, *Phys. Rev. Lett.* **70**, 1697 (1993).

[29] I.G. Kaplan and M.A. Ruvinskii, *Sov. Phys. JETP* **44**, 1127 (1976).

In: New Topics in Superconductivity Research
Editor: Barry P. Martins pp. 239-252

ISBN 1-59454-985-0
© 2006 Nova Science Publishers, Inc.

Chapter 8

Unified Explanation for the Nine Features of Inhomogeneities of Gap and Superconductivity in the High-T_c Cuprates

Fu-sui Liu and Yumin Hou*
Department of Physics, Beijing University,
Beijing 100871, China

Abstract

Recent scanning tunneling microscope (STM) experiments on Bi2212 have shed new light on the nature of superconducting state in high-T_c cuprates and have emphasized the important role played by inhomogeneities of superconductivity and energy gap in the CuO_2 plane of the high-T_c cuprates. Summarizing all related observations, we nd that there are nine features altogether for the inhomogeneities. This chapter demonstrates that the thermal perturbation leads to the uctuation of antiferromagnetic short-range coherence length (AFSRCL) in the CuO_2 plane, and further leads to the uctuation of pairing potential. The latter can cause the inhomogeneities of the gap and the superconductivity. This chapter gives a uni ed explanation for the nine features of the inhomogeneities. The physical picture of the inhomogeneities of superconductivity and gap in the CuO_2 plane is as follows. The values of the gap and the critical temperature T_c in bulk measurements are determined by the most probable value of AFSRCL. At $T = T_c$, a superconducting percolation channel is established by the locations with the most probable AFSRCL and the locations with AFSRCL larger than the most probable one. The proximity effect and pair tunneling effect exist in the locations with lower values of T_c. However, both effects are not important for the inhomogeneities. We think that the mobile $O_{p\sigma}$ holes in the CuO_2 plane are of homogeneous distribution. The gap and the superconductivity themselves are stable, and the stability does not need the help of nodal Cooper pair. This chapter also reconciles Lang *et al.*'s experimental observations with the basic concept of superconductivity.

Keywords: High-T_c cuprate, Inhomogeneity, Gap, Superconductivity

---

*E-mail address: fsliu@pku.edu.cn

1 Introduction

Pan et al. found by scanning tunneling microscope (STM) that for their 84 K T_c $Bi_2Sr_2CaCu_2O_{8+\delta}$ (Bi2212) sample, the magnitudes of the gaps in the CuO_2 plane are different in different locations, which is called by us local gaps [1]. There is a distribution of magnitudes of the gap at 4.2 K on $600\,A \times 600\,A$ area in the CuO_2 plane, with maximum value 65 meV, minimum value 25 meV, and the most probable value 40 meV [1]. That the distribution of the gap in the CuO_2 plane is inhomogeneous was also found in Refs. [2 - 9]. The observed local gap and local superconductivity have the following nine features. First, the observed cumulative counts for different magnitudes of the local gaps are different [1, 2]. The maximum and minimum values of the cumulative counts are 11 and 1 for the local gaps with 30 and 50 meV, respectively [2]. Second, the inhomogeneity of the gap in the CuO_2 plane is not stable at the same tip position in STM for long time measurement [2]. Third, when the point-contact mode was entered by producing a tip crash, dI/dV - V characteristic with a clear peak was replaced by a multi-peak structure [2]. Fourth, the values of the local gaps in the CuO_2 plane was observed in Tl2212 and Y123 as well [3, 4]. Fifth, the inhomogeneity of the gap in the CuO_2 plane exists in different hole doping level [5]. Sixth, the inhomogeneity of the gap in the CuO_2 plane exists at different temperatures [5]. Seventh, the inhomogeneity of the gap exists even above T_c, in the pseudogap range [5]. Eighth, a remnant of magnetic eld expulsion is seen in form of a transient Meissner effect at several tens of degrees above T_c, judging from the optical conductivity [6, 7]. Ninth, the CuO_2 plane breaks into distinct domains of either "good" or "bad" superconductivity [8].

Analysis of the STM data in Ref. [1] led Pan *et al.* to suggest that the observed inhomogeneous electronic structure arises from the ionic potential associated with the off-stoichiometry oxygen dopants disordered in the BiO layer [1, 10]. It is obvious that the theory in Refs. [1, 10] cannot explain that the values of the gap are not stable at the same tip position for long time measurement, and that the inhomogeneity appears in many kinds of the high-T_c cuprates without the BiO layer. Matsuda *et al.* and Lang *et al.* thought that the consensus has been that the gap inhomogeneity is a result of the in-plane charge distribution [5, 9]. However, there is no quantitative calculation based on the in-plane charge distribution. Joglekar *et al.* thought that the high-T_c cuprates are granular d-wave superconductors, studied the dynamics of Josephson coupling between such granular d-wave superconductors, focusing on the effect of nodal Cooper pairs and disorder impurities, and found that the nodal Cooper pairs give rise to a power-law Josephson coupling which leads to the stability of the superconducting phase in the CuO_2 plane [11]. Martin and Balasky thought that the inhomogeneity is induced by disordered doping, and constructed a semi-phenomenological model [12].

The basic aim of this chapter is to demonstrate that the inhomogeneities of the gap and the superconductivity come from the inhomogeneity of distribution of the antiferromagnetic short-range coherence lengths (AFSRCL) in the CuO_2 plane. In section 2 the two-local-spin-mediated interaction (TLSMI), which causes the pairing [13, 14], is introduced brie y and is extended from three-band model to four-band model. In section 3 we give T_c formulas. In section 4 we make numerical calculations to give the inhomogeneities of distributions of the gap and the superconductivity in the CuO_2 plane, and to explain the nine features of the inhomogeneities. In section 5 we give some predictions. In section 6 we give a reconciliation of the sharp contradiction between Lang *et al.*'s observations [9] and basic concept of superconductivity. In section 7 we make conclusions and give a physical picture for the inhomogeneities of the gap and the superconductivity.

2 Pairing Interaction

The Hamiltonian of Hubbard-Emery d-pσ model for the CuO$_2$ plane is given approximately by Ref. [15], which is

$$H = -\sum_{i\alpha\beta s} T_{\alpha\beta}p^+_{\alpha s}p_{\beta s} + J_K \sum_{i\alpha\beta ss'} \widehat{\mathbf{S}}_i \cdot \widehat{\sigma}_{ss'}p^+_{\alpha s}p_{\beta s'} + J\sum_{i<j} \widehat{\mathbf{S}}_i \cdot \widehat{\mathbf{S}}_j, \tag{1}$$

where the summation over α and β is for the oxygen sites around i-th Cu^{++} site, $p_{\alpha s}$ annihilates O$_{p\sigma}$ hole with spin s at site α, $\widehat{\mathbf{S}}_i$ is the local spin operator vector of Cu^{++} at site i, $\widehat{\sigma}$ Pauli matrix vector, i and j are the nearest neighbors. The main change of the four-band Hubbard-Emery d-pσ model in comparison with the three-band one in Refs. [13, 14] is that in the four-band model [15]

$$J_K = t^2(1 - \frac{x}{4})^2(B_6 + xA_7), \tag{2}$$

$$B_6 = \frac{1}{U_d - \varepsilon_p + \varepsilon_d - 2V_\circ} + \frac{1}{\varepsilon_p - \varepsilon_d}, \tag{3}$$

$$A_7 = \frac{1}{\varepsilon_p - \varepsilon_d - V_\circ} - \frac{1}{\varepsilon_p - \varepsilon_d} + \frac{1}{U_d - \varepsilon_p + \varepsilon_d - V_\circ} - \frac{1}{U_d - \varepsilon_p + \varepsilon_d - 2V_\circ}, \tag{4}$$

where t is the hopping integral between O:p_σ orbital and Cu:$d_{x^2-y^2}$ orbital, U_p, U_d and $V_\circ$ are intra- and interatomic Coulomb repulsion on O:p_σ orbital, Cu:$d_{x^2-y^2}$ orbital and between both orbitals, respectively. The expression of J still is [15]

$$J = \frac{2t^4}{(\varepsilon_p - \varepsilon_d + V_\circ)^2} \left[\frac{1}{U_d} + \frac{2}{2(\varepsilon_p - \varepsilon_d) + U_p} \right]. \tag{5}$$

Expand $p_{\alpha s}$ in k space. Here k is the wave vector in Brillouin zone of the oxygen lattice in the CuO$_2$ plane. The second term in Eq. (1) is Kondo Hamiltonian, H$_K$, which implies that the O$_{p\sigma}$ holes with k $\uparrow$ and $-$k $\downarrow$ can have interactions with the local spins of Cu^{++} at sites i and j, respectively. The third term in Eq. (1) is Heisenberg interaction, H$_H$, between the two local spins at sites i and j. The effective interaction between the two O$_{p\sigma}$ holes with k $\uparrow$ and $-$k $\downarrow$, mediated by H$_K$ $-$ H$_H$ $-$ H$_K$, is called TLSMI.

Using the extended Abrikosov's pseudo-Fermion method in Ref. [13], the expression of TLSMI

$$U_{\mathbf{kk'}} = -A(T)F_{\mathbf{kk'}} \tag{6}$$

was given in Refs. [13, 14].

$$A(T) = \frac{JJ_K^2 N''/N'w(J/T)}{T^2 + 64JJ_K^2 \sum_{\mathbf{kp}} g(\mathbf{k}, \mathbf{p})/\{1 + 32\pi^2 J_K^2[N(E_F)]^2\overline{h}(\mathbf{q})\}}, \tag{7}$$

$$g(\mathbf{k}, \mathbf{p}) = \left(\frac{1}{2N_{Cu}}\right)^2 \frac{f[\epsilon(\mathbf{k}) - E_F]\exp[i(\mathbf{k} - \mathbf{p}).(\mathbf{R}_\mu - \mathbf{R}_\nu)]F_5(\mathbf{k})F_5(\mathbf{p})}{\epsilon(\mathbf{p}) - \epsilon(\mathbf{k})}, \tag{8}$$

$$F_{\mathbf{kk'}} = \sum_{i=1}^{4} F_i(\mathbf{k})F_i(\mathbf{k'})F_5(\mathbf{k})F_5(k'), \tag{9}$$

$$F_1(\mathbf{k}) = \cos(k_x a)\cos(k_y a), \tag{10}$$

$$F_2(\mathbf{k}) = \sin(k_x a)\sin(k_y a), \tag{11}$$

$$F_3(\mathbf{k}) = \sin(k_x a)\cos(k_y a), \tag{12}$$

$$F_4(\mathbf{k}) = \cos(k_x a)\sin(k_y a), \tag{13}$$

$$F_5(\mathbf{k}) = \cos^2(k_x a/2)\cos^2(k_y a/2), \tag{14}$$

$$h(\mathbf{q}) = [\cos(q_x a/2)\cos(q_y a/2)]^4, \tag{15}$$

where $f(x)$ is Fermi distribution, $\epsilon(\mathbf{k})$ the energy of $O_{p\sigma}$ holes, E_F Fermi energy, $N(E_F)$ the density of states, N'' and N' are the total number of pair of Cu^{++} in the nearest neighbor and the total number of Cu^{++} in a cluster having AFSRCL, ξ, respectively, $w(J/T)$ is a transformation factor in the extended Abrikosov's pseudo-Fermion method [13], a is the nearest distance between two O ions in the CuO_2 plane, the curve of N''/N' vs ξ is given in Ref. [14], and is depicted here as Fig. 1. $\mathbf{R}_\mu$ and $\mathbf{R}_\nu$ are the position vectors of two Cu ions in the nearest neighbor in the CuO_2 plane. The bar above $h(\mathbf{q})$ represents average on Fermi surface. There is also the phonon-mediated interaction, V_{ep}, and the effective Coulomb interaction, V_c^*, between the two $O_{p\sigma}$ holes with $\mathbf{k} \uparrow$ and $-\mathbf{k} \downarrow$ besides TLSMI. Considering all the interactions, the BCS gap equation for the CuO_2 plane in Bi2212 and Y123 is [16]

$$\Delta'(T, \mathbf{k}) = -\sum_{\mathbf{k'}} \frac{V_{\mathbf{kk'}}\Delta'(T, \mathbf{k'})}{2m(T, \mathbf{k'})} \tanh\frac{m(T, \mathbf{k'})}{2T}, \tag{16}$$

$$m(T, \mathbf{k}) = \sqrt{[\epsilon(\mathbf{k}) - E_F]^2 + |\Delta'(T, \mathbf{k})|^2}. \tag{17}$$

$$V_{\mathbf{kk'}} = \begin{cases} V_{ep} + V_c^*, \\ \quad 0 \leq |\epsilon(\mathbf{k}) - E_F|, |\epsilon(\mathbf{k'}) - E_F| \leq E_D; \\ U_{\mathbf{kk'}}, \\ \quad 0 \leq E_F - \epsilon(\mathbf{k}), E_F - \epsilon(\mathbf{k'}) \leq E_F; \\ 0, \\ \quad \text{otherwise,} \end{cases} \tag{18}$$

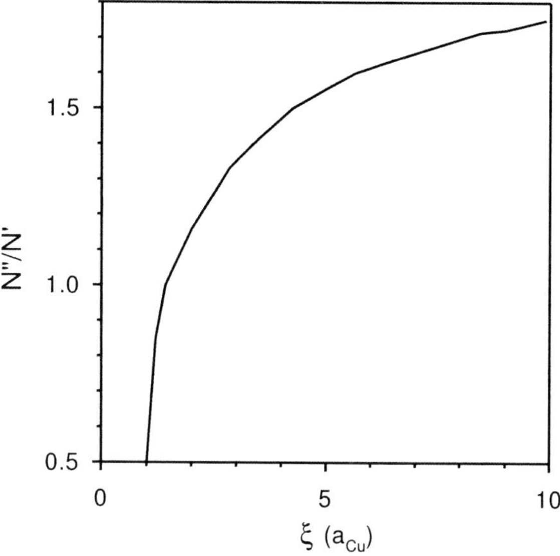

Figure 1: Theoretical curve of N''/N' vs ξ.

where $\Delta'(T, \mathbf{k})$ is a gap function in the CuO_2 plane, E_D the Debye energy. Considering that $\epsilon(\mathbf{k})$ generally is an even function, $\epsilon(\mathbf{k}) = 2.93 \cos(k_x a)\cos(k_y a)(eV)$ for Bi2212 and Y123, V_{ep} and V_c^* are constants, and considering the even or odd properties of the functions in $F_{\mathbf{k},\mathbf{k}'}$, the BCS gap equation can be separated into four decoupled equations, and has the following four solutions.

$$\Delta'_s(T, \mathbf{k}) = \Delta_1(T) + \Delta_2(T)F_1(\mathbf{k})F_5(\mathbf{k}), \tag{19}$$

$$\Delta'_d(T, \mathbf{k}) = \Delta_3(T)F_2(\mathbf{k})F_5(\mathbf{k}), \tag{20}$$

$$\Delta'_{p1}(T, \mathbf{k}) = \Delta_4(T)F_3(\mathbf{k})F_5(\mathbf{k}), \tag{21}$$

$$\Delta'_{p2}(T, \mathbf{k}) = \Delta_5(T)F_4(\mathbf{k})F_5(\mathbf{k}). \tag{22}$$

Numerical calculations show that the solution for the gap of Bi2212 and Y123 is given by Eq. (20). Note that in a coordinate system $\mathbf{k}'$ $45°$-rotating from the system $\mathbf{k}$, $\sin(k_x a)\sin(k_y a)$ becomes $[\cos(k'_x a_{Cu}) - \cos(k'_y a_{Cu})]/2$, i.e., the popular d-wave. Because $\Delta'_d(T, \mathbf{k})$ contains $F_5(\mathbf{k})$ besides $\sin(k_x)\sin(k_y a)$ we call $\Delta'_d(T, \mathbf{k})$ the composite d-wave. The a_{Cu} is the nearest distance between two Cu^{++} ions in the CuO_2 plane.

3 T_c Formula

Doniach and Inui showed that the model of high-T_c cuprates can be reduced to that of a granular superconductor [17]. According to the granular model, the "grain" is the intrinsic

Cooper pair. There is Josephson coupling energy E_J between the "grains". For a single Josephson tunneling junction, the total Hamiltonian is [18]

$$H = \frac{e^2}{2C} n^2 + E_J(1 - \cos\phi), \tag{23}$$

where e is an electron charge, C the junction capacitance, $n = N_1 - N_2$, N_1 and N_2 are the electron number in superconductors 1 and 2, respectively, $\phi = \phi_1 - \phi_2$, ϕ_1 and ϕ_2 are the phases of the cooper pairs in Superconductors 1 and 2, respectively. $[n, \phi] = -i$, so that n plays the rule of the momentum, ϕ that of the coordinate. Approximating the lower part of the cosine curve by a parabola, we get a harmonic oscillator of frequency

$$\hbar\omega = \sqrt{\frac{e^2}{C} E_J}. \tag{24}$$

The zero-point energy is $\hbar\omega/2$. When the Josephson coupling energy is equal to the zero-point energy, i. e., $E_J = \hbar\omega/2$, the phase coherence between "grains" will be damaged. Using Eq. (24), the damage condition of the phase coherence in the quantum XY-model becomes

$$E_J = \frac{e^2}{4C}. \tag{25}$$

We assume that the "grains" constitute a planar square array. The distance between two "grains" is $a_\circ$. The $a_\circ$ is

$$a_\circ^2 = \frac{a_{Cu}^2}{x/2}, \tag{26}$$

where x is the $O_{p\sigma}$ hole number in one unit cell of the CuO_2 plane. The Hamiltonian of the array is

$$H = \frac{e^2}{2} \sum_i C_{i+1,i}^{-1}(n_{i+1,i})^2 + \sum_i E_J(i+1,i)[1 - \cos(\phi_{i+1} - \phi_i)], \tag{27}$$

where $n_{i+1,i}$ is the operator that measures the number of electrons transferred between the nearest neighbor "grains", $C_{i+1,i}^{-1}$ is the inverse of the capacitance matrix, and $E_J(i+1,i)$ is the Josephson coupling energy between "grain" i and "grain" i+1. In the granular model of the high-T_c cuprates, all "grains" are the same, and Eq. (27) becomes

$$H = N\left[\frac{e^2}{2C} n^2 + E_J(1 - \cos\phi)\right], \tag{28}$$

where N is the total number of the "grains". Making comparison between Eq. (28) and Eq. (23), we know that the damage condition of the phase coherence is still given by Eq. (25) in case of the array. The E_J is given in Ref. [19], and it is

$$E_J = \frac{I_c\phi_\circ}{2\pi}, \tag{29}$$

where I_c is the maximum supercurrent of a Josephson junction, and $\phi_\circ$ is the   ux quantum. The I_c is [20, 21]

$$I_c(T) == \frac{\pi\overline{\Delta(T)}}{2eR_n}\tanh\left[\frac{\overline{\Delta(T)}}{2T}\right],\tag{30}$$

where R_n is the normal resistance at $T \approx T_c$, and $\overline{\Delta(T)}$ the average value of the gap at the Fermi surface. Ref. [21] obtained the formula of $I_c(T)$ under the condition of d-wave gap exactly. Comparing Eq. (30) with the result in Ref. [21], we know that our $I_c(T)$ formula, in which the d-wave gap is taken as average value, is a good approximation. Substituting Eq. (30) into Eq. (29) gives that E_J is a function of T and

$$E_J(T) = \frac{\pi\hbar\overline{\Delta(T)}}{4e^2 R_n}\tanh\left[\frac{\overline{\Delta(T)}}{2T}\right],\tag{31}$$

Making comparison between Eq. (1.7) in Ref. [17] and Eq. (2.1) in Ref. [22] yields

$$C = 2\epsilon_\infty a_o,\tag{32}$$

where ϵ_∞ is the high-frequency dielectric constant, typically from 1 to 5 [23]. Substituting Eq. (32) into Eq. (25) gives the damage condition of the phase coherence in quantum XY-model,

$$E_J(T) = \frac{e^2}{8\epsilon_\infty a_o}.\tag{33}$$

If Eq. (33) is set up at $T > 0$ K, then this temperature corresponds to the transition temperature T_c^{qu} determined by the damage condition of the phase coherence in quantum XY-model.

The super uid density ρ_s at 0 K is [24]

$$\rho_s(0) = B\left[E_J(0) - \frac{e^2}{8\epsilon_\infty a_o}\right]^{\frac{2}{3}}.\tag{34}$$

According to the classical phase uctuation theory proposed by Emery et $al.$ [25] and Chakraverty et $al.$ [26], the T_c equation is

$$T_c^{cl} = \frac{D\hbar^2 \rho_s(0) a'}{4m^*},\tag{35}$$

where we use T_c^{cl} to express the transition temperature determined by the damage condition, Eq. (35), of the phase coherence in classical XY-model, $D = 0.9$ in the limit that coupling between CuO_2 planes is very small, m^* is the effective mass of carrier, and a' is the average spacing between the CuO_2 planes [25]. Generally speaking, $T_c^{cl} \neq T_c^{qu}$ for any value of x. The experimentally observed value of T_c should be the smaller one between T_c^{cl} and T_c^{qu}.

4   Numerical Results

4.1   The Values of Parameters

The values of parameters appearing in the expressions of TLSMI in Eqs. (6) and (7) and in BCS gap equations, Eqs. (16), (17), and (18), can be determined by experiments and rst principle calculations. The empirical formula of T_c versus x is $T_c = T_{max}[1 - 82.6(x - 0.16)^2]$, where x is the $O_{p\sigma}$ hole number in one unit cell [27, 28]. For Bi2212 and Y123 it generally takes that $T_{max} = 93$ K. If a reference does not give the value of x, then we always use the above T_c formula to get the corresponding value of x from the observed value of T_c. The near-E_F electronic structure and Fermi surface of Bi2212 have been mapped out with angle-resolved photoemission [29, 30]. The sample in Ref. [29] is Bi2212 with $T_c = 85$ K, and $x = 0.192$. The Fermi surface can be described by $1.456[\cos(k'_x a_{Cu}) + \cos(k'_y a_{Cu})] = E_F = 0.294$ eV for Bi2212 according to the photoemission experiment [29]. For obtaining the values of E_F for different $O_{p\sigma}$ doping, we take rigid band approximation. We use ξ to express AFSRCL. $\xi = a_{Cu}/\sqrt{x}$ if $x < 0.15$ [31]. $\xi < a_{Cu}/\sqrt{x}$ if $x > 0.15$ [32]. Using the values of ξ and Fig. 1, we can obtain the values of N"/N' for any value of x. $t = 1.1$ eV, $U_d = 8$ eV, $U_p = 4$ eV, $V_o = 1$ eV, and $\epsilon_p - \epsilon_d = 2$ eV [15, 33]. The observed value of J is dependent on x. $J(x = 0) = 1333$ K [34]. $J(x = 0) = 1500 - 2500$ K [35]. $J(x = 0) = 1392$ K [36]. We take $J(x = 0) = 1600$ K. The J is a linearly decreasing function of x [34], and proportional to t^4 [15]. J can be described by $J = (1600 - 3000x)$ (K) for $x < 0.27$ [34]. The electron-phonon coupling constant $\lambda = N(E_F)V_{ep} \cong 0.25$ for $x = 0.15$ [37]. $\mu^* = N(E_F)V_c^* = 0.15$ for $x = 0.15$ [38]. The Debye energy $E_D = 60$ meV [39]. $m^* = 6m_o$ [40 - 42]. m_o is the mass of a free electron. $\rho_n = 0.001x^{-2}(10^{-5}\Omega.m)$ [43]. The R_n in Eq. (30) can be given by $R_n = \rho_n/a_o$ [44].

4.2   The Inhomogeneity

The values of parameters given in the last section are enough to solve the BCS gap equation, Eq. (16), and T_c equations, Eqs. (33) and (35). The procedures are as follows. At rst, we look for $\Delta_3(T)$ in Eq. (20). Then, we know the d-wave gap in the CuO_2 plane from Eqs. (20), (11), and (14). From the average value of $\Delta'_d(T, \mathbf{k})$ at Fermi surface, we know $\overline{\Delta(T)}$ in Eqs. (30) and (31). The T_c^{qu} and T_c^{cl} are then easy to be obtained by solving Eqs. (33) and (35), respectively. In STM experiments, the value of the gap $2\Delta(T)$ is determined by the distance of peak-to-peak in the curve of dI/dV vs V. These experimental values of gaps correspond to the maximum values of the d-wave gap in Eq. (20). Thus, to compare our theory with STM experiments, we should take $k_x a = k_y a = \pi/2$, i. e., the Cu - O bond direction. There is a one to one corresponding relation between the gap and AFSRCL ξ because N"/N' appears in Eq. (7) and $N"/N'$ is determined by AFSRCH ξ. If some location in the CuO_2 plane has uctuation of AFSRCH, ξ', due to thermal perturbation, then this location has uctuation of the gap. This is just the method, which we use to determine the inhomogeneities of the gap and the superconductivity in the CuO_2 plane of high-T_c cuprates.

Pan *et al.* found by STM that for their Bi2212 sample with $T_c = 84$ K, the distribution of magnitudes of the gaps at 4.2 K in 600 A × 600 A area of the CuO_2 plane is inhomogeneous, with maximum value 65 meV, minimum value 25 meV, and the most probable value 40 meV [1]. As we know, the antiferromagnetic short-range order is formed under thermal disturbance. Therefore, there is uctuation around the statistically most probable value for AFSRCLs. The other causes of uctuation of AFSRCL are disorder and vacancies. TLSMI is an interaction mediated by the two nearest neighbor Cu^{++} ions in the CuO_2 plane. Thus, TLSMI can be used to discuss the inhomogeneity of the high-T_c cuprates in Refs. [1 - 9].

From Eq. (7) in section 2, we know that TLSMI depends on $N"/N'$. From Fig. 1, we know that $N"/N'$ is determined by AFSRCL ξ. Therefore, we can say that TLSMI depends on AFSRCL ξ as well. In our theory, the inhomogeneity comes from the uctuation of AFSRCL ξ' at different locations in the CuO_2 plane around the most probable value ξ ($=a_{Cu}/\sqrt{x}$) [31]. Therefore, the inhomogeneity is inherent in the CuO_2 plane in our theory. However, the conjecture in Ref. [1] and the theory in Ref. [10] are that the inhomogeneity comes from the BiO layer.

From Fig. 1, the different values of ξ' have different values of $N"/N'$. From Eqs. (7) and (16) the different values of $N"/N'$ have different TLSMI, and thus different magnitudes of the gaps. From the empirical formula of T_c vs x in Refs. [27, 28], we know that $x = 0.126$ if the sample in Ref. [1] is $T_c = 84$ K. The most probable value of ξ' is $\xi = a_{Cu}/\sqrt{0.126} = 2.817\ a_{Cu}$. Correspondingly, $N"/N' = 1.33$ from Fig. 1. The corresponding most probable value of the gap at Cu-O bond direction at 0 K, $\Delta(0K)$, is 34.8 meV from BCS gap equation Eq. (16). If the uctuation values of ξ' at different locations are from 1.7 a_{Cu} to 7 a_{Cu}, then the corresponding values of $N"/N'$ are from 1.1 to 1.66. Substituting these values of $N"/N'$ into BCS gap equation yields that the uctuation values of $\Delta(0K)$ are from 20.2 meV to 58.6 meV. The experimental values of $\Delta(0K)$ in Ref. [1] are: the most probable value is 40 meV, the minimum value is 25 meV, and the maximum value is 65 meV. Note that the experimental values of $2\Delta(0K)$ in Ref. [1] come from the peak-to-peak values of the derivative conductance. Although simple, this peak-to-peak method should give a larger gap value than the true one, considering the effect of thermal smearing and/or the quasiparticle damping [45]. Therefore, we can say that our theory can explain the distribution of the values of $\Delta(4.2K)$ observed in Ref. [1].

4.3   Explanations for the Nine Features of the Inhomogeneities

The explanation for the rst feature is as follows. From solving the BCS gap equation we know that there is a one to one correspondence between the local gap and the value of ξ'. Because the value of ξ' has the most probable value ξ, the local gap also has the most probable value, and, correspondingly, the local gaps have the maximum and minimum observed cumulative counts. The 30 meV in Ref. [2] is the local gap corresponding to the maximum observed cumulative count. The 50 meV in Ref. [2] is the local gap corresponding to the minimum observed cumulative count.

Second, the value of the gap should not be stable at the same tip position in STM technology for long time measurement because the thermal perturbation sets the value of ξ' at any xed location unstable.

The explanation for the third feature is as follows. As we know, the peak on dI/dV - V characteristic corresponds to a gap. When the point-contact mode is entered by producing a tip crash, the tunneling range contains many clusters having different values of ξ', and thus dI/dV - V characteristic with a clear peak is replaced by a multi-peak structure.

Fourth, the inhomogeneity of the gap in the CuO_2 plane should be observed in many kinds of the high-T_c cuprates because the uctuation of AFSRCL ξ' is dependent on the thermal perturbation in the CuO_2 plane, and, therefore, is independent of the kinds of the high-T_c cuprates.

Fifth, the inhomogeneity of the gap in the CuO_2 plane should exist in different hole doping level. The reasons are as follows. The   uctuation of AFSRCL ξ' is dependent only on the thermal perturbation in the CuO_2 plane. Therefore, the existence of the   uctuation of AFSRCL ξ' is independent of the hole doping level.

Sixth, the inhomogeneity of the gap in the CuO_2 plane should be observed at many values of temperature. The reasons are as follows. The   uctuation of AFSRCL ξ' is dependent only on the thermal perturbation in the CuO_2 plane. Thermal perturbation exists at any temperature, even at 0 K, at which there is the zero-point oscillation of quantum mechanics. Therefore, the existence of the   uctuation of AFSRCL ξ' is independent of the temperature.

The explanation for the seventh feature is as follows. According to the observation of Ref. [5], $x = 0.11$, and $T_c = 67$ K, and the inhomogeneity of the gap in the CuO_2 plane exists at 100 K. According to our numerical calculation, the temperature T^* of opening a gap at the Fermi surface in the CuO_2 plane is 190 K. The gap at 67 K $< T <$ 190 K is called the pseudogap. The inhomogeneity of the pseudogap in the CuO_2 plane should be observed at temperature $T_c < T < T^*$ because the   uctuation of AFSRCL ξ' is dependent only on the thermal perturbation in the CuO_2 plane, and the existence of the   uctuation of AFSRCL ξ' is independent of the temperature.

The explanation for the eighth feature is as follows. The superconducting transition temperatures at some locations in the CuO_2 plane, T'_c, are different. For example, our theoretical values of T'_c for the sample in Ref [1], calculated by Eq. (35), are 34.1 K, 52.7 K, 84 K, 113.1 K, 126.3 K, 134.8 K, and 141.3 K, for the locations with $\xi' = 1.7\, a_{Cu}$ ($\Delta(0K) = 20.18$ meV), $\xi' = 2\, a_{Cu}$ ($\Delta(0K) = 24.85$ meV), $\xi' = 2.817\, a_{Cu}$ ($\Delta(0K) = 34.75$ meV), $\xi' = 4\, a_{Cu}$ ($\Delta(0K) = 46.16$ meV), $\xi' = 5\, a_{Cu}$ ($\Delta(0K) = 51.8$ meV), $\xi' = 6\, a_{Cu}$ ($\Delta(0K) = 55.47$ meV), and $\xi' = 7\, a_{Cu}$ ($\Delta(0K) = 58.57$ meV), respectively. Here, we use T_c^{cl} in Eq. (35) as T'_c because the values of T_c^{qu} in Eq. (33) are much larger than the values of T_c^{cl} in Eq. (35) for the sample in Ref. [1]. The observed value of T_c in experiment for the bulk sample should be the most probable value 84 K, which corresponds to the most probable value $\xi = 2.817\, a_{Cu}$. To explain the eighth feature, we give an example. There are still some locations with $\xi' > 4\, a_{Cu}$ in the CuO_2 plane, for which $T'_c > 113.1$ K according to our calculations for the sample in Ref. [1]. Note that the observed value of T_c for the bulk sample in Ref. [1] is 84 K. Therefore, we can say that a remnant of magnetic   eld expulsion can be seen at several tens of degrees above T_c. The remnant of magnetic   eld expulsion is given by the locations with $\xi' > 4\, a_{Cu}$ in the CuO_2 plane.

The explanation for the ninth feature is as follows. Scanning tunneling spectroscopy on cleaved Bi2212 single crystals reveals inhomogeneities on length scales of $\approx 30A$. This scale indicates that there are nearly equal AFSRCLs in the area $\approx (30A)^2$. In our theory, that the CuO_2 plane breaks into distinct domains of either "good" or "bad" superconductivity is natural because the domains can have different AFSRCL, and therefore different transition temperature T'_c. For example, $T'_c = 34.1$ K ($\xi' = 1.7 a_{Cu}$) or 134.8 K ($\xi' = 6 a_{Cu}$), which are given in the paragraph for the eighth explanation. Since there are regions with "bad" superconductivity in the CuO_2 plane, the proximity effect, which implies leakage of Cooper pair wavefunction a distance of order $\approx 15A$ (Cooper pair scale) into the "bad" superconducting regions, can occur, and therefore the "bad" superconducting regions have double gaps.

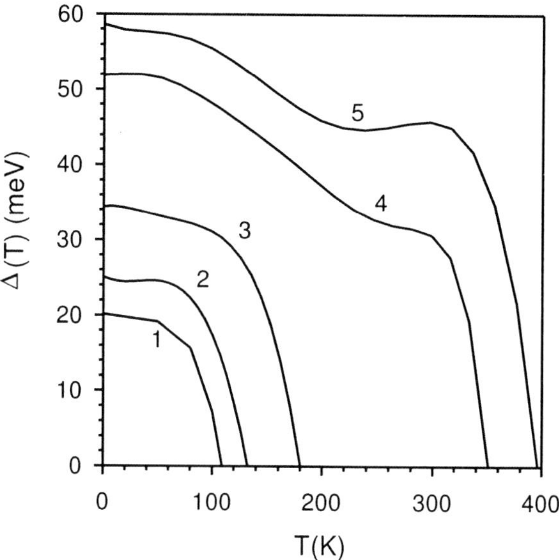

Figure 2: Theoretical curves of $\Delta(T)$ vs T. The curves 1 - 5 correspond to AFSRCLs $\xi' =$ 1.7 a_{Cu}, 2 a_{Cu}, 2.8 a_{Cu}, 5 a_{Cu}, 7 a_{Cu}, and $x = 0.126$.

5 Predictions

The dependences of the gap $\Delta(T)$ vs T at different locations in the CuO_2 plane are much different. For example, the theoretical curves 1, 2, 3, 4, and 5 in Fig. 2 correspond to ξ' = 1.7 a_{Cu}, 2 a_{Cu}, 2.817 a_{Cu}, 5 a_{Cu}, 7 a_{Cu}, and $x = 0.126$ for Bi2212 and Y123. All the ve theoretical curves in Fig. 2 have the pseudogaps, i.e., the gaps above the corresponding superconducting transition temperatures T'_c. Note that there is no discontinuty at T'_c for the ve curves. The evaluation from the pseudogap at $T > T'_c$ to the gap at $T < T'_c$ is smooth. Both the gap and the pseudogap are gap of Cooper pairs. That the people call the gap at $T > T'_c$ the pseudogap is a historical misunderstanding. The curves in Fig. 2 are dif cult to be observed in experiment because of the unstable property of AFSRCL under thermal perturbation. However, from Fig. 2 we can see that even at room temperature 300 K there are gaps larger than 30 meV in the locations with AFSRCL larger than 5 a_{Cu}. These large gaps at room temperature should be easy to be observed by long time measurement at xed location in the CuO_2 plane or by measurements at many locations in the CuO_2 plane.

6 Reconciliation of the Sharp Contradiction Between Basic Concept of Superconductivity and Lang *et al.*'s Observations

A novel atom-scale technique, designed to locally distinguish superconducting from non-superconducting regions by using quasiperticle scattering resonances at Ni impurity atoms, has recently been proposed [9]. Lang *et al.* applied this technique to carry out spectral surveys on as-grown Ni-doped Bi2212 sample [9]. Lang *et al.* found that no Ni scattering resonances are observed in any region where $\Delta > 50$ meV, Ni scattering resonances are observed in regions where $\Delta < 35$ meV, and Ni scattering resonances go to zero gradually by 50 meV. Lang *et al.*'s conclusions are that purely superconducting regions when $\Delta < 35$ meV, a mixture of superconducting and normal regions when 35 meV $< \Delta < 50$ meV, and normal regions when $\Delta > 50$ meV. As is well known, these conclusions violate basic concept on superconductivity. The gap is the bound energy of Cooper pair. Therefore, according to the basic concept, the larger the gap is, the stronger the superconductivity should be. Our numerical calculations in the paragraph for the eighth explanation accord with this basic concept. Our theory can also reconcile Lang *et al.*'s observations with the basic concept of superconductivity. The numerical calculations in the eighth explanation tell us that the longer the AFSRCL is, the larger the gap. As is well known, the longer the AFSRCL is, the stronger the antiferromagnetism. The Ni atom is magnetic. Therefore, the Ni atom is dif cult to enter the regions with stronger antiferromagnetic correlation. Our conclusions are that Ni atoms cannot enter the regions with $\Delta > 50$ meV, a few Ni atoms can enter the regions with 35meV $< \Delta < 50$ meV, and Ni atoms can easy enter the regions with $\Delta < 35$ meV.

7   Conclusions

The mobile $O_{p\sigma}$ hole distribution in the CuO_2 plane of the high-T_c cuprates is homogeneour in our theory. We demonstrate that the thermal perturbation leads to the uctuation of AFSRCL, and further leads to the uctuation of pairing potential. The latter can cause the inhomogeneities of the gap and the superconductivity in the CuO_2 plane. The values of the gap and the critical temperature T_c in bulk sample are determined by the most probable value of AFSRCL. At $T = T_c$, a superconducting percolation channel is established by the locations with the most probable AFSRCL and the locations with AFSRCL larger than the most probable one. The proximity effect and the Josephson effect exist in the locations with lower values of T'_c. However, both effects are not important for the inhomogeneities. The gap and the superconductivity themselves are stable in the CuO_2 plane, and their stability does not need the help of the nodal Cooper pair in Ref. [11].

References

[1] Pan, S. H., O'Neal, J. P., Badzey, R. L., Chamon, C., Ding, H., Engelbrecht, J. R., Elsaki, Z., Uchida, S., Gupta, A. K., Ng, K. -W., Hudson, E. W., Lang, K. M., Davis, J. C. *Nature* (London) 2001, 413, 282.

[2] Liu, Jin-Xiang, Goldman, J. C. *Phys. Rev. Lett.* 1991, 67, 2195.

[3] Wang, Lan-ping, He, Jian, Wang, Guowen *Phys. Rev.* 1989, B40, 10594.

[4] Adkins, E. A., Chandlev, C. J. *J. Phys.* 1987, C20, L1009.

[5] Matsuda, A., Fijii, T., Watanabe, T. *Physica* 2003, C388 - 389, 207.

[6] Corson, J., Mallozi, R., Orenstein, J. N., Bozovic, I. *Nature* 1999, 398, 221.

[7] Ranninger, J., Tripodi, L. *Phys. Rev.* 2003, B67, 174521.

[8] Howard, C., Fournier, P., Kapitulnik, A. *Phys. Rev.* 2001, B64, 100504.

[9] Lang, K. M., Madhavan, V., Hoffman, J. E., Hudson, E. W., Eisaki, H., Uchida, S., Davis, J. C. *Nature* (London) 2002, 415, 412.

[10] Wang, Ziqiang, Engelbrecht, J. R., Wang, Shancai, Ding, H., Pan, S. H. cond-mat/0107004, 2001.

[11] Joglekar, Y. N., Castro Neto, A. H., Balatsky, A. V. *Phys. Rev. Lett.* 2004, 92, 037004.

[12] Martin, I., Balatsky, A. V. *Physica* 2001, C357-360, 46.

[13] Liu, Fu-sui, Chen, Wan-fang *Phys. Rev.* 1998, B58, 8812.

[14] Liu, Fu-sui *Phys. Lett.* 1997, A224, 185.

[15] Matsukawa, H., Fukuyama, H. *J. Phys. Soc. Jpn.* 1989, 58, 2845.

[16] Bardeen, J., Cooper, L. N., Schrieffer, J. R. *Phys. Rev.* 1957, 108, 1175.

[17] Doniach, S., Inui, M. *Phys. Rev.* 1990, B41, 6668.

[18] Anderson, P. W. in *Lecture on the Many-Body Problem VII*, Caianiello, E. R., Ed., Academic, New York, 1964, p. 127.

[19] Hebard, A. F. in *Inhomogeneous Superconductors-1979*, Francavilla, T. L., Gubsor, D. V., Leibiwitz, J. R., Wolf, S. A., Ed., *AIP Conference Proceedings* No. 58, American Institute of Physics, New York, 1979, p. 129.

[20] Mahan, G. D. *Many-Particle Physics*, Plenum, New York, 1981, p. 819.

[21] Tanaka, Y., Kashiwaya, S. *Phys. Rev.* 1997, B56, 892.

[22] Ariosa, D., and Beck, H. *Phys. Rev.* 1991, B43, 344.

[23] Golovashkin, A. I., Kraiskaya, K. V., Shelakov, A. L. *Sov. Phys. Solid State* 1990, 32, 98.

[24] Doniach, S. *Phys. Rev.* 1981, B24, 5063 .

[25] Emery, V. J., Kivelson, S. A. *Nature* (London) 1995, 374, 434.

[26] Chakraverty, B. K., Ramakrishnan, T. V. *Physica* 1997, C282-287, 290.

[27] Williams, G. V. M., Tallon, J. L., Haines, E. M., Michalak, R., Dupree, R. *Phys. Rev. Lett.* 1997, 78, 721.

[28] Presland, M. R. *Physica* 1991, C165, 391.

[29] Dessau, D. S., Shen, Z. -X., King, D. M., Marshall, D. S., Lombardo, L. W., Dickinson, P. H., Loesser, A. G., DiCarlo, J., Park, C. -H.,Kapitulnik, A., Spicer, W. E. *Phys. Rev. Lett.* 1993, 71, 2781.

[30] Chung, Y.-D., Gromko, A. D., Dessau, D. S., Aiura, Y., Yamaguchi, Y., Oka, K., Arko, A. J., Joyce, J., Eisaki, H., Uchida, S. I., Nakamura, K., Ando, Y. *Phys. Rev. Lett.* 1999, 83, 3717.

[31] Birgeneau, P. J., Gabbe, D. R., Jensen, H. P., Kastner, M. A., Picone, P. J., Thurston, T. R., Shirane, G., Endoh, Y., Sato, M., Yamada, K., Hidaka, Y., Oda, M., Enomoto, Y., Suzuki, M., Murakumi, T. *Phys. Rev.* 1988, B38, 6614.

[32] Kitaoka, Y., Ohsugi, S., Ishida, K., Asayama, K. *Physica* 1990, C170, 189.

[33] Han, Rushan, Chew, C. W., Phua, K. K., Gan, Z. Z. *J. Phys.: Condens. Matter* 1991, 3, 8059.

[34] Oda, M., Ohguro, T., Matsuki, H., Yamada, N., Ido, M. *Phys. Rev.* 1990, B41, 2606.

[35] Batlogg, B. *Solid State Comm.* 1998, 107, 639.

[36] Lyons, K. B., Fleury, P. A., Schneemeyer, L. F., Waszczak, J. V. *Phys. Rev. Lett.* 1988, 60, 732.

[37] Timusk, T., Tanner, D. B. In *Physical Properties of High Temperature Superconductors*, Ginsberg, D. M., Ed., World Scienti c, Singapore, 1992, Vol. 3, p. 363.

[38] Schruter, M. A. *Mater. Sci. Eng.* 1993, B19, 129.

[39] Pintschovius, L. Reichardt, W. In *Physical Properties of High Temperature Superconductors*, Ginsberg. D. M., Ed., World Scienti c, Singapore, 1994, Vol. 4, p. 350.

[40] Uemura, Y. J., Luke, G. M., Sternlieb, B. J., Brewer, J. H., Carolan, J. F., Hardy, W. N., Kadono, R., Kempton, J. R., Kie , R. F., Kreitzman, S. R., Mulhern, P., Riseman, T. M., Williams, D. L., Yang, B. X., Uchida, S., Tagagi, H., Gopalakrishman, J., Slaight, A. W., Subramanian, M. A., Chien, C. L., Cieplak, M. Z., Xiao, Dang, Lee, V. Y., Statt, B. W., Stronach, C. E., Kossler, W. J., Yu, X. H. *Phys. Rev. Lett.* 1989, 62, 2317.

[41] Uemura, Y. J., Le, L. P., Luke, G. M., Sternlieb, B. J., Wu, W. D., Brewer, J. H., Riseman, T. M., Seaman, C. L., Maple, M. B., Ishikawa, M., Hinks, D. G., Jorgensen, J. D., Saito, G., Yamochi, H. *Phys. Rev. Lett.* 1991, 66, 2665.

[42] Salamon, M. B. In *Physical Properties of High Temperature Superconductors*, Ginsberg, D. M., Ed., World Scienti c, Singapore, 1989, Vol. 1, p. 40.

[43] Cooper, S. L., Gray, K. E. In *Physical Properties of High Temperature Superconductors*, Ginsberg, D. M., Ed., World Scienti c, Singapore, 1994, Vol. 4, p. 74.

[44] Abeles, B. *Phys. Rev.* 1977, B15, 2828.

[45] Hasekawa, T.,Ikuta, H., Kitazawa, K. In *Physical Properties of High Temperature Superconductors*, Ginsberg, D. M., Ed., World Scienti c, Singapore, 1992, Vol. 3, p. 533.

In: New Topics in Superconductivity Research
Editor: Barry P. Martins, pp. 253-305

ISBN: 1-59454-985-0
© 2006 Nova Science Publishers, Inc.

Chapter 9

ABOUT THE SUPERCONDUCTIVITY THEORY

R. Riera and J.L. Marín
Departamento de Investigación en Física, Universidad de Sonora
Apartado Postal 5-088, 83190 Hermosillo, Sonora, México
R. Rosas and R. Betancourt-Riera
Departamento de Física, Universidad de Sonora
Apartado Postal 1626, 83000 Hermosillo, Sonora, México

Abstract

A General Theory of Superconductivity with points of view differing from those of the BCS Theory is presented. The formation of electron pairs in a conductor material is investigated upon arriving to the critical temperature where the conductor-superconductor transition occurs. A general equation for the superconductivity is obtained based on the stable pairing of two electrons bound by a phonon for any type of superconductor material. This equation comes from a self-consistent field calculation with a screening, which is temperature dependent, showing that the total energy of the electron pairs is constant and the local energy of the paired electrons is equal to that of the phonon in the range 0 to T_C. A specific condition for the existence of the superconducting state is established, allowing the prediction of the critical temperature. The dispersion law of the elementary excitements produced by the superconductivity is obtained and correctly interpreted. The method is based on represent to the operators of Bose that characterize to phonons and to the electron-phonon interaction as a combination of products of Fermi operators corresponding to the electrons that form the pairs. The expression obtained for the critical temperature is compatible with those obtained by G.M. Eliashberg and W.L. McMillan. An expression for the bond energy of the pairs, or better known as superconductor gap, is also obtained as a function of the temperature and the critical temperature, resulting very similar to that formulated by Buckingham.

This theory is reached in the frame of self-consistent field equations for any natural or artificial solid where free electrons exist. The necessity of the electrons must be coupled by phonons for the existence of the superconducting state is also justified, arriving to a general conclusion: the superconductivity theory is based only on the theory used to carry out the electron-phonon interaction and more concretely of the phonons (harmonic or anharmonic theory, low, intermediate and high temperature). The theory is applied to the particular case of low temperature superconductors, obtaining an excellent agreement with the results of other theories (phenomenological and microscopic) as well as with experimental data.

An application of the general equation obtained for low critical temperature superconductors utilizing a phononic theory is developed. Then, we arrive to a specific expression for the bounding energy as a function of temperature. The density of states of the electron pairs is calculated and used to obtain an equation for the critical magnetic field. This result is needed to determine the electrodynamical properties. Finally, we obtain the specific heat as a function of temperature, we compare it to experimental data for Sn, and we calculate its jump at T_C for eight superconductors. We have also determined the variation of the energy gap or bond energy with the temperature of the MgB_2 superconductor and we have compared our results with another theoretical and experimental results reported in the literature, obtaining an excellent agreement with the experimental results.

I Introduction

The formation of electron pairs in a conductor material due to the electron-phonon interaction is very important in the explanation of the superconductivity theory. The attraction between electrons due to the exchange of phonons leads to the formation of a bound state of two electrons with opposite momentum and a gap in the energy spectrum [1-6].

A feature of the electron pairs excitation spectrum in superconductors is the appearance of energy gap, which it is equivalent to the bond energy of the electron pairs and it depends on the temperature, reaching their maximum value in $T = 0\,K$. There are different interpretations of the energy gap [7], however we think that in fact the name of gap energy is not the appropriate because it represents a zone of energy permitted for the electron pairs and forbidden for the individual electrons.

The corresponding energy spectrum, or law of dispersion, obtained in the BCS Theory [1-8], $\varepsilon(k) = \sqrt{E^2(k) + \Delta^2(k)}$, has not been well interpreted neither correctly used by their authors [1-8], because it is not clear the dependence with the temperature of energy gap $\Delta(k)$.

After the BCS theory was established, many works [4, 6, 8, 9] were carried out in order to make a correction to the equation that gives the critical temperature, introducing many-body techniques, such as the calculation of Green function, in order to determine the spectrum of energy. However, there are not works dedicated to calculate other physical magnitudes based in the law of dispersion, which have different points of view when are compared to the BCS theory.

In this work we have calculated, by using the self-consistent field, the energy of electron pairs correlated by the electron-phonon interaction in a superconductor material, considering the Hamiltonian of the system of electrons and phonons before and after arriving to the critical temperature. Upon arriving to the critical temperature we obtain that the kinetic energy of the free electrons with k and $-k$ is eliminated and the energy of the phonons is made equal or proportional to the energy of the electrons pairs for each temperature in the interval $[0, T_C]$. The electron-phonon interaction is worked by substituting the Bose operators of creation and annihilation of phonons as a product of Fermi operators of creation and annihilation of the electrons that are correlated, coupling the interaction of Coulomb repulsion of the electrons with the electron-phonon interaction from each of the two electrons.

Being a pair of electrons in equilibrium with a constant total energy where the effect of diminishing or increasing the temperature in the interval $(0, T_C)$ only changes the internal state of the pairs, this is to say, it only makes more or less bound the electron pairs.

On the other hand, we have obtained an expression for the critical temperature, which it is very similar to the obtained in Refs. [9] and [10]; this allows us to define a parameter, which it characterizes the particular properties of each superconductor. Finally, by using the dispersion law and this parameter, we obtain an expression for the energy gap or better called energy of bond of the pairs, which it depends on the temperature and the critical temperature as there is described by the experiments.

A well known distinctive property of the superconductors is the drop of their electric resistance to zero as they arrive to the corresponding critical temperature. It is also well known that in order to the electric current exists, it should exist the free charge carriers that can move when an electric field is applied. In pure semiconductors and dielectrics don't exist free charge carriers to zero degree Kelvin, therefore the electric current to low temperature cannot exist in these materials. Thus, it is impossible that both, conductivity and superconductivity, exist in materials where the free charge carriers do not exist; such as it occurs in the pure semiconductors and dielectrics at low temperature. If free charge carriers are introduced in these materials in some conducting phases, as is the case of the planes and chains of $Cu - O$ of the high critical temperature super conducting ceramic, then one could get superconductivity through these planes or chains.

In the Hartree-Fock theory of free electrons, the familiar set of free electron plane waves, in which each wave vector k_F occurs twice (one for each spin orientation) in the Slater determinant, gives a solution to the Hartree-Fock equation for free electrons. If the plane waves are indeed solutions, then the associated electronic charge density will be uniform. However, in the free electron gas the ions are represented by a uniform distribution of positive charge with the same density as electronic charge. Hence the electrons potential precisely cancel out the potential of the ions: $U^{ion} + U^{el} = 0$. Then the free electrons can only be dispersed by the phonons. In the solids where a free electron gas exists, the electron-phonon interaction determines the electric resistance. As in the Hartree-Fock theory of free electrons only the electron-phonon interaction is present, it is necessary that the electrons be coupled in pairs with the phonons in order to eliminate or compensate the electron-phonon interaction in the whole interval of temperature from zero until the critical temperature. This way we can consider pairs of electrons that can move with an internal state characterized by bond energy and the electric resistance can be converted to zero.

The phenomenon of superconductivity was discovered at the beginning of the past century (1911). Although the development of the microscopic theory of superconductivity involved a great deal of efforts, it was nevertheless finalized about forty years later (1957). In spite of this, the extensive investigations are still being carried out in the field of superconductivity, only its applications in engineering, and the range of topics brought under the purview of research in superconductivity is being continuously extended. New fields like the creation and study of organic and heavy-fermions superconductors also emerged around the decade of 1970. The studies of superconductors in non-equilibrium states, magnetic superconductors, organic superconductors and ceramic superconductors were initiated somewhat earlier and become of extremely importance at the moment. However, the problem

of high-temperature superconductivity in ceramic superconductors is of utmost importance
and we would even call it the number one question in superconductivity.

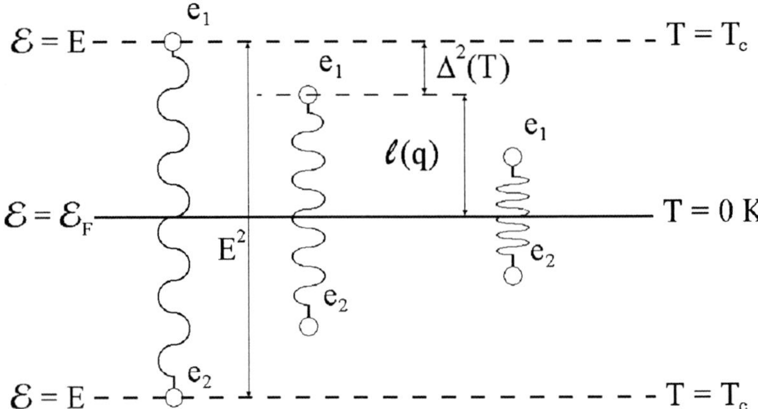

Fig. 1. A scheme of the energies that intervene in the general equation of the superconductivity, where
E is the pair total energy, which it is conserved, and $\mathcal{E}(T)$ is the electron pair energies, which are
proportional to the phonon energies, and $\Delta(T)$ is the bond energy, both vary of opposite form in order
to maintain E constant.

The starting points for the elaboration of our superconductivity theory are the following:

i)   The microscopic carriers of the superconductivity are pairs of electrons bound by
phonons (a coupled electron pair). Because the electron-phonon interaction is the
responsible of the electric resistance, it is necessary that two electrons form a
paired state in order to the electron-electron Coulomb dispersion compensates
the electron pair-phonon interaction and so eliminates the electric resistance;

ii)   The transition to the superconducting state comes from a conducting phase
where free electrons exist, allowing the formation of the pairs;

iii)   The total energy of each pair is constant and it does not depend on the wave
vector k neither on the temperature T. When the temperature is lowered,
starting from the critical temperature, the bounding energy of the pair increases
to maintain the total energy constant and vice versa; i.e., an increase or decrease
of temperature in the interval between 0 to T_C only makes a change in the
internal state of the pair in order to conserve the total energy, through a decrease
or an increase of the phonon energy, respectively (see Fig. 1);

iv)   The coupled electron pair is formed by three particles, two fermions and a boson,
then it is nor a boson, neither a fermion, this is, it is not a problem of two bodies
of identical particles but of three particles of different nature: they satisfy the
Maxwell-Boltzmann statistics;

v)   Because in the interval of temperature between $0K$ and T_C the electrons in the
pair and the phonon occupy the same energy states, the energy of the coupled

electrons pair is equal to the energy of the phonon or keep a lineal relation, through a physical constant (λ) (see Eq. (105)), which linearly depends on their wave vector $\boldsymbol{q}$ and on the temperature T. The superconductivity theory is sustained by a theory of phonons (harmonic, anharmonic, of low, intermediate and high temperature);

vi)   The density of states of coupled electrons pairs in $T = T_C$ is the same and equal to the density of states of the free electron in the conductor state at the Fermi level. The density of states of the coupled electrons pairs between $0K$ and T_C is similar to the density of states of the phonons;

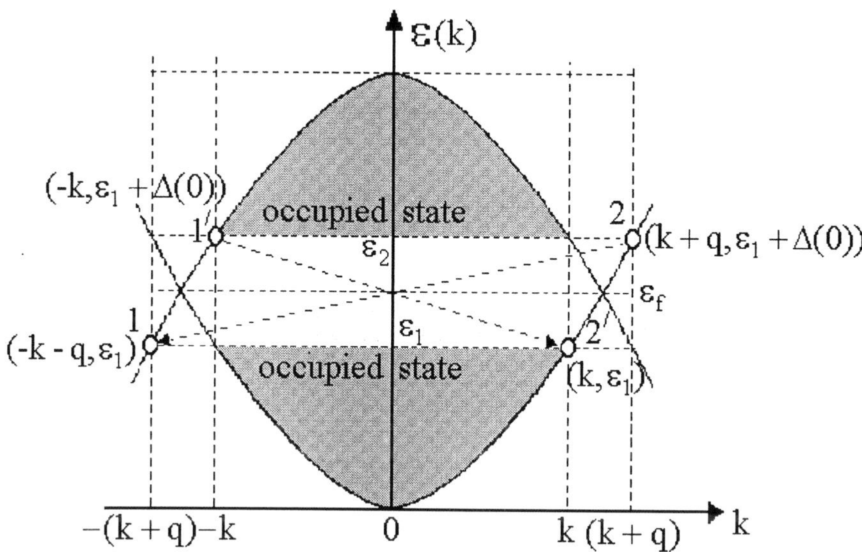

Fig. 2. A scheme of the way as the pairs are formed upon arriving to T_C and the symmetry breaking leads to an energy gap centered on the Fermi level.

Upon arriving to the critical temperature, a splitting of the conduction band occurs due to a symmetry breaking in such a way that one electron sees the other as if its mass would be negative. This symmetry breaking also leads to an energy gap (energy of the bound of the electron pairs and gap for the free electrons) (centered on the Fermi level) between the coupled electrons (see Fig.2). A transformation of the electron energies of a three-dimensional reciprocal space of wave vectors $\left(k_x, k_y, k_z\right)$ to a bi-dimensional space of electron pair energies $\left(\ell(q), \Delta(T)\right)$ has occurred (see Fig. 3).

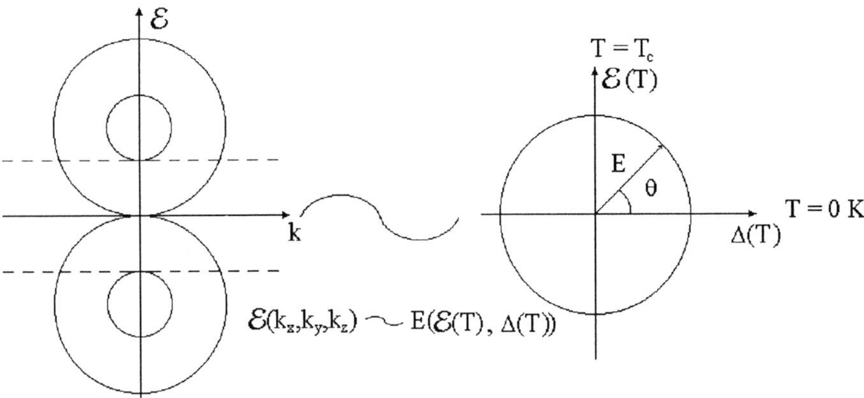

Fig. 3. The transformation of the electron energies of a three-dimensional reciprocal space of wave vectors $\left(k_x, k_y, k_z\right)$ to a bi-dimensional space of electron pair energies $\left(\ell(q), \Delta(T)\right)$.

II   Physical Model

We will consider as starting point a conductor with temperature greater than the critical temperature, $T > T_C$. The Hamiltonian of the electronic system is

$$H = K + H_0 + H_{ep},$$

(1)

where K is the kinetic energy of the electronic gas, H_0 is the free phonon energy and H_{ep} is the electron-phonon interaction, which are given by

$$K = \sum_k \varepsilon(k) C_k^+ C_k; \qquad H_0 = \sum_q \varepsilon_p(q) b_q^+ b_q$$

$$H_{ep} = \sum_{k,q} \left\{ D(-q) C_{k-q}^+ C_k b_q^+ + D(q) C_{k+q}^+ C_k b_q \right\}$$

(2)

In Eq. (1) we are considering that the crystalline potential is very small, this is to say, practically zero, which it implies that the interaction potential between the electrons is made equal but opposite sign to the potential due to the interaction of the electrons with the ions.

In this situation the electrons can be moved in a guided direction only through the conductor when a potential difference, due to the application of an external field, is created. The electric resistance felt by the electrons is only caused by their interaction with the phonons. In a pure conductor the electric resistance is due to the dispersion of the electrons caused by the electron-phonon interaction H_{ep}.

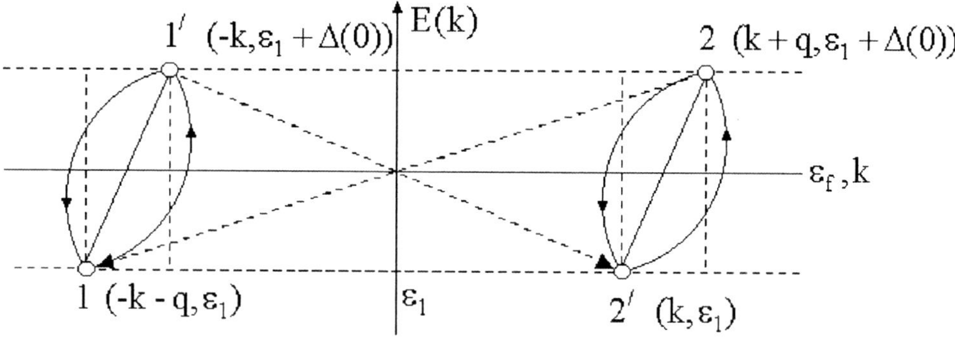

Fig. 4. A scheme which explains the way as the pairs are formed upon arriving to T_C.

Upon lowering the temperature until the critical temperature $T = T_C$, the conductor becomes superconductor and the electric resistance disappears. In order to the electron-phonon interaction not be the cause of the dispersion of the electrons and of the electric resistance, it must be used in causing another physical effect or a new elementary excitement as the formation of electron pairs. The BCS theory is based in the so-called pairs of Cooper, which are electrons pairs correlated by a real or virtual phonon, this is to say two electrons with moments k and $-k$ and spins σ and $-\sigma$ are correlated forming a new elementary excitement, using the electron-phonon interaction in order to change the internal state or bond energy of the electron pairs but not to disperse the individual electrons. We think that upon arriving to the critical temperature the following occurs:

- An unfolding of the original band energy in two subbands, one with energy $\varepsilon(k)$ and a second one of energy $\varepsilon(-k)$, of such a form that $\varepsilon(k) = -\varepsilon(-k)$ and taking the Fermi energy ε_F as the zero of energy.

- The electrons form pairs with moments k and $-k$ with spins σ and σ', such that $\sigma = -\sigma'$. The electrons are above and below the energy of Fermi.

- The electron-phonon interaction becomes electron-pair-phonon interaction coupled with the electron-electron Coulomb interaction, creating a negative constant effective potential that it doesn't depend on the total momentum k ; this makes that the total energy be conserved and it only depends on the critical temperature. The variation of the temperature from $T_C \to 0K$ only changes the internal state of the pairs, increasing their bond energy and diminishing their kinetic energy, and when the variation of the temperature is from $0K \to T_C$ the bond energy of the pairs diminishes until arrive to T_C and after that a small increase of the temperature breaks the pairs, disappearing the superconductivity.

In Figs. 2 or 4 we can see the electron pairs given by the following relations:

$$\left(-k\sigma, k\sigma'\right) \leftrightarrow \left(-k'\sigma, k'\sigma'\right)$$
$$\left(k_1\sigma, k_2\sigma'\right) \leftrightarrow \left(k_3\sigma, k_4\sigma\right).$$

Conservation of the Momentum

The wave vectors of the electrons that are paired should fulfill the following equations:

$$k_1 + k_2 = 0$$
$$k_2 - k_1 = q$$
$$k_1 - k_3 = q$$
$$k_4 - k_2 = q \tag{3}$$

From these equations we can obtain that

$$k_1 = -k_2$$
$$2k_2 = q \Rightarrow k_2 = \frac{q}{2} \Rightarrow k_1 = -\frac{q}{2}$$
$$k_2 - k_3 = 2q \Rightarrow k_3 = k_2 - 2q = -\frac{3}{2}q$$
$$-k_1 + k_4 = 2q \Rightarrow k_4 = 2q - \frac{q}{2} = \frac{3}{2}q \tag{4}$$

If we impose the following conditions

$$k_2 - k_1 = 2q \quad \text{and} \quad k_1 + k_2 = 0, \tag{5}$$

then

$$2k_2 = 2q \Rightarrow k_2 = q, \quad k_1 = -q, \quad k_3 = -2q \quad \text{and} \quad k_4 = 2q. \tag{6}$$

In general we can demand that

$$k_2 - k_1 = nq \quad n \in Z > 0$$
$$k_1 - k_2 = nq \quad n \in Z < 0 \tag{7}$$

From Figs. 2 or 4 we can also obtain that

$$\varepsilon(k) = \varepsilon(-k') \quad y \quad \varepsilon(-k) = \varepsilon(k'). \tag{8}$$

The important thing is to couple the processes of electron-e_1-phonon interaction and electron-e_2-phonon interaction with the process of Coulomb repulsion $e_1 - e_2$. This is to say, when e_1 and e_2 are under the effect of their Coulomb interaction, the e_1 emits a phonon and changes its state from $-k$ to $-k'$ and the e_2 absorbs a phonon and changes its state from k to k'. It is necessary to keep in mind that for the electron-electron interaction the pair changes from the state $(-k,k)$ to the state $(-k',k')$ and for the e_1-phonon interaction this change is from $-k$ to $-k'$, besides, the electron e_2-phonon interaction changes from k to k'. Therefore, the result is an attractive interaction between the electrons and the formation of pairs of bound electrons in equilibrium that can freely move in the crystal.

III Characteristic Equations of the Electronic Bands

From Fig. 4 we can see that the pairs are formed, upon arriving to the critical temperature, of the following form:

$$(k - k, E(T_C)) \text{ for the electrons } (k, \varepsilon(k)) \text{ and } (-k, \varepsilon(-k))$$

$$(k + q - k - q, E(T_C)) \text{ for the electrons } (k + q, \varepsilon(k,q)) \text{ and } (-k - q, \varepsilon(-k - q))$$

If in an initial instant there are electrons in $(k, \varepsilon(k))$ and $(-k, \varepsilon(-k))$, then in $(k + q, \varepsilon(k,q))$ and $(-k - q, \varepsilon(-k - q))$ there are holes, of such a form that

$$\varepsilon(k) = \varepsilon(-k - q)$$
$$\varepsilon(-k) = \varepsilon(k + q)$$

$$(9)$$

this is to say, the electron energy is equal to the hole energy.

From Fig. 4 we can obtain the equations of the electronic bands as

$$\varepsilon(k) = -\frac{\hbar^2 k^2}{2m} + \varepsilon_F$$

$$\varepsilon(-k) = \frac{\hbar^2 k^2}{2m} - \varepsilon_F$$

$$(10)$$

These equations only have in common the Fermi energy ε_F and the momentum at the Fermi surface k_F. Combining them we obtain

$$\varepsilon_F = \frac{\hbar^2 k_F^2}{2m} \quad \text{where} \quad k_F = \pm\sqrt{\frac{m}{\hbar^2}\varepsilon_F}. \tag{11}$$

We can also find that

$$-(k+q) < -k_F < -k \quad \text{and} \quad k < k_F < k+q \tag{12}$$

and then

$$k_F = k + \frac{q}{2} \tag{13}$$

IV Hamiltonian in the Superconducting State

When the conductor has reached the critical temperature the electrons form electron pairs and the Hamiltonian of the new phenomenon is

$$H = K_{e_1} + K_{e_2} + H_0 + H_{e_1 p} + H_{e_2 p} + H_{e_1 e_2}, \tag{14}$$

where K_{e_1} and K_{e_2} are the kinetic energies in the $\varepsilon(k)$ and $\varepsilon(-k)$ electronic bands, H_0 is the free phonon energy, $H_{e_1 p}$ is the electron e_1-phonon interaction, $H_{e_2 p}$ is the electron e_2-phonon interaction and $H_{e_1 e_2}$ is the electron-electron Coulomb repulsion, which are given by

$$K_{e_1} = \sum_k \varepsilon(k) C_k^+ C_k, \quad K_{e_2} = \sum_{-k} \varepsilon(-k) C_{-k}^+ C_{-k}, \quad H_0 = \sum_q \varepsilon_p(q) b_q^+ b_q$$

$$H_{e_1 p} = \sum_{-k,q} D(-q) C_{-k'}^+ C_{-k} b_q^+, \quad H_{e_1 p} = \sum_{k,q} D(q) C_{k'}^+ C_k a_q$$

$$H_{e_1 e_2} = \sum_{k,k',q} V_{-k,k} C_{k'}^+ C_{-k'}^+ C_{-k} C_k \tag{15}$$

We need that the electron-e_1-phonon interaction with the electron-e_2-phonon interaction be coupled with the electron-electron interaction in order to give a negative constant effective interaction between the two electrons, which it must be independent of $k, k', -k, -k'$, but it can depend on the phonon wave vector q, such that

$$H_{e_1 p} + H_{e_2 p} + H_{e_1 e_2} = H_{e_1 p e_2} + H_{e_1 e_2} = H_{p e_1 e_2 p}. \tag{16}$$

V Representation of the Operators of Bose as a Product of Operators of Fermi. Electron-Phonon Interaction

If the above-mentioned is true, the Bose operators b_q^+ and a_q in Eq. (15) can be expressed as a combination of products of Fermi operators C_k^+ and C_k.

We will consider the following transformation:

$$b_{k',k}^+ = \alpha_k C_{k'}^+ C_k + C_{-k'}^+ C_{k'}^+ \qquad a_{-k',-k} = \beta_k C_{-k'}^+ C_{-k} - C_{k'}^+ C_{-k'}^+$$

$$b_{k',k} = \alpha_k^* C_k^+ C_{k'} + C_{k'} C_{-k'} \qquad a_{-k',-k}^+ = \beta_k^* C_{-k}^+ C_{-k'} - C_{-k'} C_{k'} \tag{17}$$

where $\alpha_k = \alpha_k^*$ and $\beta_k = \beta_k^*$, then

$$H_{e_1 p} = \sum_{k,k',q} D(-q) C_{-k'}^+ C_{-k} \left(\alpha_k C_{k'}^+ C_k + C_{-k'}^+ C_{k'}^+ \right)$$
$$= \sum_{k,k',q} D(-q) \left\{ \alpha_k C_{-k'}^+ C_{-k} C_{k'}^+ C_k + C_{-k'}^+ C_{-k} C_{-k'}^+ C_{k'}^+ \right\} \tag{18}$$

However, the physics process given by this product of operators

$$C_{-k'}^+ C_{-k} C_{-k'}^+ C_{k'}^+ = 0, \tag{19}$$

because this process is improbable; thus Eq. (18) is

$$H_{e_1 p} = -\sum_{k,k',q} D(-q) \alpha_k C_{k'}^+ C_{-k'}^+ C_{-k} C_k . \tag{20}$$

On the other hand

$$H_{e_2 p} = \sum_{k,k',q} D(q) C_{k'}^+ C_k \left(\beta_k C_{-k'}^+ C_{-k} - C_{k'}^+ C_{-k'}^+ \right)$$
$$= \sum_{k,k',q} D(q) \left\{ \beta_k C_{k'}^+ C_k C_{-k'}^+ C_{-k} - C_{k'}^+ C_k C_{k'}^+ C_{-k'}^+ \right\} \tag{21}$$

but, by the same reason as in Eq. (19),

$$C_{k'}^+ C_k C_{k'}^+ C_{-k'}^+ = 0, \tag{22}$$

therefore

$$H_{e_2p} = -\sum_{k,k',q} D(q)\beta_k C_{k'}^+ C_{-k'}^+ C_{-k} C_k \qquad (23)$$

Now, let us see the commutation rules

$$\left[b_q^+, b_{q'}\right] = b_q^+ b_{q'} - b_{q'} b_q^+ = \delta_{qq'}$$

$$\left[C_k^+, C_{k'}\right]_+ = C_k^+ C_{k'} + C_{k'} C_k^+ = \delta_{kk'} \qquad (24)$$

$$b_{k',k}^+ b_{k,k'} = \alpha_k^2 C_{k'}^+ C_k C_k^+ C_{k'} + \alpha_k C_{k'}^+ C_k C_{k'} C_{-k'} +$$
$$+ \alpha_k^* C_{-k'}^+ C_{k'}^+ C_k^+ C_{k'} + C_{-k'}^+ C_{k'}^+ C_{k'} C_{-k'} \qquad (25)$$

$$b_{k,k'} b_{k',k}^+ = \alpha_k^2 C_k^+ C_{k'} C_{k'}^+ C_k + \alpha_k^* C_k^+ C_{k'} C_{-k'}^+ C_{k'}^+ +$$
$$+ \alpha_k C_{k'} C_{-k'} C_{k'}^+ C_k + C_{k'} C_{-k'} C_{-k'}^+ C_{k'}^+ \qquad (26)$$

therefore Eq. (25) gives us

$$b_{k',k}^+ b_{k,k'} = C_{-k'}^+ C_{k'}^+ C_{k'} C_{-k'} = n_{k'} \cdot n_{-k'} = n_q \qquad (27)$$

where n_q is the number of phonons.

Electron-Phonon Interaction

An interesting question is how to choose the constants α_k and β_k.

The theory of such constants will be presented according to the development of Luttinger and Kohn (1955) [11], about the electron-phonon interaction. Thus, let us consider the initial problem of solving the Schrödinger equation

$$H\psi = E\psi \qquad (28)$$

We propose a unitary transformation

$$\psi = e^{iS}\phi, \qquad (29)$$

where S is an hermitian operator; by substituting in the initial Schrödinger equation we obtain that

$$\tilde{H}\phi = E\phi \qquad (30)$$

Now, let us consider the transformation $\tilde{H}$ given as

$$\tilde{H} = e^{iS} H e^{-iS} ; \tag{31}$$

if S is in any sense small, the exponential functions can be expanded in the following series (we have neglected terms of third and of superior order in S):

$$\tilde{H} = \left(1 - iS - \frac{1}{2}S^2\right) H \left(1 + iS - \frac{1}{2}S^2\right) = H + i[H,S] - \frac{1}{2}[[H,S],S] , \tag{32}$$

where

$$[H,S] = HS - SH . \tag{33}$$

The matrix elements of S, in the used representation, are determined by

$$\langle nk|S|lq \rangle = \frac{i\langle nk|H_1|lq \rangle}{(E_n - E_l)} , \tag{34}$$

where

$$H = H_0 + H_1 + U . \tag{35}$$

The transformed Hamiltonian includes terms of second order in S, which can be formed noting that

$$i[H_1,S] = [[H_0,S],S] ; \tag{36}$$

then, we obtain that

$$\tilde{H} = H_0 + U + i[U,S] - \frac{1}{2}[[U,S],S] + \frac{1}{2}i[H_1,S] . \tag{37}$$

We will choose

$$H_0 = K_{e_1} + K_{e_2} + H_0 \tag{38}$$

$$H_1 = H_{e_1 p} + H_{e_2 p} \tag{39}$$

$$U = H_{e_1 e_2} \tag{40}$$

where S is chosen to satisfy the following relation:

$$i[H_0, S] = -H_1 . \tag{41}$$

We will make an approximation neglecting the commutator $[U, S]$, this is we will ignore the effect of the electron-phonon interaction on the electron-electron terms. Suppose now that eigenstates of H_0 are available, which are temporarily denoted by $|m\rangle, |n\rangle$, then

$$\langle m|S|n\rangle = \frac{i\langle m|H_1|n\rangle}{(E_m - E_n)} . \tag{42}$$

Let n_q represent the number of phonons present with wave vector q. Then, H_1 has matrix elements in which n_q changes by ± 1. Note that

$$\langle n_q - 1|b_q|n_q\rangle = n_q^{1/2} \quad \text{and} \quad \langle n_q + 1|b_q^+|n_q\rangle = (n_q + 1)^{1/2} . \tag{43}$$

The transformation operator S will have matrix elements connecting the same states:

$$\langle n_q - 1|S|n_q\rangle = i\sum_{k,k',q} D(q)C_{k'}^+ C_k \frac{n_q^{1/2}}{(\varepsilon(k') - \varepsilon(k) - \varepsilon(p))} \quad \text{and}$$

$$\langle n_q + 1|S|n_q\rangle = i\sum_{k,k',q} D(-q)C_{-k'}^+ C_{-k} \frac{(n_q + 1)^{1/2}}{(\varepsilon(-k') - \varepsilon(-k) - \varepsilon_p)} \tag{44}$$

Interest centers on matrix elements of the transformed Hamiltonian H_1 that are diagonal in the phonon occupation numbers, since it is these elements that will contain the effective electron-electron interaction. Thus,

$$\left\langle n_q \left| \frac{1}{2} i[H_1, S] \right| n_q \right\rangle = \frac{1}{2} i\sum_{\pm} \left[\langle n_q|H_1|n_q \pm 1\rangle\langle n_q \pm 1|S|n_q\rangle - \langle n_q|S|n_q \pm 1\rangle\langle n_q \pm 1|H_1|n_q\rangle \right] . \tag{45}$$

The matrix elements products are easily constructed. For example,

$$\frac{1}{2}i\langle n_q|H_1|n_q-1\rangle\langle n_q-1|S|n_q\rangle =$$

$$= -\frac{1}{2}\sum_{k,k',q} C^+_{-k'}C_{-k}C^+_{k'}C_k \left|D(q)\right|^2 \frac{n_q}{\left(\varepsilon(k')-\varepsilon(k)-\varepsilon_p\right)}$$

(46)

We have used the relation $D(-q)=D^*(q)$. After the matrix element products have been determined, the result may be summed over phonon wave vectors $\boldsymbol{q}$. The additional term in H_1 that is produced will be called H_I. Some algebra involving the change of dummy variables and use of the commutation rules yields

$$H_I = \sum_{k,k',q} \left|D(q)\right|^2 \left\{ \frac{\varepsilon_p}{\left[(\varepsilon(k)-\varepsilon(-k))^2 - \varepsilon_p^2\right]} \right\} C^+_{-k'}C^+_{k'}C_{-k}C_k$$

(47)

This interaction matrix elements can be either attractive, repulsive or any of the two and remain in equilibrium. If the states $\boldsymbol{k}, -\boldsymbol{k}$ are separated by an energy larger than ε_p, the effect is repulsive, but if the energy difference is smaller than this, an attraction is present, but if the energy difference is equal to the phonon energy then the dielectric function spreads to the infinite and is the exceptional case of the superconductivity or perfect conductor.

The developed method allows us to choose the correct form of α_k and β_k in the transformation proposal: to change from Bose operators to Fermi operators:

$$\alpha_k = \frac{D(-q)}{\varepsilon(k)-\varepsilon(-k)+\varepsilon_p}, \quad \beta_k = \frac{D(q)}{\varepsilon(-k)-\varepsilon(k)+\varepsilon_p} = \frac{-D(q)}{\varepsilon(k)-\varepsilon(-k)-\varepsilon_p},$$

(48)

then

$$H_{e_1p} + H_{e_2p} = \sum_{k,k',q} \left|D(q)\right|^2 \frac{-2\varepsilon_p}{\left(\varepsilon(k)-\varepsilon(-k)\right)^2 - \varepsilon_p^2} C^+_{k'}C^+_{-k'}C_{-k}C_k$$

(49)

VI Operators of Creation and Annihilation of Electron Pairs

Now let us define the following operators of creation and annihilation of electron pairs:

$$P^+_{k'} = C^+_{k'}C^+_{-k'}; \qquad P_{k'} = C_{-k'}C_{k'}.$$

This operator satisfies the following commutation relation

$$\left[P_k, P_{k'}^+\right] = \left(1 - n_k - n_{-k}\right)\delta_{kk'},$$

where

$$n_k = C_k^+ C_k \quad \text{and} \quad n_{-k} = C_{-k}^+ C_{-k}$$

This commutation relation is different from that for Bose particles.
Besides,

$$\left[P_k, P_{k'}\right] = 0,$$

which it is the same as for bosons.

Also we have that

$$\left\{P_k, P_{k'}\right\} = 2P_k P_{k'}\left(1 - \delta_{kk'}\right),$$

which it is different from that for Bose particles.

Therefore we can define that

$$n_{P'(k)} = P_{k'}^+ P_{k'}; \quad P_{k'}^+ = -P_{-k'}^+; \quad P_{k'} = -P_{-k'}; \quad n_{P'(-k)} = P_{-k'}^+ P_{-k'};$$

then we obtain

$$n_{p(k)} = n_k n_{-k}.$$

VII Electron-Electron Coulomb Repulsion. Effective Interaction

We argued that for many purposes the Fourier transform of the electron-electron Coulomb interaction should be screened by the electronic dielectric constant,

$$\frac{4\pi e^2}{q^2} \to \frac{4\pi e^2}{q^2 \xi^{el}} = \frac{4\pi e^2}{q^2 + k_0^2} \quad q = -k - (-k - q), \tag{50}$$

where ξ^{el} is the electronic dielectric constant, and it represents the effect of the other electrons in screening the interaction between a given pair. Besides

$$\frac{4\pi e^2}{k_0^2} = \frac{2\varepsilon_F}{3n_e}. \tag{51}$$

However, the ions also screen the interactions, and we should have used the full dielectric constant ξ instead ξ^{el}. Thus, using for ξ the following form

$$\frac{1}{\xi} = \left(\frac{1}{1+\dfrac{k_0^2}{q^2}}\right)\left(\frac{\omega^2}{\omega^2 - \omega^2(q)}\right) ,$$

(52)

we find that Eq. (50) should be replaced by

$$\frac{4\pi e^2}{q^2} \rightarrow \frac{4\pi e^2}{q^2 \xi} = \frac{4\pi e^2}{q^2 + k_0^2}\left(1 + \frac{\omega^2(q)}{\omega^2 - \omega^2(q)}\right).$$

(53)

As we can see, the effect of the ions is to multiply Eq. (50) by a correction factor that depends on frequency as well as wave vector. The frequency dependence reflects the fact that the screening action of the ions is not instantaneous, but limited by the (small on the scale of v_F) velocity of propagation of elastic waves in the lattice. As a result the part of the effective electron-electron interaction mediated by the ions is retarded.

In order to use Eq. (53) as an effective interaction between a pair of electrons, one needs to know how ω and k depend on the quantum numbers of the pairs. We know that when the effective interaction is taken to have the frequency-independent form Eq. (50), then q is to be taken as the difference in the wave vectors of the two electronic levels. By analogy, when the effective interaction is frequency-dependent, we shall take ω as the difference in the angular frequencies (i.e., the energies divided by $\hbar$) of levels. Thus, given two electrons with wave vectors $-k$ and k and energies $\varepsilon(k)$ and $\varepsilon(-k)$, we take their effective interaction to be [3, 12]

$$V_{k,-k}^{eff-ee} = \frac{1}{V}\frac{4\pi e^2}{q^2 + k_0^2}\left(\frac{[\varepsilon(k)-\varepsilon(-k)]^2}{[\varepsilon(k)-\varepsilon(-k)]^2 - \varepsilon_p^2(q)}\right); \quad \omega = \frac{\varepsilon(k)-\varepsilon(-k)}{\hbar}; \quad \varepsilon_p(q) = \hbar\omega(q)$$

(54)

There are two important qualitative features of $v_{k,-k}^{eff-ee}$:

1. The dressed phonon frequency $\omega(q)$ is of order ω_D (Debye frequency) or less. Thus, when the energies of the two electrons differ by much more than $\hbar\omega_D$, the phonon correction to their effective interaction is negligibly small. Since the range of variation of electronic energies, ε_F, is typically 10^2 to 10^3 times $\hbar\omega_D$, only electrons with energies quite close together have an interaction appreciably affected by the phonons.

2. However, when the electronic energy difference is less than $\hbar\omega_D$, the phonon contribution has the opposite sign from the electronically screened interaction, and is larger in magnitude; i.e., the sign of the effective electron-electron interaction is reversed. This phenomenon, known as "over screening" and erroneously or mistakenly it be think was thought that this effect is a crucial ingredient in the modern theory of superconductivity.

VIII   Electron Pairs-Phonon Effective Interaction

Now, let us consider the result

$$H_{pe_1e_2p} = H_{e_1p} + H_{e_2p} + H_{e_1e_2} =$$

$$\sum_{k,k',q} \left\{ |D(q)|^2 \frac{-2\varepsilon_p}{(\varepsilon(k)-\varepsilon(-k))^2 - \varepsilon_p^2} + \frac{4\pi e^2}{|k-(-k)|^2 + k_0^2} \frac{(\varepsilon(k)-\varepsilon(-k))^2}{(\varepsilon(k)-\varepsilon(-k))^2 - \varepsilon_p^2} \right\} C_{k'}^+ C_{-k'}^+ C_{-k} C_k \tag{55}$$

We will consider two cases. For the first one, we need to impose two conditions in order that superconductivity occurs; the first condition is

$$|D(q)|^2 = \frac{4\pi e^2}{|k-(-k)|^2 + k_0^2} \frac{1}{2} \varepsilon_p , \tag{56}$$

and the second condition (strong condition) is that

$$\varepsilon_p = \varepsilon(k) - \varepsilon(-k) \quad \text{or} \quad \varepsilon(k) - \varepsilon(-k) = \lambda \varepsilon_p ,$$

where λ is the parameter of electron-phonon interaction (see Eq. (105)).

For the second case the necessary conditions are:

$$\frac{2|D(q)|^2}{\varepsilon_p} \neq \frac{4\pi e^2}{|k-(-k)|^2 + k_0^2}$$

and

$$\varepsilon_p \neq \varepsilon(k) - \varepsilon(-k) ;$$

although ε_p and $\varepsilon(k) - \varepsilon(-k)$ can be related in specific forms with λ.

Let us begin considering the first case. The Hamiltonian corresponding to the imposed conditions is

$$H_{pe_1e_2p} = \sum_{k,k',q} \frac{4\pi e^2}{\left|k-(-k)\right|^2 + k_0^2} \left\{ -\frac{\varepsilon_p^2}{(\varepsilon(k)-\varepsilon(-k))^2 - \varepsilon_p^2} + \frac{(\varepsilon(k)-\varepsilon(-k))^2}{(\varepsilon(k)-\varepsilon(-k))^2 - \varepsilon_p^2} \right\} C_k^+ C_{-k'}^+ C_{-k} C_k$$

. (57)

then

$$H_{pe_1e_2p} = -\sum_{k,k',q} \frac{4\pi e^2}{\left|k-(-k)\right|^2 + k_0^2} C_{k'}^+ C_{-k'}^+ C_{-k} C_k = -\sum_{k,k',q} \frac{2|D(q)|^2}{\varepsilon_p} C_{k'}^+ C_{-k'}^+ C_{-k} C_k$$

, (58)

where we have considered that $k-(-k) = 2k = 2q$ or $k = q$.

Now making

$$\frac{4\pi e^2}{\left|k-(-k)\right|^2 + k_0^2} = V_0(q)$$

, (59)

and considering that $q = q(T)$ then

$$V_0(T) = \frac{4\pi e^2}{\left|k-(-k)\right|^2 + k_0^2} = 2\frac{|D(q)|^2}{\varepsilon_p}$$

. (60)

This last expression can be esteemed having in consideration the following relationships

$$D(q) = i\left(\frac{\hbar}{2\omega_j M}\right)^{1/2} qC_n; \qquad C_n = -\frac{2}{3}\varepsilon_F; \qquad k = q = k_F$$

$$\omega = cq; \qquad c = \frac{2\varepsilon_F}{3M} = \left(\frac{m}{3M}\right)v_F^2; \qquad \varepsilon_F = \frac{mv_F^2}{2}$$

$$g(\varepsilon_F) = \frac{mk_F}{\hbar^2\pi^2} = \frac{3}{2}\frac{n}{\varepsilon_F}; \qquad \frac{4\pi e^2}{k_0^2} = \left(\frac{\pi^2\hbar^2}{mk_F}\right)^2 = \left(\frac{1}{g(\varepsilon_F)}\right)^2$$

(61)

IX Total Hamiltonian in the Superconducting State

The Hamiltonian in the superconducting state, Eq. (14), is only formed by terms that characterize the electron pairs

$$H = \sum_k \varepsilon(k) C_k^+ C_k + \sum_{-k} \varepsilon(-k) C_{-k}^+ C_{-k} + \sum_k \left(\varepsilon(k) - \varepsilon(-k)\right) C_k^+ C_k C_{-k}^+ C_{-k}$$
$$+ \sum_{k,k',q} V_0(q) C_{k'}^+ C_{-k'}^+ C_{-k} C_k \tag{62}$$

We know that $P_k^+ P_k = N_k / 2$, where N_k is the total number of the electrons in the state k, then

$$\sum_k C_k^+ C_{-k}^+ C_{-k} C_k = \sum_k P_k^+ P_k = \frac{N}{2} \; ; \tag{63}$$

but we also know that

$$\sum_k \left(n_k + n_{-k}\right) = N , \tag{64}$$

then

$$\sum_k 2 P_k^+ P_k = \sum_k \left(n_k + n_{-k}\right) = \sum_k 2 C_k^+ C_k C_{-k}^+ C_{-k} \; ; \tag{65}$$

therefore, the third term in Eq. (62) remain as

$$\sum_k \frac{1}{2}\left(\varepsilon(k) - \varepsilon(-k)\right)\left(n_k + n_{-k}\right) = \frac{1}{2} \sum_k \left(\varepsilon(k) - \varepsilon(-k)\right)\left(C_k^+ C_k + C_{-k}^+ C_k\right) . \tag{66}$$

The first and second term of the Hamiltonian (62) are the kinetic energy of the electrons without pairing and logically should be zero when added. Then the Hamiltonian takes the following form:

$$H = \frac{1}{2} \sum_k \left(\varepsilon(k) - \varepsilon(-k)\right)\left(C_k^+ C_k + C_{-k}^+ C_{-k}\right) + \sum_{k,k',q} V_0(q) C_{k'}^+ C_{-k'}^+ C_{-k} C_k . \tag{67}$$

If we consider that

$$\left(\varepsilon(k) - \varepsilon(-k)\right) = 2 E(k) ,$$

where $E(k)$ is the energy of an electron, which is above of the Fermi level, then one can obtain

X   Diagonalization of the Pairs Hamiltonian. Law of Dispersion

We will now consider the diagonalization of the pair Hamiltonian, Eq. (81), following the method of Valatin [13]. This is quite similar to the procedure employed in the theory of antiferromagnetism.

Let us define a new operator by

$$\xi_k = \alpha_k C_k - \beta_k C_{-k}^+ \qquad\qquad \xi_{-k} = \alpha_k C_{-k} + \beta_k C_k^+$$

$$\xi_k^+ = \alpha_k C_k^+ - \beta_k C_{-k} \qquad\qquad \xi_{-k}^+ = \alpha_k C_{-k}^+ + \beta_k C_k \tag{68}$$

in which α_k and β_k are real and positive.

We will require that the ξ_k^+ and ξ_k as C_k and C_k^+, obey the fermion rules:

$$\left[\xi_{k'}^+, \xi_k\right]_+ = \xi_{k'}^+ \xi_k + \xi_k \xi_{k'}^+ = \delta_{kk'} \tag{69}$$

The bracket in Eq. (69) indicates an anti-commutator.

Let see the condition that it should be fulfilled by the constants α_k and β_k in order that the previous anti-commutation rules be valid:

$$\xi_{k'}^+ \xi_k = \alpha_k^2 C_{k'}^+ C_k - \alpha_k \beta_k C_{k'}^+ C_{-k}^+ - \beta_k \alpha_k C_{-k'} C_k + \beta_k^2 C_{-k'} C_{-k}^+$$

$$\xi_k \xi_{k'}^+ = \alpha_k^2 C_k C_{k'}^+ - \alpha_k \beta_k C_k C_{-k'} - \beta_k \alpha_k C_{-k}^+ C_{k'}^+ + \beta_k^2 C_{-k}^+ C_{-k'} \tag{70}$$

from Eq. (70), we obtain that

$$\left[\xi_{k'}^+, \xi_k\right] = \alpha_k^2 \left[C_{k'}^+, C_k\right]_+ + \beta_k^2 \left[C_{-k}^+, C_{-k'}\right]_+ , \tag{71}$$

since

$$-\alpha_k \beta_k \left(C_{k'}^+ C_{-k}^+ + C_{-k}^+ C_{k'}^+\right) = 0$$

$$-\beta_k \alpha_k \left(C_{-k'} C_k + C_k C_{-k'}\right) = 0 \tag{72}$$

and, as must occurs for electrons,

$$\left[C_{k'}^+, C_k\right]_+ = \delta_{kk'}; \quad \left[C_{-k}^+, C_{-k'}\right]_+ = \delta_{-k,-k'}; \quad \left[C_{k'}^+, C_{-k}^+\right]_+ = \left[C_{-k'}, C_k\right] = 0 . \tag{73}$$

Then, substituting Eq. (73) in Eq. (71) we obtain that

$$\alpha_k^2 + \beta_k^2 = 1 . \tag{74}$$

The transformation Eq. (68) can be easily inverted with the aid of Eq. (74):

$$C_k = \alpha_k \xi_k + \beta_k \xi_{-k}^+ \qquad\qquad C_{-k} = \alpha_k \xi_{-k} - \beta_k \xi_k^+$$

$$C_k^+ = \alpha_k \xi_k^+ + \beta_k \xi_{-k} \qquad\qquad C_{-k}^+ = \alpha_k \xi_{-k}^+ - \beta_k \xi_k \tag{75}$$

In order to substitute the operators (75) in the Hamiltonian of the superconducting pairs, Eq. (81), we look for the following products:

$$C_k^+ C_k = \alpha_k^2 \xi_k^+ \xi_k + \alpha_k \beta_k \xi_k^+ \xi_{-k}^+ + \beta_k \alpha_k \xi_{-k} \xi_k + \beta_k^2 \xi_{-k} \xi_{-k}^+$$

$$C_{-k}^+ C_{-k} = \alpha_k^2 \xi_{-k}^+ \xi_{-k} - \alpha_k \beta_k \xi_{-k}^+ \xi_k^+ - \beta_k \alpha_k \xi_k \xi_{-k} + \beta_k^2 \xi_k \xi_k^+ \tag{76}$$

and calculating the sum of the previously operators, we obtain

$$C_k^+ C_k + C_{-k}^+ C_{-k} = 2\alpha_k \beta_k \left(\xi_k^+ \xi_{-k}^+ + \xi_{-k} \xi_k \right) + \left(\alpha_k^2 - \beta_k^2 \right)\left(n_k + n_{-k} \right) + 2\beta_k^2 . \tag{77}$$

Now we will see the off-diagonal terms case; first we calculate the products

$$C_{k'}^+ C_{-k'}^+ = \alpha_{k'}^2 \xi_{k'}^+ \xi_{-k'}^+ - \alpha_{k'} \beta_{k'} \xi_{k'}^+ \xi_{k'} + \beta_{k'} \alpha_{k'} \xi_{-k'} \xi_{-k'}^+ - \beta_{k'}^2 \xi_{-k'} \xi_{k'}$$

$$C_{-k} C_k = \alpha_k^2 \xi_{-k} \xi_k - \alpha_k \beta_k \xi_k^+ \xi_k + \beta_k \alpha_k \xi_{-k} \xi_{-k}^+ - \beta_k^2 \xi_k^+ \xi_{-k} \tag{78}$$

by multiplying these two expressions we obtain how the off-diagonal term of the Hamiltonian changes:

$$C_{k'}^+ C_{-k'}^+ C_{-k} C_k =$$

$$\alpha_{k'}^2 \alpha_k^2 \xi_{k'}^+ \xi_{-k'}^+ \xi_{-k} \xi_k - \alpha_{k'}^2 \alpha_k \beta_k \xi_{k'}^+ \xi_{-k'}^+ \xi_k^+ \xi_k + \alpha_{k'}^2 \beta_k \alpha_k \xi_{k'}^+ \xi_{-k'}^+ \xi_{-k} \xi_{-k}^+$$

$$- \alpha_{k'}^2 \beta_k^2 \xi_{k'}^+ \xi_{-k'}^+ \xi_k^+ \xi_{-k} - \alpha_{k'}^2 \alpha_k \beta_k \xi_{k'}^+ \xi_{k'} \xi_{-k} \xi_k + \alpha_{k'} \beta_{k'} \alpha_k \beta_k \xi_{k'}^+ \xi_{k'} \xi_k^+ \xi_k$$

$$- \alpha_{k'} \beta_{k'} \alpha_k \beta_k \xi_{k'}^+ \xi_{k'} \xi_{-k} \xi_{-k}^+ + \alpha_{k'} \beta_{k'} \beta_k^2 \xi_{k'}^+ \xi_{k'} \xi_k^+ \xi_{-k} + \beta_{k'} \alpha_{k'} \alpha_k^2 \xi_{-k'} \xi_{-k'}^+ \xi_{-k} \xi_k$$

$$- \alpha_{k'} \beta_{k'} \alpha_k \beta_k \xi_{-k'} \xi_{-k'}^+ \xi_k^+ \xi_k + \alpha_{k'} \beta_{k'} \alpha_k \beta_k \xi_{-k'} \xi_{-k'}^+ \xi_{-k} \xi_{-k}^+ - \alpha_{k'} \beta_{k'} \beta_k^2 \xi_{-k'} \xi_{-k'}^+ \xi_k^+ \xi_{-k}$$

$$- \alpha_k^2 \beta_{k'}^2 \xi_{-k'} \xi_{k'} \xi_{-k} \xi_k + \beta_{k'}^2 \alpha_k \beta_k \xi_{-k'} \xi_{k'} \xi_k^+ \xi_k - \beta_{k'}^2 \beta_k \alpha_k \xi_{-k'} \xi_{k'} \xi_{-k} \xi_{-k}^+ - \beta_{k'}^2 \beta_k^2 \xi_{-k'} \xi_{k'} \xi_k^+ \xi_{-k} \tag{79}$$

It is convenient to define the number operator n_k for the operators ξ_k, as

$$n_k = \xi_k^+ \xi_k \quad \text{and} \quad 1 - n_k = \xi_k \xi_k^+ , \tag{80}$$

which it is typical for fermions. Therefore we obtain for the off-diagonal term

$$H = \sum_{k} E(k)\left(C_k^+ C_k + C_{-k}^+ C_{-k}\right) + \sum_{k,k',q} V_0(q) C_{k'}^+ C_{-k'}^+ C_{-k} C_k \tag{81}$$

$$
\begin{aligned}
C_{k'}^+ C_{-k'}^+ C_{-k} C_k &= \alpha_{k'}\beta_{k'}\left(\alpha_k^2 - \beta_k^2\right)\left(1 - n_{k'} - n_{-k'}\right)\left(\xi_k^+ \xi_{-k}^+ + \xi_{-k}\xi_k\right) \\
&+ \alpha_k\beta_k\left(\alpha_{k'}^2 - \beta_{k'}^2\right)\left(1 - n_k - n_{-k}\right)\left(\xi_{k'}^+ \xi_{-k'}^+ + \xi_{-k'}\xi_{k'}\right) \\
&+ \alpha_{k'}\alpha_k\beta_k\beta_{k'}\left(1 - n_{k'} - n_{-k'}\right)\left(1 - n_k - n_{-k}\right)
\end{aligned}
\tag{82}
$$

We also demand that the operators change of the following way

$$\xi_{k'}^+ \xi_{-k'}^+ = \frac{\alpha_k\beta_k}{\left(\alpha_k^2 - \beta_k^2\right)} + \xi_k^+ \xi_k^+ \quad\text{and}\quad \xi_{-k'}\xi_{k'} = \frac{\alpha_k\beta_k}{\left(\alpha_k^2 - \beta_k^2\right)} + \xi_{-k}\xi_k \;, \tag{83}$$

then

$$
\begin{aligned}
C_{k'}^+ C_{-k'}^+ C_{-k} C_k &= \alpha_{k'}\beta_{k'}\left(\alpha_k^2 - \beta_k^2\right)\left(1 - n_{k'} - n_{-k'}\right)\left(\xi_k^+ \xi_{-k}^+ + \xi_{-k}\xi_k\right) \\
&+ \alpha_k\beta_k\left(\alpha_{k'}^2 - \beta_{k'}^2\right)\left(1 - n_k - n_{-k}\right)\left(\xi_k^+ \xi_{-k}^+ + \xi_{-k}\xi_k\right) \\
&+ \frac{2\alpha_k\beta_k}{\alpha_k^2 - \beta_k^2}\left(\alpha_{k'}^2 - \beta_{k'}^2\right)\left(1 - n_k - n_{-k}\right) \\
&+ \alpha_{k'}\alpha_k\beta_k\beta_{k'}\left(1 - n_{k'} - n_{-k'}\right)\left(1 - n_k - n_{-k}\right)
\end{aligned}
\tag{84}
$$

here we have dropped out the terms of four order that are those with four operators and cannot be reduced to particle number operators.

We can manipulate the term

$$\left(1 - n_{k'} - n_{-k'}\right)\left(1 - n_k - n_{-k}\right) = \left[1 - \left(n_k + n_{-k}\right) - \left(n_{k'} + n_{-k'}\right) + \left(n_k + n_{-k}\right)\left(n_{k'} + n_{-k'}\right)\right] ,\tag{85}$$

which it can be written, after neglecting the terms of second order in n_k, as

$$\left(1 - n_{k'} - n_{-k'}\right)\left(1 - n_k - n_{-k}\right) = 1 - 2\left(n_k + n_{-k}\right) . \tag{86}$$

Now, since we must eliminate the off-diagonals terms, then

$$2E(k)\alpha_k\beta_k + \left(\alpha_k^2 - \beta_k^2\right)\sum_{k',q} V_0(q)\alpha_{k'}\beta_{k'} + \alpha_k\beta_k \sum_{k',q} V_0(q)\left(\alpha_{k'}^2 - \beta_{k'}^2\right) = 0 \;, \tag{87}$$

this equation must be resolved keeping in mind that

$$\alpha_k^2 + \beta_k^2 = 1 \; ; \tag{88}$$

we can then introduce

$$\alpha_k = \sin\theta_k \quad \text{and} \quad \beta_k = \cos\theta_k, \tag{89}$$

then

$$E(k)\sin 2\theta_k - \frac{1}{2}\cos 2\theta_k \sum_{k',q} V_0(q)\sin 2\theta_{k'} - \frac{1}{2}\sin 2\theta_k \sum_{k',q} V_0(q)\cos 2\theta_{k'} = 0. \tag{90}$$

If we make

$$\cos 2\theta_k \Delta_k = -\frac{1}{2}\cos 2\theta_k \sum_{k',q} V_0(q)\sin 2\theta_{k'} - \frac{1}{2}\sin 2\theta_k \sum_{k',q} V_0(q)\cos 2\theta_{k'}, \tag{91}$$

and the solution of Eq. (90) is of the form

$$\tan 2\theta_k = -\frac{\Delta_k}{E(k)}. \tag{92}$$

From Eq. (92), we can also obtain that

$$\sin 2\theta_k = \frac{\Delta_k}{\left(\Delta_k^2 + E^2(k)\right)^{1/2}} \quad \text{and} \quad \cos 2\theta_k = -\frac{E(k)}{\left(\Delta_k^2 + E^2(k)\right)^{1/2}}. \tag{93}$$

The energy of the electron pairs or dispersion law is related with the terms that include the expression $\left(n_k + n_{-k}\right)$:

$$\in(k) = \left(\alpha_k^2 - \beta_k^2\right)E(k) - 2\alpha_k\beta_k\left(\sum_{k',q} V_0(q)\alpha_{k'}\beta_{k'} + \frac{\alpha_k\beta_k}{\left(\alpha_k^2 - \beta_k^2\right)}\sum_{k',q} V_0(q)\left(\alpha_{k'}^2 - \beta_{k'}^2\right)\right). \tag{94}$$

Substituting α and β as given in Eq. (89)

$$\in(k) = -\cos 2\theta_k E(k) - \sin 2\theta_k\left(\frac{1}{2}\sum_{k',q} V_0(q)\sin 2\theta_{k'} + \frac{1}{2}\tan 2\theta_k \sum_{k',q} V_0(q)\cos 2\theta_{k'}\right); \tag{95}$$

the term between parentheses in Eq. (94) is what we have defined as $-\Delta_k$, thus, substituting the values of the sine and cosine functions, we obtain the dispersion law of the electron pairs

$$\in(k) = -\cos 2\theta_k E(k) + \sin 2\theta_k \Delta_k = \frac{E^2(k)}{\left(\Delta_k^2 + E^2(k)\right)^{1/2}} + \frac{\Delta_k^2}{\left(\Delta_k^2 + E^2(k)\right)^{1/2}} = \left(\Delta_k^2 + E^2(k)\right)^{1/2}$$

;(96)

this dispersion law has been obtained by many authors, however it has not been used, neither interpreted, by any of them; this is amazing because this equation contains the whole physics of the superconductivity.

Now, let us return to the equation that defines the energy gap:

$$\cos 2\theta_k \Delta_k = -\frac{1}{2}\cos 2\theta_k \sum_{k',q} V_0(q)\sin 2\theta_{k'} - \frac{1}{2}\sin 2\theta_k \sum_{k',q} V_0(q)\cos 2\theta_{k'}$$

(97)

dividing by $\cos 2\theta_k$ we have

$$\Delta_k = -\frac{1}{2}\sum_{k',q} V_0(q)\sin 2\theta_{k'} - \frac{1}{2}\tan 2\theta_k \sum_{k',q} V_0(q)\cos 2\theta_{k'} .$$

(98)

If we consider the omitted terms that contain n_k, Eq. (98) becomes

$$\Delta_k(T) = -\frac{1}{2}\sum_{k',q} V_0(q)\sin 2\theta_{k'}\left(1 - \langle n_{k'}\rangle - \langle n_{-k'}\rangle\right) - \frac{1}{2}\tan 2\theta_k\left(1 - \langle n_k\rangle - \langle n_{-k}\rangle\right)\sum_{k',q} V_0(q)\cos 2\theta_{k'}$$

(99)

Substituting the values of the sine, cosine and tangent functions [see Eqs. (92-93)] in Eq. (99), we obtain that

$$\Delta_k(T) = -\frac{1}{2}\sum_{k',q} V_0(q)\frac{\Delta_{k'}(T)}{\left(\Delta_{k'}^2 + E^2(k')\right)^{1/2}}\left(1 - \langle n_{k'}\rangle - \langle n_{-k'}\rangle\right) -$$
$$-\frac{1}{2}\frac{\Delta_k}{E(k)}\left(1 - \langle n_k\rangle - \langle n_{-k}\rangle\right)\sum_{k',q} V_0(q)\frac{E(k')}{\left(\Delta_{k'}^2 + E^2(k')\right)^{1/2}}$$

(100)

and taking

$$\langle n_k\rangle = f[E(k,T)] = \left[\exp\left(\frac{E(k,T)}{k_B T}\right) + 1\right]^{-1} ,$$

(101)

and the factors

$$1-\langle n_{k'}\rangle - \langle n_{-k'}\rangle = \tanh \frac{E(k',T)}{2k_BT} \quad \text{and} \quad 1-\langle n_{k}\rangle - \langle n_{-k}\rangle = \tanh \frac{E(k,T)}{2k_BT} \, , \quad (102)$$

then

$$\Delta_k(T) = -\frac{1}{2}\sum_{k',q} V_0(q) \frac{\Delta_{k'}(T)}{\left(\Delta_{k'}^2 + E^2(k')\right)^{1/2}} \tanh \frac{E(k',T)}{2k_BT} -$$
$$-\frac{1}{2}\frac{\Delta_k}{E(k)} \tanh \frac{E(k,T)}{2k_BT} \sum_{k',q} V_0(q) \frac{E(k')}{\left(\Delta_{k'}^2 + E^2(k')\right)^{1/2}} \quad (103)$$

but, we know that

$$V_0(q) = \frac{2D^2(q)}{\varepsilon_p} \quad \text{and} \quad \sum_q = 2 \int_0^{\hbar\omega_0} F(\varepsilon_p)d\varepsilon_p \, , \quad (104)$$

where $F(\varepsilon_p)$ is the phonon density of states; then we can define

$$\lambda = 2 \int_0^{\hbar\omega_0} \varepsilon_p^{-1} D^2(q) F(\varepsilon_p)d\varepsilon_p \, . \quad (105)$$

This magnitude is identical to the so-called parameter of electron-phonon interaction of Eliashberg, but it has dimensions of energy

If we also define that

$$E(k) = \sum_{k'} \frac{E(k')}{g(\varepsilon_F)\left(\Delta_{k'}^2 + E^2(k')\right)^{1/2}} \, , \quad (106)$$

then the equation for the gap, that is just the bond energy of the pair (necessary energy to break the pair, equivalent to destroy the superconducting state), is

$$\Delta_k(T) = -\frac{1}{2}\lambda \sum_{k'} \frac{\Delta_{k'}(T)}{\left(\Delta_{k'}^2 + E^2(k')\right)^{1/2}} \tanh \frac{E(k',T)}{2k_BT} - \lambda\Delta_k(T)g(\varepsilon_F)\tanh \frac{E(k,T)}{2k_BT} \quad (107)$$

$$\Delta_k(T)\left(1 + \lambda g \tanh \frac{E(k,T)}{2k_BT}\right) = -\frac{1}{2}\lambda \sum_{k'} \frac{\Delta_{k'}(T)}{\left(\Delta_{k'}^2 + E^2(k')\right)^{1/2}} \tanh \frac{E(k',T)}{2k_BT} \, . \quad (108)$$

$$\left(1 + \lambda g \tanh \frac{E(k,T)}{2k_B T}\right) = -\frac{1}{2}\lambda \sum_{k'} \frac{1}{\left(\Delta_{k'}^2 + E^2(k')\right)^{1/2}} \tanh \frac{E(k',T)}{2k_B T} , \tag{109}$$

where $g(\varepsilon_F)$ is the density of states of the electrons in the Fermi level, integrating the right side, we obtain

$$\left(1 + \lambda g \tanh \frac{E(k,T)}{2k_B T}\right) = g(\varepsilon_F)\lambda \int_0^{\hbar\omega_0} \tanh \frac{E(k',T)}{2k_B T} \frac{dE}{\left(\Delta^2 + E^2\right)^{1/2}} . \tag{110}$$

The solution of the previous equation evidently is in general a difficult task.

The gap parameter $\Delta(T)$ is a function that diminishes with the increase of the temperature and it is zero in T_C. Then, we can substitute $\Delta = 0$ in the previously equation and to determine T_C we make the integration of the following way

$$\frac{1 + \lambda g \tanh \dfrac{\hbar\omega_0}{2k_B T_C}}{\lambda g} = \int_0^{\hbar\omega_0} \frac{\left(\tanh \dfrac{E}{2k_B T_C}\right)}{E} dE =$$

$$= \int_0^{\hbar\omega_0/2k_B T_C} \frac{\tanh x}{x} dx \tag{111}$$

where

$$x = \frac{E}{2k_B T_C} . \tag{112}$$

Integrating by parts we obtain that

$$\frac{1 + \lambda g \tanh \dfrac{\hbar\omega_0}{2k_B T_C}}{\lambda g} =$$

$$= \ln\left(\frac{\hbar\omega_0}{2k_B T_C}\right) \tanh\left(\frac{\hbar\omega_0}{2k_B T_C}\right) - \int_0^{\hbar\omega_0/2k_B T_C} \ln x \, \mathrm{sec}\, h^2\, x \, dx \tag{113}$$

In fact, since $\hbar\omega_0/2k_B T_C$ is large, $\tanh(\hbar\omega_0/2k_B T_C) \approx 1$, and the integral could be extended to infinite, in doing so, we then have that

$$\frac{1+\lambda g}{\lambda g} = -\ln\left[A\left(\frac{\hbar\omega_0}{2k_B T_C}\right)\right],$$

(114)

where the numeric value of the constant A is given by [14]

$$\ln A = -\int_0^\infty \ln x \sec h^2 x \, dx = \gamma + \ln\left(\frac{4}{\pi}\right)$$

(115)

here, γ is the Euler constant (0.5772...); finally, we obtain

$$k_B T_C = 1.13\hbar\omega_0 \exp\left[-\frac{(1+\lambda g)}{\lambda g}\right].$$

(116)

We can observe that the Eliashberg constant $\lambda_E = \lambda g$, besides, the constant μ appearing in the equation of Eliashberg is not necessary, because it is related with the constant of electron-electron interaction, which it can be equivalent to the constant of electron-phonon interaction. This expression is exact if we take A as

$$\ln A = \int_0^{\hbar\omega_0/k_B T_C} \ln x \sec h^2 x \, dx .$$

(117)

Then, defining

$$\Delta(0) = 2\hbar\omega_0 \exp\left[-\frac{(1+\lambda g)}{\lambda g}\right],$$

(118)

we obtain the relationship

$$k_B T_C = \frac{1}{4} A 2\Delta(0) .$$

(119)

Furthermore, if the superconductivity parameter is defined as

$$\rho = \frac{4}{A},$$

(120)

where ρ is the parameter that characterizes each type of superconductor material, then

$$\rho = \frac{2\Delta(0)}{k_B T_C}.$$

(121)

Now we must look for a relationship of the gap as a function of the temperature that be agree with the experimental results. Thus, we need to return to the dispersion law of the electron pairs and interpret it correctly:

$$\epsilon^2 = \Delta^2 + E^2.$$

(122)

We should think in the following way: in the original problem the effective potential of interaction doesn't depend on k and k', besides, we also know from the experimental results that the gap or the bond energy of the electron pairs only depends on the temperature and on the critical temperature $\Delta = \Delta(T, T_C)$. Then the dispersion law is only a function of $\Delta = \Delta(T, T_C)$ if $E = E(T)$ and $\epsilon = \epsilon(T_C)$, and as the potential is constant, then the energy ϵ is constant too and only depends on T_C, thus the problem in the superconducting state is conservative.

But as $T = 0 \Rightarrow E = 0$, then $\epsilon = \Delta$ and it only depends on T_C, then $\rho k_B T_C = 2\Delta$, which it implies that ϵ is constant:

$$\epsilon = \rho k_B T_C.$$

(123)

On the other hand, E is the energy of the electron pairs for any temperature in the interval $0 < T < T_C$ and it is equal to the phonon energy, then it is logical that it depends on the temperature, $E \propto k_B T$; the constant of proportionality depends on the nature of the superconductor and therefore is ρ :

$$E = \rho k_B T.$$

With these arguments, we can obtain the equation of the gap as a function of the temperature by using the previous dispersion law:

$$\Delta^2(T) = \epsilon^2(T_C) - E^2(T) = (\rho k_B T_C)^2 - (\rho k_B T)^2$$
$$= (\rho k_B T_C)^2 \left(1 - \frac{T^2}{T_C^2}\right)$$

(124)

Finally

$$\Delta(T) = \Delta(0)\left(1 - \frac{T^2}{T_C^2}\right)^{1/2},$$
(125)

which coincides with the prediction of the experimental results in all superconductors of low temperature.

Now we will analyze the second case (most general) in which we demanded instead of Eq. (56) the following conditions:

$$\frac{2|D(q)|^2}{\varepsilon_p} \neq \frac{4\pi e^2}{|k-(-k)|^2 + k_0^2}$$

and

$$\varepsilon_p \neq \varepsilon(k) - \varepsilon(-k).$$

We will follow a similar procedure to that of McMillan in Ref. [9]. We choose

$$\Delta(E) = \Delta_0 \qquad 0 < E < \varepsilon_0$$
$$= \Delta_\infty \qquad \varepsilon_0 < E$$

where $E = \varepsilon(k) - \varepsilon(-k) = \hbar\omega$ and $\varepsilon_0 = \hbar\omega_0$ is the maximum phonon energy.

We now consider two contributions to the energy gap. From Eq. (103) and substituting $V_0(q)$ by Eq. (55), we obtain

$$\Delta_k(E) = -\frac{1}{2}\sum_{k',q}\left\{|D(q)|^2 \frac{-2\varepsilon_p}{(\varepsilon(k)-\varepsilon(-k))^2 - \varepsilon_p^2} + \frac{4\pi e^2}{|k-(-k)|^2 + k_0^2}\frac{(\varepsilon(k)-\varepsilon(-k))^2}{(\varepsilon(k)-\varepsilon(-k))^2 - \varepsilon_p^2}\right\} \times$$
$$\times\left\{\frac{\Delta_{k'}(E)}{(\Delta_{k'}^2 + E^2(k'))^{1/2}}\tanh\frac{E(k')}{2k_BT} + \frac{\Delta_k}{E(k)}\tanh\frac{E(k)}{2k_BT}\frac{\Delta_{k'}(E)}{(\Delta_{k'}^2 + E^2(k'))^{1/2}}\right\}$$
(126)

Following the procedure carried out from Eq. (105) to Eq. (111) and separating the independent calculations of the electron-phonon interaction and electron-electron Coulomb repulsion, we can obtain the contributions to the energy gap from the electron-phonon interaction for the low and high energies of the following way:

$$\Delta^1(0)=\frac{1}{1+\lambda}\left\{\Delta_0\int_0^{\varepsilon_0}\frac{dE}{E}\tanh\frac{E}{2k_BT_C}-2\int_0^{\varepsilon_0}\frac{d\varepsilon_p D^2(\varepsilon_p)F(\varepsilon_p)}{\varepsilon_p}+\Delta_\infty\int_{\varepsilon_0}^\infty\frac{dE}{E}2\int_0^{\varepsilon_0}\frac{d\varepsilon_p D^2(\varepsilon_p)F(\varepsilon_p)}{\varepsilon_p}\frac{\varepsilon_p^2}{E^2}\right\}$$

(127)

In all our calculations the second sum in Eq. (126) has been considered constant and equal to $\lambda\Delta^1$, a matter that gives the factor $(1+\lambda)$ in the denominator.

In the first term of Eq. (127), the dominant contribution to the integral in E is for small E, therefore, we can neglect E relative to ε_p in the expression $\varepsilon_p^2/(E^2-\varepsilon_p^2)$. In the second term we have neglected ε_p relative to E.

Integrating Eq. (127), we obtain

$$\Delta^1(0)=\frac{1}{1+\lambda}\left\{\lambda\Delta_0\ln\frac{\varepsilon_0}{k_BT_C}+0.5\Delta_\infty\lambda\frac{\langle\varepsilon_p^2\rangle}{\varepsilon_0^2}\right\},$$

(128)

where $\langle\varepsilon_p^2\rangle$ is an average phonon energy given by

$$\langle\varepsilon_p^2\rangle=\frac{\int_0^{\varepsilon_0}d\varepsilon_p D^2(\varepsilon_p)F(\varepsilon_p)\varepsilon_p}{\int_0^{\varepsilon_p}\frac{d\varepsilon_p}{\varepsilon_p}D^2(\varepsilon_p)F(\varepsilon_p)}.$$

(129)

At high energies the only contribution is from the Coulomb interaction. In this case $\lambda=0$ and in the expression $E^2/(E^2-\varepsilon_p^2)$ we can neglect ε_p relative to E; therefore the system behaves as if the ionic screening does not exist and it only exists the screening due to the electrons (Thomas-Fermi screening). Thus

$$\int_0^\infty\frac{dE}{E}\tanh\frac{E}{2k_BT_C}=\int_0^{\varepsilon_0}\frac{dE}{E}\tanh\frac{E}{2k_BT_C}+\int_{\varepsilon_0}^{\varepsilon_B}\frac{dE}{E}\tanh\frac{E}{2k_BT_C},$$

(130)

and integrating we obtain

$$\Delta(\infty)=-N(0)V_C\left[\Delta_0\ln\frac{\varepsilon_0}{k_BT_C}+\Delta_\infty\ln\frac{E_B}{\varepsilon_0}\right];$$

(131)

making $\Delta(\infty) = \Delta_\infty$ we find

$$\Delta(\infty) = -\frac{N(0)V_C \Delta_0}{1 + N(0)V_C \ln \dfrac{E_B}{\varepsilon_0}} \ln \frac{\varepsilon_0}{k_B T_C} , \qquad (132)$$

where we can introduce

$$\mu^* = \frac{N(0)V_C}{1 + N(0)V_C \ln \dfrac{E_B}{\varepsilon_0}} ,$$

where μ^* is the Coulomb pseudopotential of McMillan [9] and Morel and Anderson [15], E_B is the electronic bandwidth, $N(0)$ is the Fermi occupation probabilities, V_C is the matrix element of the screened Coulomb interaction averaged over the Fermi surface. The screened Coulomb interaction is described by the parameters $N(0)V_C$ and E_B, and the electron-phonon interaction by the function $D^2(\varepsilon_p)F(\varepsilon_P)$.

Then, Eq. (132) is given by

$$\Delta(\infty) = \Delta_\infty = -\mu^* \Delta_0 \ln \frac{\varepsilon_0}{k_B T_C} . \qquad (133)$$

Taking into account the Coulomb interaction at low and intermediate energies and considering that the second sum of Eq. (126) is equal to $\lambda \Delta^2(0)$, we can write

$$\Delta^2(0) = -\frac{1}{1+\lambda}\left\{ N(0)V_C \Delta_0 \ln \frac{\varepsilon_0}{k_B T_C} + N(0)V_C \Delta_\infty \ln \frac{E_B}{\varepsilon_0} \right\} .$$

Satisfying the requirement at low and high energies and considering that $\Delta(0) = \Delta_0$, we can write the equation for the total contributions of the energy gap as

$$\Delta(0) = \frac{\Delta_0 \lambda}{1+\lambda} \ln \frac{\varepsilon_0}{k_B T_C} + \frac{\Delta_\infty \lambda}{1+\lambda} \frac{\langle \varepsilon_p^2 \rangle}{\varepsilon_p^2} - \frac{N(0)V_C}{1+\lambda} \Delta_0 \ln \frac{\varepsilon_0}{k_B T_C} + \Delta_\infty \ln \frac{E_B}{\varepsilon_0} . \qquad (134)$$

Substituting Eq. (133) into Eq. (134), we find that

$$\Delta_0 = \frac{\Delta_0\left\{\lambda - \mu^* - 0.5\mu^*\lambda\dfrac{\left\langle \varepsilon_p^2\right\rangle}{\varepsilon_p^2}\right\}}{1+\lambda}\ln\frac{\varepsilon_0}{k_B T_C}.$$

From this equation we can obtain the strong-coupling formula analogous to Eq. (15) of McMillan [9]:

$$\frac{k_B T_C}{\varepsilon_0} = \exp\left[\frac{-(1+\lambda)}{\lambda - \mu^* - 0.5\lambda\mu^*\dfrac{\left\langle \varepsilon_p^2\right\rangle}{\varepsilon_p^2}}\right].$$

Remembering that $\lambda = N(0)\lambda'$, where λ' is given by Eq. (105).

XI Application of This Theory to the Low Critical Temperature Superconductivity

In this section we will use a low temperature harmonic theory of the phonons where acoustic phonons are predominant. Incorporating the parameter of superconductivity within the general expression obtained, the binding energy (energy gap) is calculated as a function of temperature. An expression for the density of states of electron pairs for this kind of superconductors is given taking into account that the energy of the local electron pair and phonon are equal. Through the use of the density of states and the energy gap we determine the relation between the critical magnetic field and the temperature. The electrodynamical properties of these superconductors are then explained and a comparison with experimental data for the specific heat and its jump at T_C is done.

For acoustic phonons the frequency ω depends directly on the moment $\boldsymbol{q}$ in such a way that $\varepsilon_p = \hbar c q$. At low temperatures $(T << \Theta_D)$ we first note that only phonons with $\hbar\omega(\boldsymbol{q})$ comparable to or lesser than $k_B T_C$ can be absorbed or emitted by electrons. In the case of absorption this is immediately obvious, since these are the only phonons present in appreciable number. It is also true in the case of emission that, in order to emit a phonon, an electron must be far enough above the Fermi level and the final electron level (whose energy is lower by the quantity $\hbar\omega(\boldsymbol{q})$) to be unoccupied; since levels are occupied only to within order $k_B T$ above ε_F, and unoccupied only to within order $k_B T$ below of ε_F, only phonons with energies $\hbar\omega(\boldsymbol{q})$ of the order of $k_B T$ can be emitted.

Well below of Debye temperature, the condition $\hbar\omega(q) \leq k_B T_C$ requires q to be small compared with k_D. In this regime ω is of the order of cq so the wave vectors q of the phonons are of the order of $k_B T_C / \hbar c$ or less. Thus within the surface of phonons that the conservation laws allow to be absorbed or emitted, only a piece of linear dimensions proportional to T, and hence of area proportional to T^2, can actually participate. We conclude that the number of phonons that can scatter an electron declines as T^2 becomes well below the T_C.

As we have already previously seen, upon arriving to the critical temperature the electron pair energy E is constant and equal to the phonon energy for q_{max}. Then, let make

$$E = 2\varepsilon(k) = \ell(q_{max}) = \varepsilon_p(q_{max}) = \rho k_B T_C, \tag{135}$$

where ρ is the parameter of the superconductivity, which it is characteristic of each superconductor material and is determined from the experiments, measuring the bond energy of the electron pairs $2\Delta(T = 0K)$ and the critical temperature.

However, for $T = 0K$ we also know that $\ell(0) = 0$ since $q \approx 0$, then $E = 2\Delta(0) = \rho k_B T_C$.

Therefore, for any temperature in the interval $0 < T < T_C$, $2\varepsilon(k) = \ell(q) = \rho k_B T$. Taking into account the equation

$$E^2 = \ell^2(q) + (2\Delta(T))^2, \tag{136}$$

we obtain

$$(2\Delta(T))^2 = (\rho k_B T_C)^2 - (\rho k_B T)^2. \tag{137}$$

From here we obtain the general equation for the low temperature superconductors

$$\Delta(T) = \frac{1}{2}\rho k_B T_C \left(1 - \frac{T^2}{T_C^2}\right)^{1/2} \qquad \text{or} \qquad \Delta(T) = \Delta(0)\left(1 - \frac{T^2}{T_C^2}\right)^{1/2}. \tag{138}$$

Eq. (138) allows us to calculate the bond energy of the electron pairs as a function of the temperature. All the physical magnitudes of superconductors that vary with the temperature are related with this energy. Notice that for $T = T_C$, $\Delta(T_C) = 0$ and for $T = 0$, $\Delta(0) = 0$. The form suggested by Buckingham [16] is

$$\Delta(T) = 3.2 k_B T_C \left(1 - \frac{T}{T_C}\right)^{1/2},$$

(139)

which it is empiric and it is not valid in $T = 0K$. The BCS [2] theory doesn't arrive to an explicit relationship for the bond energy of the electron pairs.

XII Density of States

Some of the one-electron or lattice properties, like the specific heat and the magnetic field, are of the form

$$Q = \frac{2}{V} \sum_{n\vec{k}} Q_n(\boldsymbol{k}) = 2 \sum_n \int \frac{d\boldsymbol{k}}{(2\pi)^3} Q_n(\boldsymbol{k})$$

or

$$Q = \frac{1}{V} \sum_{k,s} Q(\omega_s(\boldsymbol{q})) = \sum_s \int \frac{d\boldsymbol{q}}{(2\pi)^3} Q(\omega_s(\boldsymbol{q})),$$

(140)

where, for each n, the sum is over all the allowed $\boldsymbol{k}$ given (physically distinct levels) and S is the phonon branch. It is often convenient to reduce such quantities to energy or frequency integrals, introducing a density of levels or normal modes with infinitesimal energies or frequencies ranging between ε and $\varepsilon + d\varepsilon$ or ω and $\omega + d\omega$

$$\int d\varepsilon g(\varepsilon) Q(\varepsilon) \quad \text{or} \quad \int d\omega g(\omega) Q(\omega).$$

(141)

Comparing (140) and (141), we find that density of states is given by

$$g(\varepsilon) = \sum_n \int \frac{d\boldsymbol{k}}{4\pi^3} \delta(\varepsilon - \varepsilon_n(\boldsymbol{k})) \quad \text{or} \quad g(\omega) = \sum_s \int \frac{d\boldsymbol{q}}{(2\pi)^3} \delta(\omega - \omega_s(\boldsymbol{q})).$$

(142)

In the case of electron pairs the density of states corresponding to the total energy E is constant, but that corresponding to the energy of the electron pair $\ell(\boldsymbol{q})$ is variable and considering that this energy is equal to the phonon energy and it depends linearly on $\boldsymbol{q}$ (notice that it does not depend on the square of $\boldsymbol{q}$), then we have:

i) In the first case, where $E = const$ and $\ell(\boldsymbol{q}_{max}) = E$, the density of states of the electron pair does not depend on $\boldsymbol{k}$, then

$$g(E) = \int \frac{dk}{(2\pi)^3} \delta\!\left(E - E_{q_{max}}\right) = n\delta\!\left(E - E_{q_{max}}\right) ,$$

(143)

carrying out the integration in spherical coordinates in the reciprocal space of wave vectors, we obtain

$$n = \frac{k_F^3}{3\pi^2} = \left(\frac{2}{3}\right)\varepsilon_F \, g(\varepsilon_F) ,$$

(144)

with

$$g(\varepsilon_F) = \frac{mk_F}{\hbar^2 \pi^2} ,$$

(145)

but as E is constant, then $\varepsilon_F = E$ and

$$n = \left(\frac{2}{3}\right)\rho k_B T_C \, g(\varepsilon_F) = \left(\frac{2}{3}\right) E g(\varepsilon_F) .$$

(146)

ii) In the second case $\ell(q)$ depends linearly on q and it varies from $\ell(q = 0) = 0$ in

$$T = 0K \text{ until } \left(E^2 - \ell^2(T)\right)^{1/2} ,$$

(147)

for any temperature $0 < T < T_C$ or $0 < q < q_{max}$, taking into account that $\ell(q \text{ or } T) = \varepsilon_p = \hbar c q$; then, the integral of Eq. (143) is

$$g(\ell(T)) = \frac{1}{2\pi^2} \frac{\ell^2(T)}{\hbar^3 c^3} ,$$

(148)

where

$$\hbar^3 c^3 = \frac{q_F^3 \hbar^3 c^3}{q_F^3} = \left(\frac{\varepsilon_F^3}{3\pi^2}\right)\left(\frac{3\pi^2}{q_F^2}\right) = \frac{\varepsilon_F^3}{3n\pi^2} ,$$

(149)

and substituting the value of n, we obtain

$$g(\ell(T)) = \frac{1}{2\pi^2} \frac{\ell^2(T)}{\hbar^3 c^3} = \frac{\ell^2(T)}{\varepsilon_F^2} g(\varepsilon_F)$$

(150)

Notice that when $\ell(T) = \varepsilon_F$, then $g(\ell) = g(\varepsilon_F)$; finally we can write

$$g(\ell(T)) = \frac{\ell^2(T)}{E^2} g(\varepsilon_F)$$

(151)

which it is different to the obtained in the BCS [2] theory (see Eq. (3.26) of Ref. [2]).

XIII Critical Magnetic Field

The critical magnetic field for a bulk superconductor material of unit volume is calculated by equating the magnetic energy with the average bond energy of the electron pairs:

$$\frac{H_C^2(T)}{8\pi} = \int \varepsilon g(\varepsilon) d\varepsilon$$

(152)

For $T = 0K$, $\varepsilon = E$ and $g(\varepsilon) = g(\varepsilon_F)$, the critical magnetic field can be calculated as

$$\frac{H_C^2(0)}{8\pi} = \int_0^E \varepsilon g(\varepsilon_F) d\varepsilon = g(\varepsilon_F) \frac{E^2}{2} = g(\varepsilon_F) \frac{\rho k_B T_C}{2}$$

(153)

obtaining

$$H_C(0) = (4\pi g(\varepsilon_F))^{1/2} \Delta(0)$$

(154)

In order to calculate the critical magnetic field as a function of the temperature we use Eq. (152) in the following form

$$\frac{H_C^2(T)}{8\pi} = \int \ell(T) g(\ell(T)) d\ell$$

(155)

substituting the expression for the density of states given by Eq. (151), we obtain

$$\frac{H_C^2(T)}{8\pi} = \int \ell(T) \frac{\ell^2(T)}{E^2} g(\varepsilon_F) d\ell = 2g(\varepsilon_F) \frac{\ell^4(T)}{4E^2} \Bigg|_0^{(E^2-\ell^2)^{1/2}}$$

(156)

and this way we have obtained that

$$\frac{H_C^2(T)}{8\pi} == 2g(\varepsilon_F)\frac{\Delta^4(T)}{4E^2}.$$

(157)

Substituting Eq. (138) in Eq. (157), we obtain

$$H_C(T) = \sqrt{4\pi g(\varepsilon_F)}\left(\frac{\rho}{2}k_B T_C\right)\left[1 - \frac{T^2}{T_C^2}\right] = H_C(0)\left(1 - \frac{T^2}{T_C^2}\right).$$

(158)

This expression is valid in the whole interval of temperature $0 \le T \le T_C$ and it coincides with the empirical law obtained from the experiments; in this case the BCS theory deduces an expression for the critical magnetic field only in $T = 0K$.

XIV Electrodynamical Properties

The electrodynamical properties of the superconductors are closely related or very related with the critical magnetic field considered in Eq. (158). Taking into account the Maxwell equation

$$\nabla \times \boldsymbol{H} = \left(\frac{4\pi}{c}\right)\boldsymbol{J},$$

(159)

and applying the rotational in both members and considering that $\nabla \cdot \boldsymbol{H} = 0$, we obtain

$$\nabla^2 \boldsymbol{H} = -\left(\frac{4\pi}{c}\right)\nabla \times \boldsymbol{J}.$$

(160)

Now if we consider that an electric force of the form $dp/dt = eE$ is generated, and considering that $\boldsymbol{J} = nev$, we obtain

$$\left(\frac{m}{ne^2}\right)\left(\frac{d\boldsymbol{J}}{dt}\right) = \boldsymbol{E};$$

(161)

then, applying the rotational in both members and considering the Maxwell equation

$$\nabla \times \boldsymbol{E} = -\left(\frac{1}{c}\right)\left(\frac{d\boldsymbol{H}}{dt}\right),$$

(162)

we obtain that

$$\left(\frac{cm}{ne^2}\right)\nabla\times\boldsymbol{J}=-\boldsymbol{H}$$

Now if we make

$$\Lambda=\frac{4\pi\lambda_L^2}{c^2}=\left(\frac{m}{ne^2}\right), \tag{163}$$

and substituting in the previous equation we obtain one of the London equations [17]

$$\boldsymbol{H}=-c\nabla\times\Lambda\boldsymbol{J}, \tag{164}$$

which, when it is combined with

$$\nabla^2\boldsymbol{H}=-\left(\frac{4\pi}{c}\right)\nabla\times\boldsymbol{J},$$

leads to the second London equation [17]

$$\nabla^2\boldsymbol{H}=\frac{\boldsymbol{H}}{\lambda_L^2}. \tag{165}$$

This implies that a magnetic field is exponentially screened from the interior of a sample and only can penetrate the length λ_L, this is the Meissner effect. Thus, the parameter λ_L is operationally defined as a penetration depth.

From the equation

$$\left(\frac{cm}{ne^2}\right)\nabla\times\boldsymbol{J}=-\boldsymbol{H}, \tag{166}$$

and by using the relation $\boldsymbol{H}=\nabla\times\boldsymbol{A}$ we obtain

$$\boldsymbol{J}=-\left(\frac{ne^2}{mc}\right)\boldsymbol{A}=-\left(\frac{1}{\Lambda c}\right)\boldsymbol{A}, \tag{167}$$

this is the diamagnetic density of current, which is valid for $T=0K$; therefore we can write it of the following form

$$J(0) = -\frac{1}{c\Lambda(0)} A(0) = J_D(0) \quad . \tag{168}$$

In order to calculate the paramagnetic density of current we use Eq. (158) and substituting the equation

$$H = -c\nabla \times \Lambda J \, , \tag{169}$$

we obtain

$$-c\Lambda\nabla \times J(T) = H_c(0)\left(1 - \frac{T^2}{T_C^2}\right) = -c\Lambda(0)\left(1 - \frac{T^2}{T_C^2}\right)\nabla \times A(0) \, , \tag{170}$$

where

$$\Lambda(T) = \Lambda(0)\left(1 - \frac{T^2}{T_C^2}\right) \quad . \tag{171}$$

This equation coincides with the equation (5.25) of the paper of BCS theory [2]

$$\Lambda(T) = \Lambda(0)\left(1 + \frac{\beta}{\Delta(T)} \frac{d\Delta(T)}{d\beta}\right) \, , \tag{172}$$

with $\beta = \frac{1}{k_B T}$, if we use Eq. (138) for the electron pairs bond energy $\Delta(T)$.

XV Critical Currents

The paramagnetic current is

$$J_p(T) = \frac{1}{c\Lambda(T)} A(0) \quad . \tag{173}$$

Notice that the paramagnetic density of current contains the diamagnetic density of current, in such a form that the total induced density of current is

$$J = J_D + J_p = -\frac{1}{c\Lambda(0)} A + \frac{1}{c\Lambda(T)} A = -\frac{1}{c\Lambda(T)} A \, , \tag{174}$$

then, using Eq. (163) we obtain Eq. (5.26) of the paper of BCS theory [2]:

$$J = J_D + J_p = -\frac{\Lambda(0)}{\Lambda(T)}\frac{ne^2}{mc}A(0).$$

(175)

XVI Penetration Lengths

Now, considering the equation

$$\nabla^2 H = \frac{H}{\lambda_L^2},$$

(176)

and substituting $H_C(T)$, we obtain

$$\nabla^2 H_C(T) = \frac{1}{\lambda_L^2(0)}\left(1-\frac{T^2}{T_C^2}\right)H_C(0);$$

(177)

then making

$$\frac{1}{\lambda_L^2(T)} = \frac{1}{\lambda_L^2(0)}\left(1-\frac{T^2}{T_C^2}\right),$$

(178)

we obtain that $\lambda_L(T)$ is given by the following expression:

$$\lambda_L(T) = \lambda_L(0)\left(1-\frac{T^2}{T_C^2}\right)^{-1/2},$$

(179)

which it is in correspondence with our equation for $\Lambda(T)$ and with $\lambda_L(T)$ from the BCS theory; however it does not coincide with the empirical law

$$\lambda_L(T) = \lambda_L(0)\left(1-\frac{T^4}{T_C^4}\right)^{-1/2}.$$

(180)

XVII Coherence Distance

Taking the relationship between the phonon moment q and the coherence distance ξ_0

$$\xi_0 = \frac{1}{\pi q},$$

(181)

and knowing that in $T = 0K$, $2\Delta = \rho k_B T_C = \hbar c q_{max}$, then

$$q = \frac{\rho k_B T_C}{\hbar c},$$

(182)

and after substituting in Eq. (181) we obtain

$$\xi_0(0) = \frac{1}{\pi}\frac{\hbar c}{\Delta(0)} = \frac{1}{\pi}\frac{1}{\rho}\frac{\hbar c}{k_B T_C}.$$

(183)

In the BCS theory $1/\rho\pi = 0.18$, but Faber and Pippard [18] obtained a value of 0.15 and Glover and Tinkhan [19] obtained a value of 0.27. In order to calculate the dependence on the temperature of $\xi_0(T)$ we take into account the equation $2\Delta(T) = \hbar c q(T)$, where

$$q(T) = \frac{\rho k_B T_C}{\hbar c}\left(1 - \frac{T^2}{T_C^2}\right)^{-1/2},$$

(184)

obtaining finally that

$$\xi_0(T) = \frac{\hbar c}{\pi\rho k_B T_C}\left(1 - \frac{T^2}{T_C^2}\right)^{-1/2}.$$

(185)

Taking into account Eq. (180) for $\lambda_L(T)$ we obtain that

$$\kappa = \frac{\lambda_L(T)}{\xi_0(T)} = \frac{\hbar c \lambda_L(0)}{\pi\Delta(0)},$$

(186)

the ratio of the two characteristic lengths defined in the Ginzburg-Landau [20] theory, which it is independent of the temperature, but it is characteristic for each superconductor.

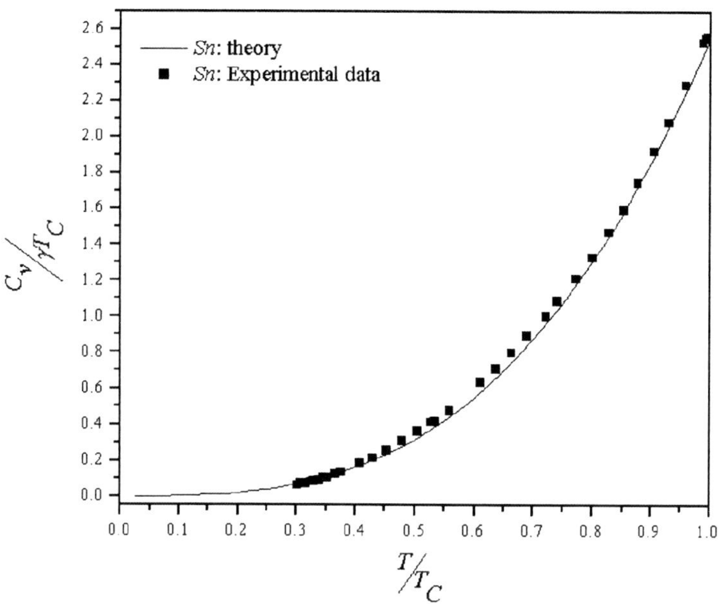

Fig. 5. Ratio c_v/c_v^e vs T/T_C for the specific heat as a function of temperature and comparison with experimental data for Sn.

XVIII   Specific Heat

Upon arriving to T_C, we can continue with our starting points in the general theory of the superconductivity, where the electron pairs energy is equal to the phonon energy. We can say that physically the specific heat of the free electrons becomes into the specific heat of the electron pairs and then into the specific heat of the phonons; therefore we should make a theory of the specific heat very similar to that of phonons one, but using the Maxwell-Boltzmann distribution function and introducing the superconductivity parameter. The low temperature specific heat of the phonons in the quantum theory of the harmonic solid is given by

$$c_v = \frac{\partial}{\partial T} \sum_s \int \frac{d\mathbf{q}}{(2\pi)^3} \frac{\hbar \omega_s(\mathbf{q})}{e^{\frac{\hbar \omega_s(\mathbf{q})}{k_B T}} - 1}.$$

(187)

As we can see, at very low temperature modes with $\hbar \omega_s(\mathbf{q}) \gg k_B T$ will contribute negligibly to the specific heat, since the integrand will exponentially vanish. However, because $\omega_s(\mathbf{q}) \to 0$ as $\mathbf{q} \to 0$ in the three acoustic branches, these conditions will fail to be satisfied by acoustic modes of sufficiently long-wavelength, no matter how low is the temperature. These modes (and only these) will continue contributing appreciably to the specific heat.

Bearing this in mind, we can make the following simplifications in Eq. (187), all of which result in a vanishing small fractional error, in the zero-temperature limit: i) Even if the crystal has a polyatomic basis, we can ignore the optical modes in the sum over s, since their frequencies are bounded below; ii) We can replace the dispersion relation $\omega = \omega_s(q)$ for the three acoustic branches $k_B T/\hbar$ because they are substantially lesser than those frequencies at which the acoustic dispersion curves begin to differ appreciably from their long-wavelength linear forms; iii) We will substitute the Bose-Einstein distribution function for the Maxwell-Boltzmann distribution function; thus at low temperature for the superconductor electron pairs Eq. (187) may be simplified to

$$c_v = \frac{\partial}{\partial T} \sum_s \int \frac{dq}{(2\pi)^3} \hbar\omega_s(q) e^{-\frac{\hbar\omega_s(q)}{k_B T}} \quad , \tag{188}$$

where the integral is over the q values in the interval $0 < q < q_{max}$, which it corresponds to the temperature interval $0 < T < T_C$. We evaluate the integral in spherical coordinates, writing $dq = q^2 dq d\Omega$ and making the sum on S, we obtain

$$c_v = \frac{3}{2\pi^2} \frac{\partial}{\partial T} \int \hbar c q^3 e^{-\frac{\hbar c q}{k_B T}} dq \quad . \tag{189}$$

If we make the change of variables $\ell = \hbar c q$ and considering that $\partial/\partial T = (\partial\ell/\partial T)(\partial/\partial\ell)$ and $\ell = \rho k_B T$ we obtain

$$c_v = \frac{3}{2\pi^2} \frac{1}{\hbar^3 c^3} \rho k_B T \frac{d}{d\ell} \int_0^{\ell(T)} \ell^3 e^{-\frac{\ell}{k_B T}} d\ell \quad ; \tag{190}$$

previously we have obtained that $\hbar^3 c^3 = E^2/2\pi^2 g(\varepsilon_F)$ and considering that $E = \rho k_B T_C$ and $\ell = \rho k_B T$, we obtain

$$c_v = 3g(\varepsilon_F) k_B^2 \rho^2 \left(\frac{T}{T_C}\right)^3 (1 - e^{-\rho}) \quad , \tag{191}$$

and as the electronic specific heat in $T \geq T_C$ is

$$c_v^e(T = T_C) = \frac{2\pi^2}{3} k_B^2 g(\varepsilon_F) T_C \quad , \tag{192}$$

taking into account that $\rho = 2\Delta(0)/k_B T_C$ we finally obtain for the specific heat in the superconducting state the following expression:

$$\frac{c_v}{c_v^e(T_C)} = \frac{2}{\pi^2}\left(\frac{T}{T_C}\right)^3 \left(\frac{2\Delta(0)}{k_B T_C}\right)^2 \left(1 - e^{-\frac{2\Delta(0)}{k_B T_C}}\right). \tag{193}$$

This expression is valid in the whole interval of temperature $0 < T \leq T_C$ and it describes the jump of the specific heat in $T = T_C$, taking into account the particular characteristic of each superconductor material through the parameter ρ of the superconductivity. This expression can be written as $c_v/c_v^e(T_C) = \zeta(T/T_C)^3$, where

$$\zeta = \left(\frac{2}{\pi^2}\right)\rho^2\left(1 - e^{-\rho}\right), \tag{194}$$

which it can be called a *characteristic parameter of the specific heat*.

The ratio c_v/c_v^e vs T/T_C is plotted in Fig. 5 (using Eq. (193)) and it is compared with experimental values for tin; we used $\rho = 3.5 \pm 0.1$ reported in Ref. [21]. The agreement of our expression is rather good. In Table 1 we show measured values of the ratio $\left.\left(c_v - c_v^e\right)/c_v^e\right|_{T_C}$ reported in Ref. [21] and calculated for our equation.

Table 1. We show measured values of the ratio $\left.\left(c_v - c_v^e\right)/c_v^e\right|_{T_C}$ reported in Ref. [21] and calculated for our equation. We have considered the values listed of the ρ parameter of Ref. [21], which they have an uncertainty $\pm.1$. The values in bold and underlined are in good agreement with the experiment data, considering the uncertainty in ρ.

ELEMENTS	SUPERCONDUCTIVITY PARAMETERS $\rho = \dfrac{2\Delta(0)}{k_B T_C} \pm 0.1$	EXPERIMENT DATA $\left.\left(c_v - c_v^e\right)/c_v^e\right|_{T_C}$	THEORETICAL DATA $\left.\left(c_v - c_v^e\right)/c_v^e\right|_{T_C}$		
			-0.1	ρ	+0.1
Al	3.4	1.4	1.12	1.26	**1.41**
Nb	3.8	1.9		**1.86**	
Pb	4.3	2.7		**2.69**	
Sn	3.5	1.6	1.26	1.41	**1.55**
V	3.4	1.5	1.12	1.26	**1.41**
Ta	3.6	1.6	1.41	**1.55**	1.70
Tl	3.6	1.5	1.41	**1.55**	1.70
In	3.6	1.7	1.41	1.55	**1.70**

XIX MgB$_2$ Superconductivity

The recent remarkable discovery of superconductivity at $T_C \approx 39\,K$ in Magnesium Diboride, MgB_2, has arose great interest in the scientific community [22-32]. MgB_2 has attracted tremendous attention since the discovery of superconductivity [33], because this simple intermetallic compound shows a remarkably high transition temperature $T_C \approx 39\,K$.

For MgB_2, the symmetry of the Cooper pairs has been confirmed to be of a spin-singlet type from a nuclear magnetic resonance study [34]. The Boron isotope effect measurement [35, 36] indicates phonon-mediated superconductivity. The critical temperature T_C of MgB_2 lies on or beyond the estimated upper limit of T_C for phonon-mediated superconductivity in the one-band model [37].

Band structure calculations [38-41] predict the coexistence of two-dimensional covalent in-plane (hole-like $B - 2p_{x,y}$ or σ band) and three-dimensional metallic-like interlayer (electron-like $B - 2p_z$ band or π band) conducting bands at the Fermi level $\left(\varepsilon_F\right)$ for a peculiar feature of MgB_2. For explaining its large T_C within the BCS scheme [39-41], a strong interaction of high frequency phonons originating in the in-plane Boron vibrations with electronic states at ε_F is a plausible scenario. However, the conclusions from band calculations are not consistent with one another. Kong $et\ al.$ [41] conclude that MgB_2 seems to be an intermediate-coupling phonon-assisted s wave superconductor, since they can explain the value of T_C using the obtained electron-phonon interaction constant $\lambda = 0.87$ and the Coulomb pseudopotential $\mu^* = 0.14$. Other studies have reported that the value of T_C cannot be reproduced if one uses the obtained $\lambda = 0.65 - 1$ and the commonly accepted values for μ^* [39-42], suggesting the need for a theory beyond the ordinary BCS.

It is plausible to suggest that the superconductivity in MgB_2 is described by a multi (two)-band model including the intraband and interband electron-phonon and Coulomb couplings. The two-band models for the superconductivity have been known for a long time [43, 44], and the corresponding developments for the high-T_C cuprate superconductors; see for example Refs. [45. 46]. In some papers various two-band schemes for MgB_2 are discussed [47-49]. According to Ref. [47], in the clean limit, MgB_2 should have two very different order parameters, which in turn should affect the thermodynamic properties in the superconductivity state. Using the Allen-Dynes approximation formulas for T_C [50], they found that to have $T_C = 40\,K$, a Coulomb pseudopotential of $\mu^* = 0.13$ is needed.

Using the fitting parameters in the two-band model with the repulsive interband transfer of intraband pairs between effective σ and π-bands, together with the σ-intraband electron-phonon attraction and Coulomb interaction, the temperature dependences of specific heat and superconducting gaps (for experimental data on gaps see Refs. [51-57]) are calculated [58].

In Ref. [48] the nesting properties in MgB_2 were calculated, indicating strong interband interaction. In Ref. [59] the specific heat capacity was calculated in MgB_2 from the spectral Eliashberg function $\alpha^2(\omega)F(\omega)$ first in the one-band model using the isotropic $\alpha^2(\omega)F(\omega)$ as given by Kong et al. [41]. Then they calculated the heat capacity in a two-band model by reducing the Eliashberg functions $\alpha_{ij}^2(\omega)F_{ij}(\omega)$ appropriate for the four Fermi surface sheets into four Eliashberg functions corresponding to an effective two-band model with a σ and π-band only. In Ref. [59] it is shown that the two-band model describes the temperature dependence of the specific heat [52, 60] better than the one-band model.

A quantitative first-principles calculation of T_C in M_gB_2 including the full variation of the electron-phonon interaction on the Fermi surface and the anharmonicity of the phonons is given in Refs. [61, 62]. For M_gB_2 the anisotropic k and ω dependent Eliashberg equation is solved. It is shown that the anisotropy (the electronic state dependence) of the electron-phonon interaction on the Fermi surface is strong enough to rise T_C to $39\,K$ even though the interaction is weakened by the anharmonicity of the phonons as compared to the harmonic case. In Refs. [61, 62], an ab initio calculation of the superconducting gaps in M_gB_2 and their effects on the specific heat are presented.

The compound M_gB_2 has structural features analogous to the graphite intercalation compounds (GIC). In Refs. [63-69] the superconductivity in $GIC - sC_xK$ $(x = 6.8)$ and C_xNa $(x = 2, 3, 4)$ was explained by the interband electron-phonon coupling. Large values of the interband interaction λ_{12} were obtained in Refs. [48, 60].

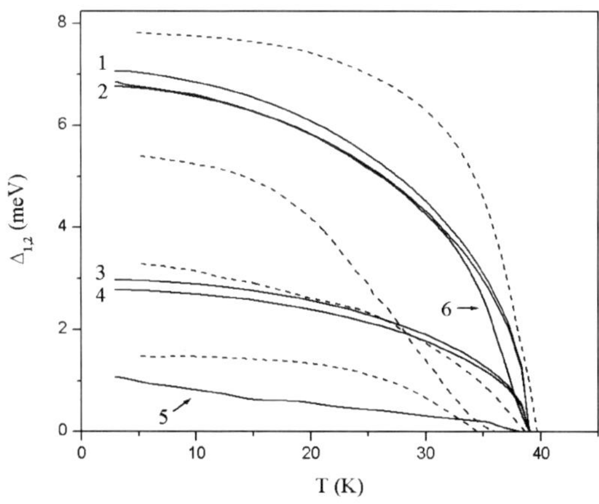

Fig. 6. Theoretical and experimental dependences of superconducting gap on temperature for the MgB_2 superconductor. The curves 1, 2, 3 and 4 are our results for $\Delta_1(0) = 7.1$ and $6.8\,MeV$, and $\Delta_2(0) = 2.8$ and $3.0\,MeV$. The other results are obtained from Fig. 2 of Ref. [30], where the curves 5 and 6 are theoretical results and the other are experimental results.

The very recent experimental study of the anisotropic superconductor MgB_2 using a combination of scanning tunneling microscopy and spectroscopy reveals two distinct energy gaps at $\Delta_1 = 7.1\,meV$ and $\Delta_2 = 2.3\,meV$ at $4.2\,K$ [70]. The π band superconducting gap, $\Delta_2 = 2.2\,meV$, was also recently observed by scanning tunneling spectroscopy [71]. Many results show that conventional phonon-mediated electron pairing theory can explain superconductivity in MgB_2 when both the anisotropy of the nonlinear electron-phonon interaction and the anharmonicity of the phonons are properly taken into account [72-75].

From Eq. (138), the variation of the energy gaps as a function of the temperature is plotted in curves 1, 2, 3 and 4 of Figs. 6 and 7. We have compared our results with those theoretical and experimental results showed in the Figs. 1 of Ref. [22] and 2 of Ref. [30], obtaining an excellent agreement with the experimental data, even better than the theoretical results (curves 5 and 6) reported in these references.

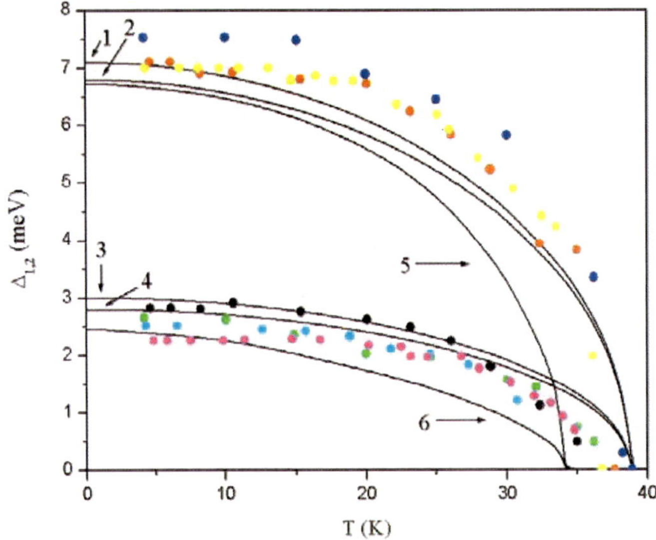

Fig. 7. Same as in previous figure but the results are compared with the obtained from Fig. 1 of Ref. [22], where the curves 5 and 6 are theoretical results and the other are experimental results.

Our curves were plotted taking the values of the gaps in $T = 0K$ as $\Delta_1(0) = 6.8\,MeV$ and $\Delta_2(0) = 3.0\,MeV$ of the Ref. [22] and $\Delta_1(0) = 7.1\,MeV$ and $\Delta_2(0) = 2.8\,MeV$ of the Ref. [23, 24].

XX Conclusion

We can conclude that upon arriving to the critical temperature a breakup of the symmetry has occurred with the splitting and reversing of the conduction band of the electrons and a canonical transformation of the three-dimensional space of the wave vectors $\left(k_x, k_y, k_z\right)$ to the bi-dimensional space of the energy $\left(\ell(T), \Delta(T)\right)$ (see Fig. 3).

From Fig. 3 we can see how the angle of dispersion of the electron pair varies between $0K$ and T_C:

$$\sin\theta = \frac{\Delta(T)}{\sqrt{\ell^2(\boldsymbol{q})+\Delta^2(T)}}; \quad \cos\theta = \frac{\ell(\boldsymbol{q})}{\sqrt{\ell^2(\boldsymbol{q})+\Delta^2(T)}}; \quad \theta = \arctan\frac{\Delta(T)}{\ell(\boldsymbol{q})}.$$

We think that the problems of the various theories of superconductivity we have presented here-in, are due to an incorrect physical interpretation of this phenomenon and of the energies appearing in the general equation of superconductivity. Besides, those theories don't keep in mind that the electron pair energies are equal to the phonon energies.

The superconductivity can only occur in materials where free carriers exist, namely, such as the conductors. The superconductivity in pure dielectrics cannot exist, because in those materials the free electric charges don't exist, so that, for superconductivity takes place it should occur a conductor-superconductor phase transition. Due to this condition, if we want to obtain a superconductor with high critical temperatures, then we need an electron-phonon coupling constant strong enough to have electron pairs correlated with small coherence distance; this is to say very near electrons. In order to obtain strong coupling between the electrons and phonons we need materials like dielectrics, but we know that in those materials there are not free electric carriers, therefore, we need to introduce conductor or metallic phases inside them with the purpose of free carriers can move through that phases as it happens in the high T_c superconductor ceramics. Thus, the ideal superconductor should have a big electron-phonon coupling constant, which it is characteristic of the dielectrics, and conductor phases where the free charges exist and they can form the electron pairs.

The relation $2\Delta(0) = 3.5\,k_B T_C$ imposed in the BCS theory [2] forbids the existence of superconductivity in high T_C. We think this relation is important and it should be rewritten in the following form: $2\Delta = \rho\,k_B T_C$, where ρ is our superconducting parameter, which it depends on each type of superconductor and it can be experimentally measured. For the superconductors characterized by high T_C this parameter is big.

It is physically acceptable that an electron pair is a new quasi-particle, which it is formed by two electrons and one phonon occupying the same states, but without losing any of their own identity, the electrons continue being fermions and the phonon a boson; that is the reason because we use the statistics of Maxwell-Boltzmann.

If the hypothesis of the formation of electron pairs is experimentally proven in any type of superconductivity, then the electron-electron Coulomb repulsion should compensate any interaction responsible for the electric resistance, which in this case it is considered as a different type of interaction from the electron-phonon one.

Until here we can conclude that the theory of superconductivity leads to a theory of the phonons (harmonic, anharmonic, of low, intermediate or high temperature).

The difference between the calculations carried out by Eliashberg and McMillan, in which they use a technique of many bodies, and our results are insignificant; there only exists a small difference in the equation that gives the critical temperature, however the dispersion law is always the same. A correct interpretation of the terms of the dispersion law and a correct use for low temperature allow us to obtain an expression of the energy gaps as a function of temperature that satisfies the experimental data so much for the superconductors of low critical temperature as the MgB_2 superconductor, such as we have shown in our results.

Now, let us suppose that there are only two physically distinguishable sources of scattering (for example, scattering by electron-phonon interaction and scattering by electron-electron Coulomb repulsion) as it is the case. If the presence of one mechanism does not alter the way in which the other mechanism functions, then the total collision rate W will be given by the sum of the collision rates due to the separate mechanisms $W = W^{ee} + W^{ep}$. In the relaxation-time approximation this immediately implies that

$$\frac{1}{\tau} = \frac{1}{\tau^{ee}} + \frac{1}{\tau^{ep}}.$$

If, additionally, we assume a k-independent relaxation time for each mechanism, then, since the resistivity is proportional to $1/\tau$, we will have

$$\rho = \frac{m}{ne^2\tau} = \frac{m}{ne^2\tau^{ee}} + \frac{m}{ne^2\tau^{ep}} = \rho^{ee} + \rho^{ep}.$$

This asserts that the resistivity in the presence of several distinct scattering mechanisms is simply the sum of the resistivities that one would have if each mechanism were present alone.

The isotopic effect in the low temperature superconductors, where the phonon theory is harmonic, can be described as

$$\omega\ (q) = \left(\frac{4\pi n_e Z^2 e^2}{\xi(q)}\right)\left(\frac{1}{M}\right) = \left(\frac{1}{\hbar}\rho k_B T_C\right)^2;$$

then we can write

$$T_C = \gamma\left(\frac{1}{\sqrt{M}}\right),$$

with

$$\gamma = \sqrt{\frac{4\pi n_e}{\xi(q)}}\left(\frac{Ze\hbar}{\rho k_B}\right).$$

The isotopic coefficient is 0.5 for these superconductors. It is necessary to notice that γ is characteristic of each material. This permits that the isotopic effect can be screened according to γ, for some superconductors.

References

[1] Cooper, L. N. *Phys. Rev.* 1956, 104, 1189-1190.

[2] Bardeen, J.; Cooper, L. N.; Schrieffer, J.R. *Phys. Rev.* 1957, 106, 162-164.

[3] Bogoliubov, N. N. *Soviet Phys. JETP* 1958, 7, 41-46.

[4] Gor'kov, L. P. *Soviet Phys. JETP* 1958, 7, 505-508.

[5] Bardeen, J.; Cooper, L. N.; Schrieffer, J. R. *Phys. Rev.* 1957, 108, 1175-1204.

[6] Migdal, A. B. *Soviet Phys. JETP* 1958, 7, 996-1001.

[7] Eliashberg, G. M. *Soviet Phys. JETP* 1960, 12, 1000-1002.

[8] Eliashberg, G. M. *Soviet Phys. JETP* 1960, 11, 696-702.

[9] McMillan, W. L. *Phys. Rev.* 1967, 167, 331-344.

[10] Eliashberg, G. M. *Soviet Phys. JETP* 1963, 16, 780-781.

[11] Luttinger, J. M.; Kohn, W. *Phys. Rev.* 1955, 97, 869-883.

[12] Frohlich, H. *Phys. Rev.* 1950, 79, 845-856.

[13] Valatin, D. G. *Nuovo Cimento* 1958, 7, 843-857.

[14] Rickayzen, G. *Theory of Superconductivity*; Wiley: New York, NY, 1965.

[15] Morel, P.; Anderson, P. W. *Phys. Rev.* 1962, 125, 1263-1271.

[16] Buckinham, M. J. *Phys. Rev.* 1956, 101, 1431-1432.

[17] London, F. *Proc. Roy. Soc.* 1935, *A*152, 25.

[18] Pippard, A. B. In *Advances in Electronics and Electron Physics*; Marton, L.; Ed.; Academic Press, Inc.: New York, NY, 1954; Vol. 6, pp. 1-45.

[19] Glover, R. E.; Tinkham, M. *Phys. Rev.* 1956, 104, 844-845.

[20] Ginzburg, V. L.; Landau, L. D. *Zh. Eksp. Teor. Fiz.* 1950, 20, 1064-1082.

[21] Meservey, R.; Schwartz, B. B. In *Superconductivity*; Parks, R. D., Ed.; Marcel Dekker: New York, NY, 1969; pp. 117-191.

[22] Konsin, P.; Sorkin, B. *Supercond. Sci. Techol.* 2004, 17, 1472-1476.

[23] Wang, Y.; Ancilotto, F.; Toigo, F.; Wang, K. *Physics Letters A*. 2003, 317, 483-488.

[24] Punpocha, S.; Hoonsawat, R.; Tang, I. M. *Solid State Comm.* 2004, 129, 151-156.

[25] Mishonov, T. M.; Klenov, S.I.; Penev, E. S. *Phys. Rev. B* 2005, 71, 024520.

[26] Zehetmayer, M.; Eisterer, M.; Jun, J.; Kazakov, S. M.; Karpinski, J.; Weber, H. W. *Phys. Rev. B* 2004, 70, 214516.

[27] Cappelluti, E.; Pietronero, L. *Phys. Stat. Sol. (b)* 2005, 242, 133-150.

[28] Khare, N.; Singh, D. P.; Gupta, A. K.; Aswal, D. K.; Sen, S.; Gupta, S. K.; Gupta, L. C. *Supercond. Sci. Technol.* 2004, 17, 1372-1375.

[29] Mucha, J.; Pekata, M.; Szydtowska, J.; Gadomski, W.; Akimitsu, J.; Fagnard, J. F.; Vanderbemden, P.; Cloots, R.; Ausloos, M. *Supercond. Sci. Technol.* 2003, 16, 1167-1172.

[30] Floris, A.; Profeta, G.; Lathiotakis, N. N.; Lüders, M.; Marques, M. A. L.; Franchini, C.; Gross, E. K. U.; Continenza, A.; Massidda, S. *Phys. Rev. Lett.* 2005, 94, 037004.

[31] Kortus, J.; Dolgov, O. V.; Kremer, R. K. *Phys. Rev. Lett.* 2005, 94, 027002.

[32] Bianconi, A.; Agrestini, S.; Campi, G.; Filippi, M.; Saini, N. L. *Current Applied Physics* 2005, 5, 254-258.

[33] Nagamatsu, J.; Nakagawa, N.; Muranaka, T.; Zenitani, Y.; Akimitsu, J. *Nature* 2001, 410, 63-64.

[34] Kotegawa, H.; Ishida, K.; Kitaoka, Y.; Muranaka, T.; Akimitsu, J. *Phys. Rev. Lett.* 2001, 87, 127001.

[35] Bud'ko, S. L.; Lapertot, G.; Petrovic, C.; Cunningham, C. E.; Anderson, N.; Canfield, P. C. *Phys. Rev. Lett.* 2001, 86, 1877-1880.

[36] Hinks, D. G.; Claus, H.; Jorgensen, J. D. Nature 2001, 411, 457-460.

[37] Ginzburg, V. L. *J. Supercond.* 2000, 13, 665-677.

[38] Belashchenko, K. D.; van Schilfgaarde, M.; Antropov, V. P. *Phys. Rev. B* 2001, 64, 092503.

[39] Kortus, J.; Mazin, I. I.; Belashchenko, K. D.; Antropov, V. P.; Boyer, L. L. *Phys. Rev. Lett.* 2001, 86, 4656-4659.

[40] An, J. M.; Pickett, W. E. *Phys. Rev. Lett.* 2001, 86, 4366-4369.

[41] Kong, Y.; Dolgov, O. V.; Jepsen, O.; Andersen, O. K. *Phys. Rev. B* 2001, 64, 020501.

[42] Bohnen, K. P.; Heid, R.; Renker, B. *Phys. Rev. Lett.* 2001, 86, 5771-5774.

[43] Suhl, H.; Matthias, B. T.; Walker, L. R. *Phys. Rev. Lett.* 1959, 3, 552.

[44] Moskalenko, V. A. *Fiz. Met. Metalloved.* 1959, 8, 503.

[45] Konsin, P.; Kristoffel, N.; Örd, T. *Phys. Lett. A* 1988, 129, 339-342.

[46] Kristoffel, N.; Konsin, P.; Örd, T. *Riv. Nuovo Cimento* 1994, 17, 1-41.

[47] Liu, A. Y.; Mazin, I. I.; Kortus, J. *Phys. Rev. Lett.* 2001, 87, 087005.

[48] Hase, I.; Yamaji, K. *J. Phys. Soc. Japan* 2002, 71, 371-371.

[49] Örd, T.; Kristoffel, N. *Physica C* 2002, 370, 17-20.

[50] Allen, P. B.; Dynes, R. G. *Phys. Rev. B* 1975, 12, 905-922.

[51] Wang, Y. X.; Plackowski, T.; Junod, A. *Physica C* 2001, 355, 179-193.

[52] Bouquet, F.; Fisher, R. A.; Phillips, N. E.; Hinks, D. G.; Jorgensen, J. O. *Phys. Rev. Lett.* 2001, 87, 047001.

[53] Yang, H. D.; Lin, J. Y.; Li, H. H.; Hsu, F. H.; Liu, C. J.; Li, S. C.; Yu, R. C.; Jin, C. Q. *Phys. Rev. Lett.* 2001, 87, 167003.

[54] Szabó, P.; Samuely, P.; Kacmarcik, J.; Klein, T.; Marcus, J.; Fruchart, D.; Miraglia, S.; Marcenat, C.; Jansen, A. G. M. *Phys. Rev. Lett.* 2001, 87, 137005.

[55] Giubileo, F.; Roditchev, D.; Sacks, W.; Lamy, R.; Thanh, D. X.; Klein, J.; Miraglia, S.; Fruchart, D.; Marcus, J.; Monod, P. *Phys. Rev. Lett.* 2001, 87, 177008.

[56] Chen, X. K.; Konstantinovic, M. J.; Irwin, J. C.; Lawrie, D. D.; Franck, J. P. *Phys. Rev. Lett.* 2001, 87, 157002.

[57] Tsuda, S.; Yokoya, T.; Kiss, T.; Tacano, Y.; Togano, K.; Kito, H.; Ihara, H.; Shin, S. *Phys. Rev. Lett.* 2001, 87, 177006.

[58] Kristoffel, N.; Örd, T.; Rägo, K. *Int. J. Mod. Phys. B* 2002, 16, 1585-1589.

[59] Golubov, A. A.; Kortus, J.; Dolgov, O. V.; Jepsen, O.; Kong, Y.; Andersen, O. K.; Gibson, B. J.; Ahn, K.; Kremer, R. K. *J. Phys.: Condens. Matter* 2002, 14, 1353-1360.

[60] Kremer, R. K.; Gibson, B. J.; Ahn, K. 2001. Preprint cond-mat/0102432. Preprint at arXiv.org.

[61] Choi, H. J.; Roundy, D.; Sun, H.; Cohen, M. L.; Louie, S. G. *Phys. Rev. B* 2002, 66, 020513.

[62] Choi, H. J.; Roundy, D.; Sun, H.; Cohen, M. L.; Louie, S. G. *Nature* 2002, 418, 758-760.

[63] Al-Jishi, R. *Phys. Rev. B* 1983, 28, 112-116.

[64] Jishi, R. A.; Dresselhaus, M. S. *Phys. Rev. B* 1992 45, 12465-12469.

[65] Jishi, R. A.; Dresselhaus, M. S.; Chaiken, A. *Phys. Rev. B* 1991, 44, 10248-10255.

[66] Konsin, P. *Phys. Status Solidi b* 1995, 189, 185-191.

[67] Konsin, P.; Örd, T. *Physica C* 1992, 191, 469-476.

[68] Konsin, P.; Sorkin, B.; Ausloos, M. *Supercond. Sci. Technol.* 1998, 11, 1-3.

[69] Tanaka, Y. *Phys. Rev. Lett.* 2002, 88, 017002.

[70] Iavarone, M.; Karapetrov, G.; Koshelev, A. E.; Kwok, W. K.; Crabtree, G. W.; Hinks, D. G.; Kang, W. N.; Choi, E. M.; Kim, H. J.; Kim, H-J.; Lee, S. I. *Phys. Rev. Lett.* 2002, 89, 187002.

[71] Eskildsen, M. R.; Kugler, M.; Tanaka, S.; Jun, J.; Kazakov, S. M.; Karpinski, J.; Fischer, O. *Phys. Rev. Lett.* 2002, 89, 187003.

[72] Serquis, A.; Zhu, Y. T.; Peterson, E. J.; Coulter, J. Y.; Peterson, D. E.; Mueller, F. M. *Appl. Phys. Lett.* 2001, 79, 4399-4401.

[73] Yildirim, T.; Gülseren, O.; Lynn, J. W.; Brown, C. M.; Udovic, T. J.; Huang, Q.; Rogado, N.; Regan, K. A.; Hayward, M. A.; Slusky, J. S.; He, T.; Haas, M. K.; Khalifah, P.; Inumaru, K.; Cava, R. J. *Phys. Rev. Lett.* 2001, 87, 037001.

[74] Hlinka, J.; Gregora, I.; Pokorný, J.; Plecenik, A.; Kús, P.; Satrapinsky, L.; Benacka, *Phys. Rev. B* 2001, 64, 140503.

[75] Angst, M.; Puzniak, R.; Wisniewski, A.; Jun, J.; Kazakov, S. M.; Karpinski, J.; Roos, J.; Keller, H. *Phys. Rev. Lett.* 2002, 88, 167004.

INDEX

D

H

I

J

K

L

oxygen, ix, 45, 47, 49, 52, 53, 54, 55, 56, 57, 58, 59, 60, 61, 62, 93, 197, 202, 203, 204, 205, 206, 208, 212, 213, 216, 219, 220, 240, 241
oxygen absorption, 45

P

pairing, xii, 172, 179, 239, 240, 250, 253, 272, 300
parameter, xi, 8, 16, 17, 26, 27, 48, 52, 67, 86, 90, 91, 92, 122, 126, 138, 157, 159, 160, 161, 162, 164, 166, 168, 169, 172, 174, 175, 176, 178, 180, 181, 182, 184, 188, 189, 196, 201, 202, 203, 204, 205, 206, 207, 208, 209, 210, 211, 212, 215, 216, 217, 218, 255, 270, 278, 279, 280, 285, 286, 291, 295, 297, 301
particles, 104, 159, 169, 175, 181, 188, 227, 231, 235, 256, 268
passive, xiv
percolation, xii, 239, 250
periodicity, 178
permeability, 141
perspective, 14
phase boundaries, 46, 47
phase diagram, xvii, 54, 61, 91, 98
phase transformation, xv
phase transitions, ix, xvii, 45, 46, 55, 67
phenomenology, 17
phonons, viii, ix, xiii, 2, 5, 7, 9, 10, 11, 12, 19, 20, 24, 25, 27, 28, 64, 76, 197, 224, 253, 254, 255, 256, 257, 258, 264, 266, 269, 285, 286, 295, 298, 299, 300, 301
photoemission, 77, 81, 83, 89, 91, 99, 100, 246
photons, 198
physical properties, ix, xiv, 45, 73, 104, 232
physics, vii, xvi, xvii, 3, 9, 14, 25, 74, 189, 226, 232, 263, 277
piezoelectric properties, 62
plane waves, 255
plastic deformation, 45, 46, 55
plasticity, 55
PM, 122, 123, 124, 125, 126, 128, 140
polarized light microscopy, 53
polymers, xiii
poor, 60, 62, 129
positron, 196, 197, 198, 199, 200, 201, 202, 203, 204, 205, 206, 208, 209, 211, 212, 213, 214, 215, 216, 218, 219
positrons, 196, 197, 199, 201, 203, 204, 206, 208, 209, 211, 212, 213, 214, 215, 216, 219
power, vii, viii, xiii, xiv, 2, 3, 4, 6, 8, 9, 12, 13, 19, 21, 24, 25, 26, 27, 46, 66, 67, 90, 185, 240
prediction, xiii, 253, 282
preparation, xiii, 29, 48

pressure, xi, xvi, xvii, 47, 158, 187, 188, 189, 190, 191, 192, 193
principle, 116, 170, 183, 189, 224, 235, 246
probability, 14, 208, 213
probe, xiv, xvii, 196, 203, 204, 209, 213, 217
production, 46
program, 7, 116
propagation, ix, 6, 45, 46, 269
proportionality, 281
pulse, xvi, 111

Q

quadratic curve, xi, 158, 187, 192
quanta, 178, 179
quantization, xi, 157, 172, 174, 178, 179, 224, 225, 232
quantum dots, 5, 6, 12, 23, 25
quantum fluctuations, xvii
quantum mechanics, 248
quantum theory, 295
quantum well, 4, 5, 6
quarks, 227
quartz, 48
quasiparticles, xii, xvi, xvii, 24, 25, 223, 225, 232, 233

R

radiation, vii, viii, 1, 2, 3, 4, 5, 6, 7, 8, 25, 27, 196, 197, 198, 199, 200, 205, 206, 217
radius, 65, 175
Raman spectra, 60
range, vii, viii, ix, xii, xiii, 1, 2, 3, 4, 5, 10, 12, 45, 46, 48, 49, 50, 51, 52, 53, 54, 55, 56, 58, 60, 61, 62, 63, 67, 91, 93, 96, 124, 127, 188, 200, 201, 203, 204, 205, 212, 213, 214, 215, 217, 219, 220, 239, 240, 246, 247, 253, 255, 269
reality, 102
recall, xvi, 62, 74
reciprocity, 167
recombination, 25
reconcile, 250
reconciliation, 241
recrystallization, 52
redistribution, 56, 59, 60, 61, 62, 196, 202, 206, 207, 208, 209, 211, 217
reduction, 59, 152
reflection, 5
refugees, 29
regression, 57
regression equation, 57

superconductor, vii, viii, ix, x, xi, xii, xiii, xiv, xvi,
     xvii, 2, 3, 4, 5, 7, 20, 25, 26, 28, 45, 64, 86, 87,
     92, 107, 108, 109, 111, 114, 115, 122, 125, 129,
     134, 135, 136, 140, 141, 142, 144, 145, 150, 151,
     152, 157, 158, 159, 164, 165, 166, 167, 168, 169,
     170, 174, 180, 181, 187, 188, 190, 191, 192, 193,
     203, 208, 210, 211, 213, 214, 215, 217, 243, 253,
     254, 255, 259, 280, 281, 286, 289, 294, 296, 297,
     298, 299, 300, 301, 302
superfluid, xvi, 164, 245
superlattice, xii, 12, 23, 223, 233, 234, 236
supply, vii, 1
surface layer, 46
surface region, 66
susceptibility, xiv, 49, 50, 54, 62, 63, 66, 74, 89, 95
symbols, 7, 8, 15, 16, 90, 99, 126, 128, 228, 234, 235
symmetry, 60, 61, 64, 80, 123, 140, 160, 161, 168,
     172, 174, 175, 177, 179, 188, 224, 233, 257, 298,
     300
systems, xii, xiv, 7, 24, 75, 87, 93, 102, 135, 140,
     168, 193, 196, 204, 223, 224, 227, 232

T

targets, 20
technology, x, xiii, xvi, 4, 10, 107, 108, 109, 152,
     158, 247
television, 152
temperature, vii, viii, ix, x, xi, xii, xiii, xiv, xvi, xvii,
     1, 2, 3, 4, 5, 6, 7, 8, 9, 11, 12, 13, 14, 16, 19, 20,
     21, 22, 24, 25, 26, 27, 28, 29, 45, 46, 47, 48, 49,
     50, 51, 52, 53, 54, 55, 56, 57, 58, 60, 61, 62, 63,
     64, 73, 74, 76, 77, 78, 81, 82, 84, 85, 86, 87, 88,
     89, 90, 91, 92, 95, 96, 97, 98, 99, 100, 101, 102,
     104, 106, 107, 108, 109, 119, 120, 121, 122, 131,
     140, 157, 158, 159, 164, 165, 167, 168, 169, 172,
     173, 174, 175, 179, 180, 181, 184, 187, 188, 189,
     190, 191, 192, 193, 196, 202, 203, 204, 205, 206,
     207, 208, 209, 210, 211, 212, 213, 214, 215, 216,
     217, 218, 219, 220, 224, 239, 245, 248, 250, 253,
     254, 255, 256, 257, 258, 259, 261, 262, 279, 281,
     282, 285, 286, 288, 289, 290, 294, 295, 296, 297,
     298, 299, 300, 301, 302
temperature dependence, 24, 52, 77, 102, 119, 120,
     203, 204, 205, 214, 298, 299
tensile strength, 113
theory, xiii, xvii, 12, 14, 62, 80, 92, 93, 95, 128, 159,
     169, 172, 188, 224, 240, 245, 246, 247, 250, 253,
     254, 255, 256, 257, 264, 270, 273, 285, 287, 289,
     295, 298, 300, 301, 302
thermal activation, 62
thermal properties, 169, 173, 180
thermal resistance, 9, 10, 27

thermal stability, xv
thermal treatment, 54
thermalization, 24
thermodynamic properties, xi, 81, 157, 168, 298
thermodynamics, 20, 170, 183, 184, 189
thin films, xiv, xvi, xvii, 5, 57, 59, 99
three-dimensional space, 300
threshold, 64
time, vii, viii, xiv, 2, 5, 6, 7, 8, 10, 16, 18, 19, 23, 24,
     25, 27, 28, 48, 53, 67, 90, 97, 98, 111, 113, 123,
     129, 135, 136, 138, 145, 146, 158, 159, 160, 161,
     173, 175, 179, 188, 200, 240, 247, 249, 298, 302
tin, 297
topology, 74
total energy, xiii, 253, 255, 256, 287
tracking, 142
training, 29
transducer, 46
transformation, 204, 225, 233, 234, 242, 257, 258,
     263, 264, 265, 266, 267, 274, 300
transition, x, xi, xii, xvii, 3, 4, 7, 14, 26, 47, 52, 61,
     64, 73, 74, 89, 101, 103, 104, 105, 141, 157, 168,
     169, 172, 173, 192, 196, 203, 205, 206, 207, 208,
     209, 210, 211, 213, 214, 215, 216, 217, 219, 245,
     248, 253, 256, 298, 301
transition temperature, xvii, 105, 192, 196, 206, 207,
     209, 211, 213, 214, 215, 216, 245, 248, 298
transitions, ix, xvii, 47
translation, 163, 233
transmission, vii, xiii, 5, 10, 85
transport, viii, ix, x, xiv, 2, 8, 10, 14, 15, 16, 17, 18,
     19, 23, 45, 50, 64, 65, 66, 67, 73, 92, 96, 104, 144,
     150, 157, 158, 164, 165, 166
transport processes, 14
transportation, x, 107, 108, 111, 133, 140, 152
trend, 204, 212, 214
tunneling, viii, xii, 2, 25, 84, 91, 92, 96, 166, 172,
     239, 244, 247, 248, 300
tunneling effect, xii, 166, 239
twinning, 46, 52
twins, 62, 64

U

ultrasound, 46, 62
uncertainty, 297
uniform, 24, 109, 114, 116, 124, 129, 130, 140, 144,
     206, 255
universality, 47, 173, 193
universities, vii